Studies in Logic

Logic and Argumentation

Volume 110

The Cognitive Dimension of Social Argumentation

Proceedings of the 4th European Conference on Argumentation

Volume II

Volume 101
The Logic of Partitions. With Two Major Applications
David Ellerman

Volume 102
Bounded Reasoning Volume 1: Classical Propositional Logic
Marcello D'Agostino, Dov Gabbay, Costanza Larese, Sanjay Modgil

Volume 103
The Fertile Debate. Affective Exploration of a Controversy
Claire Polo

Volume 104
Argument, Sex and Logic
Dov Gabbay, Gadi Rozenberg and Lydia Rivlin

Volume 105
Logic as a Tool. A Guide to Formal Logical Reasoning
Valentin Goranko

Volume 106
New Directions in Term Logic
George Englebretsen, ed

Volume 107
Non-commutative Algebras. Pseudo-BCK Algebreas versus m-pseudo-BCK Algebras
Afrodita Iorgulescu

Volume 108
Semitopology: decentralised collaborative action via topology, algebra, and logic
Murdoch J. Gabbay

Volume 109
The Cognitive Dimension of Social Argumentation. Proceedings of the 4th European Conference on Argumentation, Volume I. Fabio Paglieri, Alessandro Ansani and Marco Marini, eds.

Volume 110
The Cognitive Dimension of Social Argumentation. Proceedings of the 4th European Conference on Argumentation, Volume II. Fabio Paglieri, Alessandro Ansani and Marco Marini, eds.

Volume 111
The Cognitive Dimension of Social Argumentation. Proceedings of the 4th European Conference on Argumentation, Volume III. Fabio Paglieri, Alessandro Ansani and Marco Marini, eds.

Studies in Logic Series Editor
Dov Gabbay   dov.gabbay@kcl.ac.uk

The Cognitive Dimension of Social Argumentation

Proceedings of the 4th European Conference on Argumentation

Volume II

Edited by

Fabio Paglieri

Alessandro Ansani

and

Marco Marini

© Individual author and College Publications, 2024
All rights reserved.

ISBN 978-1-84890-472-9

College Publications
Scientific Director: Dov Gabbay
Managing Director: Jane Spurr

http://www.collegepublications.co.uk

Cover prepared by Laraine Welch

All rights reserved. No part of this publication may be reproduced, stored in a retrieval system or transmitted in any form, or by any means, electronic, mechanical, photocopying, recording or otherwise without prior permission, in writing, from the publisher.

The Cognitive Dimension of Social Argumentation

Proceedings of the 4th European Conference on Argumentation

Volume 2

Edited by
Fabio Paglieri
Alessandro Ansani
Marco Marini

TABLE OF CONTENTS

INTRODUCTION V
FABIO PAGLIERI, ALESSANDRO ANSANI, MARCO MARINI

META-ARGUMENT AND SPECIOUS ALLEGATIONS OF 1
FALLACY
SCOTT AIKIN & JOHN CASEY

COLLABORATIVE OPPOSITIONALITY, JUDGMENT 15
REVISION, AND CRITICAL THINKING EDUCATION
SHARON BAILIN & MARK BATTERSBY

META-ARGUMENT AND PARA-ARGUMENT 29
JOHN CASEY & SCOTT AIKIN

INCONCLUSIVE ARGUMENT APPRAISAL: DIFFICULTIES IN 41
CONCLUDING ARGUMENT EVALUATIONS AND
EVALUATING ARGUMENTS WITHOUT CONCLUSIONS
DANIEL COHEN

THE ROLE OF IMPLICITNESS IN PERSUASIVE 51
ARGUMENTATION: LOOKING AT DIFFERENT TEXT
GENRES
CLAUDIA COPPOLA, DORIANA CIMMINO, FEDERICA
COMINETTI, GIULIA GIUNTA, GIORGIA MANNAIOLI, VIVIANA
MASIA & EDOARDO LOMBARDI VALLAURI

NOT TO PLAY DEVIL'S ADVOCATE, BUT... 67
MARÍA INÉS CORBALÁN & GIULIA TERZIAN

VISUAL ARGUMENTATION IN THE FRAMEWORK OF THE 81
DUAL-INFERENCE SYSTEM
HÉDI VIRÁG CSORDÁS & ALEXANDRA KARAKAS

INSINUATED VS. ASSERTED AD HOMINEM: AN 93
EXPERIMENTAL APPROACH TO THEIR RHETORICAL
EFFECTIVENESS

Daniel de Oliveira Fernandes, Steve Oswald &
Pascal Gygax

Cognitive Machine Argumentation 105
Emmanuelle Dietz, Antonis Kakas & Adamos Koumi

Sinister interest as argumentational vice. 119
Bentham's Handbook of Political Fallacies and
Virtue Argumentation Theory
Iovan Drehe

Two Points for a Feminist View on Adversariality 125
Lucija Duda

Minimal Argumentation: a Research Program 139
Michel Dufour

Frames and Inferences 151
Isabela Fairclough

The reasonableness of fallacy accusations: An 165
exploratory study
José Ángel Gascón

The Argument's the Thing ... 179
Geoffrey C. Goddu

Two Subtypes of Illocutionary Acts of Arguing 191
Amalia Haro Marchal

Forget the Toulmin Scheme, Remember the 203
Epicheireme!
Mika Hietanen

Cultivating Normative Terrains in Information 217
Ecologies
Beth Innocenti

QUALITATIVE AND QUANTITATIVE EVIDENCE FOR LINGUISTIC AND DISCURSIVE FEATURES OF REPHRASE KONRAD KILJAN & MARCIN KOSZOWY	229
CONNECTIONS BETWEEN AGE AND INTERPERSONAL ARGUING IN UKRAINE, WITH SPECULATIONS ABOUT WAR'S EFFECTS IRYNA KHOMENKO, CRISTIÁN SANTIBÁNEZ & DALE HAMPLE	243
DISAGREEMENT ON REDDIT: AN EMPIRICAL STUDY ZLATA KIKTEVA & ANNETTE HAUTLI-JANISZ	259
THE CLASSIFICATION AND RECONSTRUCTION OF AUDITORY ARGUMENTS GABRIJELA KIŠIČEK	273
GAMING THE COGNITIVE PRINCIPLE OF RELEVANCE IN SOCIAL MEDIA MANFRED KRAUS	289
EVOCATION OF RELEVANT QUESTIONS: HOW DOES THIS WORK? LEONARD KUPŚ & MARIUSZ URBAŃSKI	301
THE PARABLES OF JESUS AS ANALOGICAL ARGUMENTATION – A CASE IN POINT: TEN BRIDESMAIDS NIILO LAHTI	313
EMPLOYING ARGUMENT MINING FOR REASON-CHECKING JOHN LAWRENCE & JACKY VISSER	328
NORMAL VS. INTENSE SCRUTINY: DISTINCT MODES OF CRITICAL THINKING? LAWRENCE LENGBEYER	340
AUDIENCE: A CENTRAL CONCEPT IN SOCIAL ARGUMENTATION	353

Jiaxing Li

**On the Reducibility and the Irreducibility of
Analogical Arguments**   367
Yan-Lin Liao

**Resolving Open-Textured Rules with Templated
Interpretive Arguments**   379
John Licato, Logan Fields & Zaid Marji

INTRODUCTION

FABIO PAGLIERI
Istituto di Scienze e Tecnologie della Cognizione, Consiglio Nazionale delle Ricerche (ISTC-CNR), Roma, Italy
fabio.paglieri@istc.cnr.it

ALESSANDRO ANSANI
Centre of Excellence in Music, Mind, Body and Brain, University of Jyväskylä, Finland
alessandro.a.ansani@jyu.fi

MARCO MARINI
Istituto di Scienze e Tecnologie della Cognizione, Consiglio Nazionale delle Ricerche (ISTC-CNR), Roma, Italy
marco.marini@istc.cnr.it

The European Conference on Argumentation (ECA) is an academic initiative launched in 2013 by a group of Europe-based argumentation scholars, with the aim of inaugurating a series of biennial international conferences on this thriving area of studies (for details, see https://ecargument.org/), to complement other large-scale events on similar or related topics: the conference of the International Society for the Study of Argumentation (ISSA), the conference of the Ontario Society for the Study of Argumentation (OSSA), the conference on Computational Models of Argument (COMMA), the Rhetoric in Society conference and, more recently, the events organized by the Argumentation Network of the Americas (ANA). The first edition of ECA took place in Lisbon (PT) in 2015, followed by a second one in Fribourg (CH) in 2017 and a third one in Groningen (NL) in 2019: then the Covid-19 pandemic struck and the next edition of ECA had to be postponed to 2022, when it took place in Rome, from September 28 to September 30.

The conference lasted two days and a half, with a very intense and diverse programme, including 3 keynote talks, 1 plenary panel, 16 long papers with invited commentators, and as many as 118 regular papers (14 of which were presented as part of 3 thematic panels). Most of these contributions are collected in written form in these three volumes, as follows:

- *Volume 1* includes 2 of the 3 keynotes presented at the conference, authored by Catarina Dutilh Novaes and Harvey Siegel, followed by 16 long papers, ordered alphabetically by first author's surname: the majority of those (9 out of 16) are accompanied by their respective commentary.
- *Volume 2* includes 30 regular papers, ordered alphabetically by first author's surname, from Aikin & Casey to Licato et al.
- *Volume 3* includes the remaining 30 regular papers, again ordered alphabetically by first author's surname, from Liga to Zemplén & Tanács.

Even though these proceedings cover only a selection of what was presented at the ECA conference in Rome, they provide a faithful approximation of the breadth and depth of the ongoing discussion in argumentation scholarship. They also attest the interdisciplinary character of this field: this has been the hallmark of argumentation studies since their inception, yet the disciplines brought to bear on this subject matter have steadily increased over the years; whereas philosophy and linguistics were always partners in this endeavor, nowadays they are supported also by computer science and experimental psychology, as well as communication and media studies in a broader sense. At the same time, the study of argument from a philosophical perspective is no longer regarded as a specialistic niche for philosophers with a grudge against deductive logic as a model of human reasoning, but it is taking back its place as a central concern for philosophical inquiry in general, as it was at the dawn of the discipline (Aristotle's work is an obvious example).

These are welcome developments in the natural evolution of argumentation studies, which the ECA initiative has always intended to promote and nourish: thus, we expect to see more of the same in future editions of ECA, starting with ECA 2025 in Warsaw, Poland, on "Argumentation in the digital society".

META-ARGUMENT AND SPECIOUS ALLEGATIONS OF FALLACY

SCOTT AIKIN
Vanderbilt University
scott.aikin@vanderbilt.edu

JOHN CASEY
Northeastern Illinois University
j-casey1@neiu.edu

Abstract
Accusations of fallacy are specious when they either inaccurately represent the target argument or are incorrect evaluations of their argumentative success. The burden of successfully making such a case is a meta-argumentative burden, and there are not only inferences about the target argument that hang on these meta-arguments but inferences about the debate more widely and the arguers themselves that are often made. Consequently, accusations of fallacy are not only meta-argumentative moves bearing on particular arguments, but they provide meta-evidence about the broader argumentative context. So specious allegations of fallacy are sites for not only focused error about a particular argument but also broader errors about arguers and the ongoing exchange.

1. Some stage-setting

In a cartoon by artist and activist Sabrina Symington (in *Life of Bria Comics*), a mild-mannered fellow, Derek Beardguy, reads about the straw man fallacy. This knowledge turns him into an internet supervillain, STRAWMAN. He possesses, with this knowledge, "amazing argumentative powers that allow him to win ANY debate with the wave of his hand." In the next pane, he waves off a critic with his reply, "Ha! That's just a flimsy STRAW MAN of my real, ingenious argument and not even worth refuting further!" This strategy is announced to have made him "invincible." The trouble, of course, is that his critic replies, "Actually, I was quoting you directly...," but all of that is ignored.[1]

Assuming that this comic reflects an accurate impression of how deploying the concepts of fallacy can derail instead of re-rail critical discussion, a few questions are worth asking. First, how does this strategy make Derek "invincible"? Second, it seems clear that there are both errors of correct fallacy-identification and argument correction here. So, for example, if the critic was actually quoting Derek, it seems his charge of straw man is less plausible. But there are consequent errors downstream for the critical discussion, since straw man accusations also paint pictures of the interlocutor and the quality of the discussion. Finally, there are questions as to whether errors on this level are theoretically unique, that is, whether there are *meta-argumentative fallacies*.

We will argue here that there are meta-argumentative fallacies, fallacies we can commit only when we reason about reasons or argue about arguments. Straw man arguments are exemplary of this kind of error (since one can straw man only when one reasons about other reasons or other views, so it's a meta-argument in and of itself).[2] But fallacy-accusation generally is explicitly meta-argumentative, so it's a place where these kinds of unique errors can occur.[3] We will argue here that there are two kinds of meta-argumentative error one can make in fallacy-accusation. First, there are the core errors one can make in accusation of fallacy, that of erring in representing or in criticizing another's reasons. These are familiar elements of fallacy-accusation and argument re-railing, that one bear a burden of proof for the fallacy-charge to stick and for the interlocutor to change their tack. Second, there are more global errors downstream from these charges – those that paint one's interlocutor as incompetent or argumentatively vicious or that depict the broader discussion as badly run and so invalid or, further, being more evidence that one is right. Given that these are broader argumentative moves, these global inferences have been under-theorized.

---

[1] The comic is archived at the following address: http://www.thealmightyguru.com/Wiki/index.php?title=File:Life_of_Bria_-_2017-02-20_-_Straw_Man_and_the_Freedom_Forum.png

[2] See Aikin and Casey (2022a) for a full discussion of this point. Straw-manning is unique as a fallacy, generally, because it takes place as an error of representing and criticizing reasoning, instead of an error in representing what one is reasoning about.

[3] See Aikin and Casey (forthcoming a and forthcoming b) for an overview of how meta-argument is a unique site for argument content with unique forms of error.

2. Two burdens of proof for fallacy-charges

Those who accuse others of committing fallacies must take on a burden of proof. They must (a) represent the reasoning they criticize and (b) show that the reasoning fits the form of an identifiable fallacy type or show how the argument fails some criterion for acceptability. So, for example, consider Arjun who argues that:

Since all communists hold that workers should be treated well and Carter holds that workers should be treated well, it follows that Carter is a communist.

It would be appropriate for Swathi to charge Arjun with a fallacy, that of asserting from the consequent. Swathi would reconstruct Arjun's argument as follows:

1. All communists hold workers should be treated well.
2. *So,* if Carter is a communist, he holds that workers should be treated well.
3. Carter does hold that workers should be treated well.
4. *So,* Carter's a communist.

Insofar as this accurately represents Arjun's reasoning, the problem is that steps 2 and 3 don't imply 4, because there are counter-examples to the form that the reasoning takes (for example, liberals are not communists, but hold that workers should be treated well). So, Swathi can argue something along these lines:

That's like reasoning that since all dogs have hair, Garfield is a dog because he has hair. Or that because all cubes have many sides, the Pyramids are cubes because they have many sides. That's a fallacy, friend. Called 'asserting the consequent.'

Of course, the examples may vary with audience-familiarity with Jim Davis cartoons or world heritage sites, and so they could be about numbers, planets, or famous pop singers. So long as the cases are accessible as counter-examples (true premises, false conclusion) and clearly mirror the form of the reasoning criticized, Swathi has lived up to her burden of proof for fallacy identification.[4]

So, fallacy-accusation, if dialectically appropriate, has a dual burden of proof – that of representational accuracy and critical manifestness. You've got to be right about what the core argument was, and you've got to be right about what's wrong with it in a way accessible to your interlocutor.

---

[4] See, for models of meta-argument with charges of fallacy or points of dialectical order Krabbe (2003), van Eemeren and Houtlosser (2009), Innocenti (2011), Balin and Battersby (2011), and Castro (2022). For analyses of how these meta-discussions can be badly run, see Linker (2014) and Innocenti (2022).

So, when we negotiate these burdens of proof, we are arguing about arguments. These are meta-arguments.

Meta-arguments are, to start, arguments about arguments. We use meta-arguments to clarify our first-order arguments, show their limits, explain how they show what they show, and who they are supposed to reach. We also use meta-arguments to evaluate arguments, so we ask whether this is a valid argument or whether it meets a degree of scrutiny that arguments in these situations must meet, and we identify who has the burden of proof and why. Finally, we use meta-arguments to assess the overall evidential situation with the issues over which we are deliberating, and so we ask: have we closed the issue, do we need more evidence, is further conversation with this group worthwhile, are we getting anywhere on this question, and is this argument worth the trouble?[5]

Fallacy-accusation opens a meta-argumentative stage in a critical discussion – in particular, it is a matter of evaluation of our first-order arguments. Our point here is that specious allegations of fallacy are meta-argumentative errors in the sense that they are errors of argument-evaluation. But these errors can prompt further meta-argumentative errors about the more global argumentative circumstance. Two examples of evaluative errors along the dual burdens of proof for fallacy-accusation will be occasions for more global meta-argumentative conclusions.

Again, the two fronts of the burden of proof for fallacy-accusation are that of correctly representing and then correctly criticizing the target argument, so there may be two types of error for this task. A TYPE 1 error is in argument reconstruction, namely, that the argument criticized is not the argument given. A TYPE 2 error is in argument evaluation, namely, that the argument is criticized for being deficient when it is in fact not deficient in the way identified. We will now turn to these two types of error, then then we will show how these can both occasion broader meta-argumentative error.

3. Type 1 error – Coulter's *ad hominem* charge

Ann Coulter is a famous American conservative political pundit and culture critic. The thesis of her 2002 book *Slander* is that *"ad hominem* attack is the liberal's idea of political debate" (2002: 12). The book has

---

5 The literature on meta-argument is growing, starting with the contributions from Cohen (2001) and Finocchiaro (2007 and 2013) being exemplary. Further developments of norms of meta-argument and sites for error are Linker (2014), Aikin (2020), Innocenti (2022), Godden (2022), and Aikin and Casey (2022a; 2022b; and forthcoming a and forthcoming b).

detailed accounts of many, many, liberal critics charging conservatives with racism, flirtations with totalitarianism, incuriousness, religious bigotry, and so on. A regular target for criticism from liberals at the time was then-President, George W. Bush. Coulter quotes many liberals directly, noting the regularity with which many on the left just call the President and his supporters "dumb" or "religious nutcases." She concludes:

> The "you're stupid" riposte is part of the larger liberal tactic of refusing to engage ideas [...] A vicious personal attack, they believe, constitutes clever counter-argument (2002: 53)

Now, conceding the evidence Coulter's book has amassed of liberals calling conservatives bad names ranging from 'dumb' to 'fanatic' to 'Nazi' to 'Christofacist,' the charge of *ad hominem* argument as the target reconstruction is underdetermined. There are two problems. The first is that the name-calling may not be the argument that Coulter thinks it is. The *ad hominem* interpretation requires that the name-calling function as a *premise*, so something along the following lines:

Premise: Conservatives are vicious
Conclusion: Conservative are wrong

That's the frame for Coulter's reconstruction, but that liberals are committed to the claim that conservatives are vicious doesn't mean this is a premise, but it could just as well be a conclusion. So, for example, take the fact that some conservatives may opine about many things (in the post 9/11 environment, particularly) that Muslims value and may pronounce on Islam's theology and political trajectory, but they yet cannot tell the difference between Shia and Sunni practitioners and states. This is a good reason to infer that this person is a religious bigot. So, it is a kind of *ad hominem* argument in the sense that it is against a person, but it is not *from* the person's moral character to the quality of their views. Rather, it's the other way around – they have ridiculous views and their intellectual character being bad is a plausible explanation for the terrible quality of their views. So, the fact of widespread character abuse is not necessarily evidence of the structure of a widespread form of fallacy, but it can be the result of widespread assessment of intellectual character.

Second, conceding (for the sake of argument) that Coulter is right that there are, in fact, many *ad hominem* abusive argumentative fallacies committed by the American left, one more problem still hangs for her case. Coulter takes these instances of egregiously bad argument as reflective of the best that liberals can do in argument on these matters. She says:

> Progress cannot be made on serious issues because one side is making argument and the other side is throwing eggs [...] Pevarication and denigration are the hallmarks of liberal argument. Logic is not their métier (2002: 3).

So, Coulter's attention to liberal *ad hominem* is not just because it is widespread, but because (as her case implies and she explicitly concludes above) it is the best that the liberals have done. The problem is that this is simply not the case. The liberal argument is not an *ad hominem* against conservatives, but a case for progressive values and a case of protecting liberties.[6] So, even if Coulter has accurately represented the liberals she criticizes, she has misrepresented the opposition's case by attending only to their worst arguments and takes them to be representative of the overall liberal case. This is a broad form of the straw man fallacy, one best termed the *weak man fallacy*.[7] So, it still fails the representational requirement for fallacy accusation (by assiduously following it for specially selected cases purporting to be representative), because it fails to represent the reasons of higher quality the general opposition makes. Even if Coulter is fastidious about properly documenting and directly quoting her examples, she still has misrepresented her opponent's case by only attending to their worst versions.

4. Type 2 error: Hillard's charge of question-begging

Mallory Ortberg was the lead writer for Slate's "Dear Prudence" advice column. Slate is an online magazine for progressive readers, so it regularly runs columns about labor politics, global warming, gender issues, and race relations. The "Dear Prudence" advice column, then, features advices for liberal to left-inclined readers navigating a country with many fraught issues: how to handle relatives who are anti-trans, friends who oppose gay marriage, and what to say about yourself on dating apps in overwhelmingly conservative states. Graham Hillard writes a culture column at National Review, which is a magazine for cultural and fiscal conservatives. NR runs stories on how government regulation stifles business growth and has opinion pieces on how (Christian) religion should be more prominent in daily and political life. Hillard reported in his culture column on Ortberg's "Dear Prudence" advice in Slate. In his report, Hillard tells us that Ortberg's advice to progressive readers who face scorn from their conservative family members and neighbors never has them pause to reconsider their own cultural orientation. There is never an argument for trans rights or that gay marriage is acceptable. Rather, these commitments are just assumed, and the advice is how to pursue

---

[6] Let's just say that Rawls's *On Justice* is not very plausibly read as a long *ad hominem* abusive argument.

[7] See Aikin and Casey (2022a) for a taxonomy of forms of straw man fallacy,

those goals or how to handle living among people who don't accept these values. The same, Hillard observes, is the case for most every wedge issue Ortberg's column addresses – the advice assumes that the culturally left are correct on these issues, so the advice is just about how to manage relations with people who are so wrong on these truths. It, he holds, casts a deeply distorted picture.

> The problem with these cubes of P.C. baloney [...] is that their cumulative effect is to move acceptable discourse (indeed, acceptable thought) ever leftward. *Because Ortberg makes pronouncements rather than arguments* when discussing the latest trends in gender and sexuality, the casual reader could be forgiven for believing that the argument has already happened somewhere, that the left won, and the only remaining thing to do is climb on board (2017, emphasis added)

Hillard is charging Ortberg with begging the question in assuming the correctness of these values and only reasoning from (instead of to) them. The charge is that of presuming one is right on a matter of controversy and proceeding as though the matter is settled. With matters on which there is disagreement, this argumentative omission is objectionable, and this is why the question-begging tag is applied.

Hillard has correctly identified an argument omission, given the controversy. But the problem is that charges of question-begging are audience-indexed. One may introduce premises in appropriate dialectical order only if one's audience finds them acceptable. If one's audience finds them acceptable, those premises are dialectically permitted (but they may yet require other support, but the problem of question-begging is no longer the problem).[8] So, for example, if A is trying to convince two audience members B and C to accept p, A may have an arsenal of arguments at their disposal: M, N, and O. Let's say B accepts premises only of M (and rejects those of both N and O), and C accepts premises only of O (and rejects M and N). A would beg the question with both A and B by giving N, and A begs the question with B in giving O and with C by giving M. The simple fact of the matter is that the issue at hand for Otberg's column are not that of resolving the controversies of the day but that of maintaining many relationships that those controversies of the day make complicated. The advice is about how to live by one's principles when they are controversial. Seen from this perspective, Ortberg's column is full of arguments: about the importance of consent, clarity with sexual partners about what one expects both in and outside of the bedroom, the need for boundaries with

---

8 The requirements of truth and epistemic justification are more demanding and, it should be said I, express aspirations for argument beyond pragmatic coordination. For the case for the more demanding requirements, see Kasser and Cohen (2002), Freeman (2005), and Godden (2015).

one's relatives, and so on. And none (quite so obviously) beg the question for an audience in need of resolution on those issues, but they do for those who reject the premises that there is noting obviously wrong with being gay or that reproductive rights are worthwhile. And given that it is an advice column in a liberal-to-left-leaning magazine, its primary audience is the former, not the latter. And so Ortberg's argument does not beg the question as properly assessed, as there is no argumentative lacuna in the case, because these questions, these deliberations, are not at issue for this audience in this case.[9] Hillard's error is that he's assumed that he and those of his political orientation are the primary audience for all of these communications, but he's (and they are) not. So expecting that "Dear Prudence" answer his objections to progressive lifestyle choices is out of order for the argumentative context of the column, and so his objection that the argument is short on argument is in error. In particular, it is a meta-argumentative error.[10]

5. Meta-argument and meta-evidence

Meta-arguments can clarify arguments, and they can then be used to assess the overall dialectical situation on the issue the arguments address. And when there are erroneous meta-arguments, there are downstream errors of collecting meta-evidence about the overall state of the evidence on the matter. Meta-evidence is evidence about evidence; it may show: that the available evidence is incomplete or complete, that it is inconsistent or consistent, that there are defeaters for the evidence or not, and that the evidence supports one option or not.[11] For example, consider the following cases:

> An expert's testimony that p is evidence that p, because it is evidence that there is evidence that p and that someone with knowledge and expertise has assessed the range of available evidence regarding p and has concluded that it supports p.

---

9 This is a pragmatic model for the analysis for begging the question. See for prominent defenses of this model: Rescher 1977, Walton 1989, and van Eemeren and Grootendorst 2004.

10 One way to capture the thought is that Hillard's column, because it is for right-leaning audiences, contains no explicit argument for conservative values, but rather proceeds from them. It does not thereby follow that his case begs the question, too. Or consider the simple fact that we (the authors) assume here that fallacies are argumentative moves that we should not commit, but we have not explicitly argued for that, have we?

11 For accounts of how critical dialogue doesn't just aggregate but displays meta-evidence, see Christensen (2009) and Feldman (2008; 2009).

Evidence that one's instruments (perhaps, for locating a lost phone) are malfunctioning is evidence about one's evidence from the instruments, because it is evidence that the evidence from the instruments is unreliable or even misleading.

Meta-arguments are instances of collecting and evaluating meta-evidence on a matter of deliberation.[12] So, if one were to take the expert testimony of another to be meta-evidence (and so, then, evidence) on an issue, criticisms of their reasoning would function as meta-evidence on their reliability. That the expert testifies that p is meta evidence, again, that the evidence supports p; however, as we look at the various reasons that the expert may give and see fallacies, we have new evidence that undercuts our assessment of the expert as reliable. So we have evidence that the evidence provided by their reports is weak, perhaps even defeated.

Call the following line of reasoning, from fallacy-detection to assessments of the overall dialectical situation, *a meta-argument loop*. The rough outline of the reasoning runs that one takes some arguments given by one's interlocutor and represents it for evaluation. On that representation, one detects a fallacy, and (again assuming the argument is representative of the interlocutor's general argument quality), we have reason to infer that the interlocutor's overall case is not good and that the exchange, if not well-run, is explained by the bad arguments proffered by the interlocutor. Again, assuming that these reasons given are representative, there is good reason to infer that there is no reasonable opposition on the matter. So, not only are the potential rebutting defeaters arising from *prima facie* peer disagreement eliminated, but if there are not good reasons against one's views, there are reasons to think the reasons for one's view are quite good.[13] That is, if the mendacious, ill-informed, or rationally incompetent are the only opponents, then it is good reason to think one is on the side of the issue is unproblematically right and good. This loop can be performed either on the basis of the reasons one's interlocutors give or on the basis of the character they show with their reasons. Either way, the meta-argumentative loop runs that because the opposition has no good case and have run critical dialogue poorly, one has

---

12 See Levy (2019), Green (2019), Aspietia (2020), and Aikin and Casey (2022b) for accounts of how argumentative meta-evidence can be either misleading or illuminating, depending on the circumstances.

13 A side note is that these charges and the meta-evidence is presented to a third party or onlooking audience and not the target for the crticitism. Rather, the audience for the meta-argument is already on the speaker's side on the issue. See Lewinski (2019), Aikin and Talisse (2019; 2020), and Aikin and Casey (2022a).

reason to not only think one's view is not truly challenged by theirs, but that it is in fact indirectly improved by their failures.[14]

Returning to our Fallacy-Accusation cases, Ann Coulter makes an inference to the best explanation from the persistence of liberal *ad hominem* argument: that liberals *know* that they have no real argument, but they nevertheless cling to their views and get ugly with anyone who sees through their charade. That liberals seem to be leaders of the best thinking on the issues is more a function of their power and skill at bullying conservatives:

> [Liberals] self-consciously hold themselves outside of the argument and make snippy personal comments about anyone who is actually talking about something [...] If it were true that conservatives were racists, sexists, homophobes, and fascists, shouldn't their arguments be easier to deconstruct? [...] Why the evasion? (2002: 13)

Coulter's charge of *ad hominem*, either as identifying the real argument given or by weak-manning the entire side by the bad arguments of a few, is taken as meta-argumentative evidence about the overall dialectical situation. The picture painted is that liberals actually flee a real exchange of ideas, and that conservative views are unscathed by liberal lines of argument. Further, whatever apparent expertise our intellectual standing liberals may have is undone by attending to the bad arguments they give and their evasion of substantive argumentative engagement with those who disagree. Finally, assuming evasion implies that the one evading exchange does so expecting defeat, we see that there is, by detecting the evasion, meta-evidence that improves the conservative side. In short: even liberals see how comparatively strong the conservative case is – that's why they avoid argument, and when they do argue, it's all character-assassination.

Graham Hillard argues in a similar meta-argumentative loop. He takes it that the conservative point of view is not only natural but obviously right, so there must be profound distortions in place for someone to maintain their progressive commitments and liberal politics. And for there to be apparent expertise by liberal writers. A grand dissimulation must have been put in place. This is what he points to when he says that this is the effect of making pronouncements rather than arguments – the left

---

[14] These cases are analogues to shifts in peer disagreement cases with higher order evidence. For example, consider the difference between people having a slight difference of opinion on a relatively simple math question and those with vast differences. In the latter case, there can be good reason to take the opponent's case is very weak, given what seems an obvious error. Kelly (2010) and Kornblith (2010) argue along similar lines, since there are bounds for what would count as a reasonable disagreement – those outside the bounds are no longer epistemic peers on the issue, so will offer no undercutting reasons in their disagreements.

avoids making arguments with the right because to do so would concede that there is controversy. But if that is conceded, the conservative view would get air time, and conservatives being right and obviously so would complicate things for the progressives. So the left avoids exchange, and the silence on the controversy, then, is misleading meta-evidence about the argument – that it happened already and the left won handily. The normalizing effect this has, then, is to shift the burden of proof onto conservatives. This is exactly backwards, thinks Hillard, given his assessment of the evidence:

> Regular people [...] know instinctively and by hard experience that to live as the sexual Left preaches is to enter a world of confusion, heartbreak, and deep, abiding dissatisfaction [...] *Dear Prudence* is dangerous because it does precisely what advice columns have always done: It shapes readers' sense of what is proper, what is expected, and what is owed. That, in doing so, it sneaks ahead of popular opinion while implicitly presenting itself as mainstream is, of course, exactly the point (2017)

Hillard's meta-argumentative loop runs that the question-begging argument from Ortberg is not only a studied avoidance of fair exchange but a long-term tactic of shifting perspectives on what the defaults are for critical discussion. So, at the end of this, requirements that conservatives have a burden of proof instead of allowances of presumption are evidence that the dialogue has been debased. In the end, Hillard's case makes conservatives more sure they are right, and what seems to be their ideas not even appearing in liberal discussions evidence that they are actually winning the debate.

6. Lessons

There are two lessons here. The first is a familiar one: namely, that fallacy-accusation opens a meta-argument. The accuser is the one with a dual burden of proof – that of representational accuracy and appropriate critical assessment. Some arguers can avoid the burden of proof altogether (as we saw with STRAWMAN) and those who shoulder it can fail along either front. The second lesson is that the narrow meta-argumentative context of evaluating an argument as a fallacy opens a wider meta-argumentative discussion of the overall state of dialectical play between the two interlocutors or the two sides of the debate generally. Allegations of fallacy, then, not only have narrow meta-argumentative import, but they have broad import, too. So, errors on the meta-level of particular arguments are then treated as meta-evidence about the state of the evidence, the performance of others in the argument, and what is really implied by the evidence. Specious allegations of fallacy have, then, not only local but potentially global meta-argumentative consequences.

References

Aikin, S. (2020). The Owl of Minerva Problem. *Southwest Philosophy Review*, 36, 12-22.
Aikin, S. and Casey, J. (2022a). *Straw Man Arguments*. London: Bloomsbury.
Aikin, S. and Casey, J. (2022b). Bothsiderism. *Argumentation*, 36, 249-268.
Aikin, S. and Casey, J. (forthcoming a). The Free Speech Fallacy as Meta-Argumentative Error. *Argumentation*.
Aikin, S. and Casey, J. (forthcoming b). Fallacies of Meta-argumentation. *Philosophy and Rhetoric*.
Aikin, S. and Talisse, R. (2019). *Why We Argue*. New York: Routledge
Aikin, S. and Talisse, R. (2020). *Political Argument in a Polarized Age*. London: Polity.
Aspietia, A. A. B. (2020). Whataboutisms and Inconsistency. *Argumentation*, 34, 433-97.
Balin, S. and Battersby, M. (2015). Fallacy Identification in a Dialectical Approach to Teaching Critical Thinking. *Inquiry: Critical Thinking Across the Disciplines*, 30, 9-16.
Castro, D. (2022). Argumentation in Suboptimal Settings. *Argumentation*, 36, 393-414.
Christensen, D. (2009). Disagreement as Evidence. *Philosophy Compass*, 4/5, 756-767.
Coulter, A. (2002). *Slander*. New York: Three Rivers Press.
Feldman, R. (2009). Evidentialism, Higher-Order Evidence, and Disagreement. *Episteme*, 6, 294-312.
Feldman, R. (2008). Epistemological Puzzles about Disagreement, in S. Hetherington (Ed.), *Epistemology Futures* (pp. 215-36). Oxford: Oxford University Press.
Finocchiaro, M. (2007). Arguments, Meta-Arguments, and Meta-Dialogues. *Argumentation*, 21, 253-268.
Finocchiaro, M. (2013). *Meta-Argumentation*. London: College Press.
Freeman, J. (2005). *Acceptable Premises*. Cambridge: Cambridge University Press.
Godden, D. (2015). Argumentation, Rationality, and Psychology of Reasoning. *Informal Logic*, 35, 135-166.
Godden, D. (2022). Getting out in front of the Owl of Minerva Problem. *Argumentation*, 36, 35-60.
Green, J. (2019). Metacognition as an Epistemic Virtue. *Southwest Philosophy Review*, 35, 117-129.
Hillard, G. (2017). Bad Advice: The Exquisite Political Correctness of *Slate's* 'Dear Prudence'. *National Review*. July 15. https://www.nationalreview.com/2017/07/advice-column-dear-prudence-slate-political-correctness-social-issues-leftward-reader-comments/
Innocenti, B. (2011). Countering Questionable Tactics by Crying Foul. *Argumentation and Advocacy*, 47, 178-88.
Innocenti, B. (2022). Demanding a Halt to Meta-Discussion. *Argumentation*, 36, 345-364.

Johnson, R. (2000). *Manifest Rationality*. Mahwah, NJ: Lawrence Erlbaum Publishers.

Kasser, J. and Cohen, D. H. (2002). Putnam, Truth, and Informal Logic. *Philosophica,* 69, 85-109.

Kelly, T. (2010). Peer Disagreement and Higher Order Evidence, in R. Feldman and T. Warfield (Eds.). *Disagreement* (pp. 111-171). Oxford: Oxford University Press.

Kornblith, H. (2010). Belief in the Face of Controversy,. in R. Feldman and T. Warfield (Eds.) *Disagreement* (pp. 29-52). Oxford: Oxford University Press.

Krabbe, E. (2003). Metadialogues," in van Eemeren (Ed.), *Anyone Who Has a View* (pp. 83-90). Kluwer.

Levy, N. (2019). No-platforming and Higher-Order Evidence. *Journal of the American Philosophical Association,* 5, 487-502.

Lewiński, M. (2019). Argumentative Discussion: The Rationality of What? *Topoi,* 38, 645-658.

Linker, M. (2014). Epistemic Privilege and Expertise in the Context of Meta-Debate. *Argumentation,* 28, 67-84.

Rescher, N. (1977). *Dialectics*. Albany: State University of New York Press.

van Eemeren, F. H. and Grootendorst, R. (2004). *A Systematic Theory of Argumentation*. Cambridge: Cambridge University Press.

van Eemeren, F. H. and Houtlosser, P. (2009). How Should One Respond to Fallacious Moves? *Argumentation and Advocacy,* 45, 198-206.

Walton, D. (1989). *Informal Logic*. Cambridge: Cambridge University Press.

COLLABORATIVE OPPOSITIONALITY, JUDGMENT REVISION, AND CRITICAL THINKING EDUCATION

SHARON BAILIN
Faculty of Education, Simon Fraser University, Vancouver Canada
bailin@sfu.ca

MARK BATTERSBY
Department of Philosophy, Capilano University, Vancouver Canada
mbattersby@criticalinquirygroup.com

Abstract

A primary goal of critical thinking education is teaching students to make reasoned judgments, which implies modifying one's judgments if the reasons are not adequate to support them. There is, however, evidence that a willingness to revise one's judgments when warranted is difficult to achieve. In this paper, we demonstrate how an inquiry approach to critical thinking education involving collaborative oppositionality can foster this virtue.

1. Introduction

A primary goal of critical thinking education is teaching students to make reasoned judgments. Although the role of sound argumentation in achieving this goal is widely recognized, insufficient attention has been paid to the role of modifying one's judgments when warranted by the evidence and arguments. Such a willingness to revise one's judgments in the face of compelling evidence, although central to epistemic improvement, has proven difficult to achieve. In this paper, we demonstrate how an inquiry approach to critical thinking education involving collaborative oppositionality can help to overcome the resistance to warranted judgment revision and can foster the virtue of willingness to revise one's judgments in the face of compelling evidence.

2. Judgment revision

Although this issue is sometimes characterized in terms of belief change, we find this description problematic. There are numerous conceptual, epistemic, and ethical questions surrounding the notion of belief change, including issues of how beliefs are acquired and change or are changed, whether our beliefs are voluntary, and whether it is ethical to try to change other's beliefs (Casey, 2020). Moreover, the term belief change may evoke a kind of wholesale and sudden conversion. Our conception of critical thinking centres not on belief but rather on judgment. We characterize the goal of critical thinking in terms of coming to reasoned judgments (Bailin & Battersby, 2016). An important feature of judgments is that we *make* judgments. And making judgments is an active process amenable to educational intervention. In addition, we have chosen to talk about revision rather than change to emphasize that the modification of judgments can have degrees and take place over time. We would argue that the focus of critical thinking education is not on trying to change students' beliefs. Rather, it is directed at the process by which students make and revise judgments, including guiding them in employing a judicious evaluation of competing positions and arguments to arrive at and modify their judgments.

3. Centrality of judgment revision

In focusing on judgment revision, we are referring to *warranted* judgment revision -- the revision of one's judgments when justified by the weight of evidence and arguments. Constantly modifying one's judgments without the judicious weighing of a range of arguments is not an epistemic virtue. Indeed, having what Aikin has called "a community of doxastic weathervanes, constantly changing their orientation depending on the last gust of argumentative wind" (Aikin, 2008, p. 571) would not be a desirable state of affairs. Moreover, changing one's judgments in the face of poor reasons is an epistemic vice.

Warranted judgment revision is implicit in how critical thinking is commonly conceptualized, as making judgments on the basis of reasons implies modifying one's judgments if the reasons are not adequate to support them and evidence and arguments point to better justified judgments. Many prominent conceptions of critical thinking explicitly emphasize the importance of warranted judgment revision, for example Lipman's description of critical thinking as self-correcting (Lipman, 2003) and Siegel's conception of critical thinking as involving not just reasoning well but being appropriately *moved* by reasons (Siegel, 1988). The willingness to modify one's own position is also considered an important virtue by many virtue argumentation theorists (e.g., Cohen, 2005; Aberdein, 2010). The literature on open-mindedness also points to a willingness to revise one's judgments when warranted as an important critical thinking virtue (Cohen, 2009; Hare, 2003).

The importance of warranted judgment revision is also implicit in arguments regarding the value of adversarial argumentation. One of the primary arguments in support of adversariality is that the strongest arguments will be put forth and the best view will prevail (Zarefsky, 2012). Yet the potential for such epistemic improvement arising from the rigorous confrontation of opposing views will be realised only if those holding less justified views are willing to revise their judgments (Bailin & Battersby, 2020).

Warranted judgment revision is also central to Mercier's and Sperber's argumentative theory of reasoning (Mercier & Sperber, 2017). According to this theory, the success of argumentation depends not only on the division of argumentative labour, but also on the idea that people will change their views based on the best arguments (Mercier, Boudry et al., 2017, p. 5). Moreover, one of the conditions of successful group deliberation, according to Mercier et al., is that participants feel free to *revise their own opinions* (Mercier, Boudry et al. 2017, p. 5).

4. Resistance to warranted judgment revision

Considerable research suggests, however, that beliefs and judgments, once formed, have remarkable tenacity (Lord, Ross, & Lepper, 1979). Moreover, people tend to resist revising their judgments even in the face of discrediting evidence (Jennings, Lepper, & Ross, 1981) and often become more entrenched in their original views when defending them against opposing arguments (Sloman & Fernbach, 2017).

4.1. Epistemic and cognitive factors

Among the factors contributing to resistance to judgment revision is a lack of awareness of the full range of argumentation on various sides of an issue, including objections and responses. Information silos provide information and arguments that reinforce existing beliefs and limit exposure to opposing views and evidence, as well as to objections to and counterarguments against certain positions. Moreover, the versions of opposing views that are seen in these contexts are often straw-person versions. The lack of exposure to credible versions of the opposing views and credible accounts of the whole debate and best arguments means that an individual may not see the arguments and evidence that should lead to judgment revision.

In order to revise judgments when faced with compelling evidence and arguments, one also needs to be in a position to recognize compelling evidence and arguments. Thus, obstacles to belief revision include the lack of understanding of the relevant epistemic criteria for evaluating the accuracy and force of evidence and reasons as well as a lack of understanding of the process for comparatively evaluating, weighing, and balancing opposing arguments.

Depending on the judgments involved, the lack of knowledge of the criteria for judging causal claims, for evaluating statistical arguments, and the role of consensus in validating scientific claims can prevent proper assessment and appreciation of the force of counterarguments and claims.

Given that so much of what we know depends on trusted sources, the understanding of source credibility and expertise is also crucial for identifying claims that require revision of one's judgments. To have a reasonable trust in scientific information and scientific experts requires understanding the crucial role that scientific consensus, peer review, peer criticism, experimental replication, and statistical inference play in the establishment and advancement of scientific claims. Failure to understand how these processes contribute to the self-correcting nature of science can lead people to deny the significance of scientific claims and refuse to acknowledge when they should provide a basis for judgment revision. It can also provide an excuse for not revising one's view in light of changing scientific views and recommendations.

Many complex issues can only be decided by comparatively weighing and balancing opposing arguments. This process requires a willingness and ability to access the opposing views and an understanding of the epistemic criteria for assessing the weight of opposing arguments. Moreover, lack of understanding of the process for comparatively evaluating, weighing, and balancing opposing arguments reduces the likelihood of people moving away from polarized positions on such complex issues to more appropriately nuanced judgments.

Myside bias can also contribute to resistance to judgment revision, leading to a biased evaluation of the weaknesses of one's own position and the strengths of opposing positions (Stanovich, 2011).

4.2. Psychological, social, and emotional factors

Such epistemic factors do not, however, account for the resistance to judgment revision which occurs even among those who have epistemic competence and access to credible opposing arguments (Kahan, 2013). Here, other factors come into play, factors of a psycho-social, emotional, and attitudinal nature.

One of these involves defensive biases. People tend to identify with their beliefs and so perceive challenges to their beliefs as threatening to their sense of self-worth. They are, as a consequence, motivated to protect their beliefs as a way of protecting their feelings of adequacy (Cohen, Bastardi, Sherman, McGoey & Ross, 2007) and so to resist judgments that challenge these beliefs.

Another manifestation of motivated reasoning is cultural cognition, which involves individuals maintaining a strong commitment to the particular judgments of their group as a way of expressing their group identity and solidarity. These individuals will tend to evaluate information in a selective pattern that reinforces their group's worldview and resist information and evidence which challenge the judgments of the group (Kahan, 2013). Modifying any of these judgments would entail considerable social and emotional costs.

4.3. Attitudinal factors

In addition to the preceding, there is set of factors affecting the willingness to revise one's judgments which we are calling attitudinal.

One kind of attitudinal factor relates to how the enterprise of coming to or revising judgments is understood. The goal of the enterprise is, at its core, an epistemic one - arriving at judgments best justified by available evidence and arguments. Yet "our practices of transacting reasons with one another" (Godden, 2021, p. 845), i.e., the activity of argumentation, is often framed in adversarial terms - as involving a confrontation between arguers arguing opposing positions (Govier, 1999) who have as their goal to win the argument. This win-lose framing disincentivizes judgment revision as such revision would involve acknowledging weaknesses in one's own argument and would thus be paramount to losing. Thus, the commitment to winning may eclipse the goal of coming to a reasoned judgment, undermining a willingness to concede to the strongest reasons and to revise one's judgments if warranted by the arguments (Bailin & Battersby, 2020).

The framing of the enterprise as a contest between two opposing positions may also result in the differences between positions being overemphasized and positions being stated with an unjustified level of confidence (Govier, 2020). Thus, the kind of in-between positions that would involve some modification of both positions might not emerge and the possibility for a more nuanced judgment that takes into consideration the strengths and weaknesses of the different positions and is held with a justified level of confidence is less likely to occur.

Another factor inhibiting judgment revision is excessive certainty, both intellectual and moral. When individuals are utterly certain of the rightness of their position, the likelihood of their being open to a serious consideration of opposing views and to modifying their own view is greatly diminished. Such certainty can likely be explained by a number of factors. One is the cognitive bias of overconfidence in which a person's subjective confidence in their judgments is greater than the objective accuracy of those judgments (Moore & Healy, 2008). Another is the common human tendency to feel discomfort with uncertainty (Dugas, Buhr, & Ladouceur, 2004) and so to express more confidence than is warranted by the situation. Another possible factor is epistemic. A lack of understanding of the fallible nature of inquiry and the evolving, self-correcting nature of knowledge can lead to a feeling of pride in maintaining one's judgments regardless of new evidence. It can also lead to criticisms of justified revisions of judgments by others and to the view that, therefore, their judgments cannot be relied upon.

The willingness to revise one's judgments may also be hampered by a lack of what Cohen has called *reason responsiveness* (Cohen, 2019). Although someone is exposed to arguments which are in opposition to their own view, they may not really listen to or hear those arguments. Thus, they will not be in a position to respond appropriately, if at all, to those arguments nor to modify their own view in response, if required. Being reason-responsive requires a willingness to listen to opposing views. The most egregious problem is when one refuses to listen at all. But one can also listen in an adversarial manner which does not promote reason-responsiveness, for example, listening only to object, to ignore, to call out, or to exploit (Davis & Godden, 2021). Really listening requires trying to understand why those making the opposing arguments hold the views that they do (Beatty, 1999). It is listening in this sense which is a prerequisite to the kind of serious consideration of opposing views which can lead to judgment revision (Cohen, 2005).

5. The role of critical thinking education in countering resistance

Given the centrality to critical thinking of revising one's judgments when warranted, addressing resistance to warranted judgment revision is an important task for critical thinking education. Critical thinking education is a particularly apt venue for such an endeavour as it is doubtless easier to foster a willingness to revise one's judgments among students, who are likely less committed to certain views, than among adults, who may already have strong positions.

What, then, does critical thinking education need to do? Fostering epistemic competence is certainly an important task as students need to be in a position to recognize compelling evidence and arguments that require a revision to their current judgments. But such competence must include learning a variety of criteria and norms that go beyond the usual norms of argument evaluation typically taught in critical thinking courses, including criteria for evaluating a variety of types of judgments (e.g., causal claims, statistical arguments) as well as criteria for evaluating the credibility of sources.

Since warranted judgment revision requires exposure to, and examination of, the full range of argumentation on various sides of an issue, students also need to learn to search outside their information silos for credible versions of the competing arguments and also to acquire competence at comparatively evaluating these various arguments and adjudicating the debate. Crucially, they need to recognize when the process of inquiry supports judgments that are contrary to their existing views, and judgment revision is required. Moreover, an understanding of the process for comparatively evaluating, weighing, and balancing opposing arguments increases the likelihood of students moving away from polarized positions on such complex issues to more appropriately nuanced judgments.

Given that resistance to judgment revision occurs even among those who have epistemic competence and access to credible opposing arguments, it is vital for education in critical thinking to address some of the additional factors which give rise to such resistance. One of these is the adversarial framing of argumentation, which construes the goal of argumentation as winning and thus frames changing one's view as losing. It is important for critical thinking instruction to promote an understanding of the project of critical thinking as an epistemic one with the common goal of epistemic improvement. Students need to come to understand that the exchange of reasons, in this context, is not a zero-sum game with winners and losers but rather that all parties are better off if we make better judgments. Such an understanding can help to foster a more collaborative, inquiry approach to critical thinking, with the willingness to revise one's judgments as warranted seen as an integral aspect of the process.

Critical thinking education also needs to foster an understanding of the fallible nature of inquiry and the evolving nature of knowledge. A recognition that our judgments constitute a moment in the ongoing process of inquiry implies that judgments are not final and need to be revisited and often revised, and that judgment revision often takes place over time. Thus, critical thinking education needs to foster in students the habit of ongoing reflection on their judgments and an understanding that revising one's judgments when warranted is not a sign of lack of knowledge or intellectual weakness. Rather, the willingness to revise judgments is how our knowledge progresses. Understanding this is vital for countering overconfidence and the negative assessment of judgment revision with respect to both the judgments of others and to one's own judgments. Such an understanding may also help to counter defensive bias as one comes to realize that being wrong is a normal part of inquiry.

Addressing the psychological and social obstacles should also be an essential aspect of critical thinking education, although it is one that is rarely a focus of critical thinking instruction. Critical thinking involves not only an understanding of how to evaluate competing arguments, but also a commitment to base one's judgments and actions on inquiry, what Siegel (1988) has called the critical spirit and we have called the spirit of inquiry (Bailin & Battersby, 2016), as well as concomitant virtues which are grounded in that commitment, including fair-mindedness, a willingness to listen, a willingness to follow arguments where they lead, and, importantly, intellectual humility. These virtues are central to judgment making and judgment revision. Thus, critical thinking education must develop the competencies and understandings described above in the context of settings, structures, and relationships which foster these virtues.

6. Dialectical inquiry and collaborative oppositionality

The approach to critical thinking education which we believe can best address these challenges is dialectical inquiry (Bailin & Battersby, 2016). The goal of dialectical inquiry is to come to a reasoned judgment on complex issues. This is accomplished through students working collaboratively to comparatively evaluate opposing views within a collaborative framework - what we call collaborative oppositionality (Bailin & Battersby, 2020) - and in the context of a community of inquiry.

This process ensures an exposure to a range of positions and opposing arguments, including those which might warrant a revision to current judgments. Such an exposure can also serve to counter myside bias. A useful heuristic in this regard is a dialectical argument table which represents the debate on the issue, including the arguments pro and con as well as objections and responses. Moreover, students must research the conflicting arguments that have actually been presented and not rely on possibly straw-person versions that may have been offered by opponents of the view.

Through learning epistemic norms as well as considering the context of the debate, students are in a position to come to a reasoned judgment through a comparative evaluation of the relative strengths of the various arguments in the overall case. Students also learn that the best justified judgment may not consist in a total acceptance of one or other of the opposing sides of the debate as framed but, rather, involve a modification, qualification, or combination of the positions.

In coming to their judgment, there is also a requirement that students specify the level of confidence warranted by the overall adjudication and weighing of the evidence and competing arguments (Bailin & Battersby, 2015a). This requirement helps students to learn to calibrate their views and counters the tendency to overconfidence.

A central aspect of collaborative oppositionality is that, while it involves a confrontation of opposing views, students work collaboratively to come to reasoned judgments, a goal which is emphasized throughout. Thus, the epistemic gains that can arise from an evaluation of conflicting views can be achieved without the negative aspects of adversariality, i.e., without undermining a willingness to concede to the strongest reasons and to revise one's judgments if warranted. Collaborative oppositionality fosters an epistemic orientation, emphasizing the goal of arriving at the best judgments.

It is important to point out that our view, focusing on collaborative oppositionality, differs substantially from a cooperative view of argumentation of the kind attributed to us by a number theorists (e.g., Stevens & Cohen, 2019; Godden, 2021; Govier, 2021). Stevens (2015) characterizes the cooperative view, exemplified by Foss & Griffin (1995), as asking arguers to learn from each other's perspectives, improve each other's claims and arguments, and develop the weak points in the other's arguments. The aim, according to Stevens, is not to test claims but rather to enable others to develop their views further. Collaborative oppositionality, in contrast, *does* focus on testing claims through a confrontation and comparative evaluation of conflicting positions and arguments. Thus, referring to our view as an example of a cooperative view is inaccurate. We agree that some aspects of Foss and Griffin's approach, e.g., really listening to and attempting to understand the view of someone with whom you disagree, has an important role to play in pedagogy. It is listening in this sense which is a prerequisite to the kind of serious consideration of opposing views which can lead to judgment revision (Cohen, 2005). Thus, we view this kind of listening as a prerequisite to or aspect of testing. We do not view it as an alternative. Evaluating claims and arguments and making judgments are central to collaborative oppositionality.

In addition, the competencies and understandings described above must be developed in the context of settings, structures, and relationships which foster the spirit and virtues of inquiry. Such a culture of inquiry can be achieved through the creation of a community of inquiry in the classroom (Lipman, 2003). This is a community which is centred on the practice of inquiry and instantiates its norms and in which the epistemic goals of argumentation and the essentially collaborative nature of the enterprise are emphasized. Such a community is characterized by and values open-minded and fair- minded exchanges, rigorous but respectful critique, following arguments where they lead, and revising one's judgments when justified by the evidence and arguments (Bailin & Battersby, 2015, 2016).

Creating such a community requires a classroom centred on student interaction and collaborative inquiry. Students frequently engage in group interaction, discussing, questioning, challenging, and critiquing. They engage in collaborative inquiries, using a collaborative oppositionality approach to jointly research, evaluate, debate, and come to a joint judgment. They also engage in individual inquiries, conducting the inquiry in stages using the techniques of collaborative oppositionality, working in groups to get critique from peers and the instructor at each stage (Bailin & Battersby, 2017).

The use of groups is a central aspect of the approach. There is considerable evidence that well-functioning groups can more reliably come to well-reasoned judgments than can individuals on their own (Mercier & Landmore, 2012). By expanding the range of arguments to be considered, group inquiry can expose students to new arguments that may require a modification to their current views. Member of groups can also often compensate for the cognitive biases of others in the group, for example myside bias, which can interfere with warranted judgment revision (Sunstein, 2006). These epistemic gains appear to be limited to groups which are heterogeneous[1] and which reflect the ethos of rational inquiry and instantiate its norms and practices. In addition, participants must feel free to express their views, to critique the views of others, and to *revise their own opinions* (Mercier, Boudry et al. 2017, p. 5). The virtue of willingness to revise one's views where warranted is thus both required for, and can be further cultivated through, the critical interchange involved in collaborative inquiry.[2]

It is also important, in fostering an understanding of inquiry as an epistemic enterprise, to engage students in a discussion about the nature of collaborative inquiry and about appropriate expectations among students and between students and the instructor when engaging in collaborative inquiry (Bailin & Battersby, 2017). Another important aspect is instructor modelling of the virtues of inquiry, including an openness to altering their own position if warranted.

A community of inquiry can also play a role in countering the psychological and social obstacles listed previously. Such a community can mitigate defensive biases in that it is a community in which value is placed not on maintaining a commitment to a particular view but rather on being reasonable and in which students feel free to and encouraged to revise their views when warranted. A community of inquiry can also help to address the challenges posed by cultural cognition by creating a community of affiliation as an alternative to or counter-balance to one's cultural community. In a community of inquiry, group identity is constituted not by a commitment to specific beliefs and judgments but rather by adherence to the norms of rational inquiry.

---

[1] For strategies to inject further diversity of views into groups discussion, see Bailin & Battersby (2015b, 2020).
[2] For additional pedagogical strategies, see Bailin & Battersby (2020).

7. Conclusion

Although our inquiry approach to critical thinking education was developed with a focus on critical thinking courses, we believe that it is applicable throughout the curriculum and across all educational levels (Bailin & Battersby, 2021; Battersby & Bailin, 2015; Battersby & Bailin, 2021). Competence in making reasoned judgments is central in most areas as is an understanding of the nature and goals of inquiry. But, crucially, education in all contexts should aim to foster the spirit and virtues of inquiry, including a commitment to making judgments on the basis of reasons, and, importantly, to revising these judgments when warranted.

References

Aberdein, A. (2010). Virtue in argument. *Argumentation*, 24(2), 165–179.
Aikin, S. (2008). Holding one's own. *Argumentation*, 22, 571-584.
Bailin S. & Battersby, M. (2015a). Argumentation, degrees of confidence, and the communication of uncertainty. In F.H. van Eemeren & B. Garssen (Eds.), *Reflections on Theoretical Issues in Argumentation Theory* (pp. 71-82). Springer.
Bailin S., & Battersby M. (2015b). Fostering the virtues of inquiry. *Topoi* 35(2), 367-374.
Bailin, S. & Battersby, M. (2016). *Reason in the balance: An inquiry approach to critical thinking* (2nd ed.). Cambridge, Mass: Hackett.
Bailin, S. & Battersby, M. (2017). What should I believe? Teaching critical thinking for reasoned judgment. *Teaching Philosophy*, 40(3), 275-295.
Battersby, M. & Bailin, S. (2017). Teaching critical thinking as inquiry. In M. Davies & R. Barnet (Eds.), *Palgrave handbook of critical thinking in higher education* (pp. 123-138) New York: Palgrave.
Bailin, S. & Battersby, M. (2020). Is there a role for adversariality in teaching critical thinking? *Topoi* 40(5), 951-961.
Bailin, S. & Battersby, M. (2021). Infusing critical inquiry in the elementary curriculum. Presentation at the Southwestern Alberta Teachers' Association Conference, Feb. 19, 2021. https://www.youtube.com/watch?v=73j8tBzmVC4
Battersby, M. & Bailin, S. (2021). Infusing critical inquiry in the secondary curriculum. Presentation at the Southwestern Alberta Teachers' Association Conference, Feb. 18, 2021.
Beatty, J. (1999). Good listening. *Educational Theory*, 49, 281–298.
Casey, J. (2020). Adversariality and argumentation. *Informal Logic*, 40(1), 77–108.
Cohen, D. (2005). Arguments that backfire. In D. Hitchcock (Ed.), *The uses of argument: Proceedings of the 6th Ontario Society for the Study of Argumentation (OSSA) Conference* (58–65). Hamilton ON: OSSA.
Cohen, D. (2009). Keeping an open mind and having a sense of proportion as virtues in argumentation. *Cogency* 1(2), 49-64.
Cohen, D. (2019). Argumentative virtues as conduits for reasons causal efficacy: Why the practice of reasons giving requires that we practice hearing reasons. *Topoi* 38(4), 711–718.

Cohen, G. L., Bastardi, A., Sherman, D. K., McGoey, M., Hsu, L., & Ross, L. (2007). Bridging the partisan divide: Self-affirmation reduces ideological closed-mindedness and inflexibility in negotiation. *Journal of Personality and Social Psychology, 93,* 415-430.

Davis, J. & Godden, D. (2021). Adversarial listening in argumentation, *Topoi* 40(5): 925–937.

Dugas, M. J., Buhr, K., & Ladouceur, R. (2004). The role of intolerance of uncertainty in the etiology and maintenance of generalized anxiety disorder. In R. G. Heimberg, C. L. Turk, & D. S. Mennin (Eds.), *Generalized anxiety disorder: Advances in research and practice* (pp. 143–163). New York: Guilford Press.

Foss, S. K. & Griffin, C. L. (1995). Beyond persuasion: A proposal for an invitational rhetoric. *Communications Monographs,* 62(1), 2–18.

Godden, D. (2012). Rethinking the debriefing paradigm: The rationality of belief perseverance. *Logos & Episteme,* 3(1), 51-74.

Govier, T. (1999). *The philosophy of argument.* Newport News: Vale Press.

Govier, T. (2020). Opposition and polarization. In J.A. Blair & C. Tindale (Eds)., *Rigour and reason: Essays in honour of Hans Wilhelm Hansen* (pp. 87 – 110). Windsor: Windsor Studies in Argumentation.

Govier, T. (2021). Reflections on minimal adversariality, *Informal Logic,* 41(4), 523-537.

Godden, D. (2021). The compliment of rational opposition: Disagreement, adversariality, and disputation. *Topoi,* 40(5), 845–858.

Hare, W. (2003). Is it good to be open-minded? *International Journal of Applied Philosophy,* 17(1), 73–87.

Jenings, D. L., Leper, M. R., & Ross, L. (1981). Persistence of impressions of personal persuasiveness: Perseverance of erroneous self-assessments outside the debriefing paradigm. *Personality and Social Psychology Bulletin,* 7, 257-263.

Kahan, D. M. (2013). Ideology, motivated reasoning, and cognitive reflection. *Judgment and Decision Making,* 8(4), 407-424.

Lipman M. (2003). *Thinking in education* (2nd ed.). Cambridge: Cambridge University Press.

Lord, C. G., Ross, L., & Lepper, M. R. (1979). Biased assimilation and attitude polarization: The effects of prior theories on subsequently considered evidence. *Journal of Personality and Social Psychology,* 37(11), 2098-2109.

Mercier, H. (2016). The argumentative theory: Predictions and empirical evidence. *Trends in Cognitive Sciences,* 20(9), 689-700.

Mercier, H., Boudry, M., Paglieri, F., Trouche, E. (2017). Natural-born arguers: Teaching how to make the best of our reasoning abilities. *Educational Psychologist* 52(1), 1–16.

Mercier, H. & Landemore, H. (2012). Reasoning is for arguing: Understanding the successes and failures of deliberation. *Political Psychology,* 33(2), 243-258.

Mercier, H. & Sperber, D. (2017). *The enigma of reason.* Cambridge, Mass: Harvard University Press.

Moore, D. A., & Healy, P. J. (2008). The trouble with overconfidence. *Psychological Review,* 115(2), 502–517.

Siegel, H. (1988). *Educating reason: Rationality, critical thinking, and education.* New York: Routledge.

Sloman, S. & Fernbach, P. (2017). *The knowledge illusion: Why we never think alone.* Penguin.

Stanovich, K. (2011). *Rationality and the reflective mind.* Oxford: Oxford University Press.
Stevens, K. (2016). The virtuous arguer: One person, four roles, *Topoi,* 35(2), 375–383.
Stevens, K. & Cohen, D. (2019). The attraction of the ideal has no traction on the real: On adversariality and roles in argument. *Argumentation and Advocacy,* 55(1), 1–23.
Sunstein, C. R. (2006). *Infotopia: How many minds produce knowledge.* Oxford: Oxford University Press.
Zarefsky, D. (2012). A challenge and an opportunity for argumentation studies. *Argumentation and Advocacy,* 48(3), 175-178.

META-ARGUMENT AND PARA-ARGUMENT

JOHN CASEY
Northeastern Illinois University
j-casey1@neiu.edu

SCOTT AIKIN
Vanderbilt University

Abstract

Sometimes we argue about cats, or about whether there is a largest prime number. Other times we argue about arguments. When we do this, we engage in meta-arguments. There are many different ways to meta-argue. For example, we meta-argue when we call attention failures in ground-level arguments by employing the meta-argumentative concepts of burden of proof, or fallacies. In this case, our meta-arguing has bearing on ground-level arguing. We engage in meta-argument also when we argue about arguments qua arguments, as Dan Cohen (2001) has observed. We may, for instance, invoke a meta-argumentative principle to resist a troublesome conclusion of what appears to be an otherwise good argument (MARGA = Meta-Argument for Resisting Good Arguments). Take a complicated mathematical proof that 1=0. The steps may exceed the ability of the addressee to assess, so they invoke a principle of meta-rationality to resist what seems like a false conclusion. In this paper, we consider a variation on this form of meta-argument. We call it the meta-argument for making bad arguments (MAMBA = Meta-Arguments for Making Bad Arguments). MARGA-style reasons and their variations are principles invoked by addressees, by contrast, MAMBA reasons are invoked by arguers. MAMBA reasoning comes in various versions, but we identify two: strong and weak. Strong MAMBA licenses one to make what we will call para-argumentative moves. These are non-argumentative communicative acts relevant to the argument. Weak MAMBA is a principle allowing one to make merely bad arguments. Examples of para-argumentative communication include sarcastic abuse, ridicule, ad baculum threats; examples of weak MAMBA include merely fallacious (e.g., straw men) or weak arguments directed at certain addressees. In addition to identifying the forms, we assess the costs and opportunities of employing MAMBA reasoning..

1. Introduction

Sometimes we argue about cats or about whether there is a largest prime number. It often happens that in making progress on how many cats to buy or whether there's a largest prime number, we have to settle questions *about* how we're settling the question, about arguments in other words. These further questions might include who has the burden of proof, or whether someone has committed, say, a fallacy of appeal to unqualified authority. Broadly speaking, to have recourse to the burden of proof or to the language of fallacies is to engage in meta-argument, argument about arguments. It's very alarming, but perhaps not surprising, how much of our everyday arguing is actually arguing about arguing. After all, argument is constitutively normative: an argument is an argument to the extent that it follows, or purports to follow, a particular set of norms. It's not just a bunch of sentences with an illative word; it's an allegation that a norm of some kind applies, a norm that requires the addressee to do something (and that the speaker alleges to have followed). To this extent, argument is meta-argumentative in a weak sense (Aikin and Casey 2022b).

But argument is not only meta because it has to have recourse to its own norms to achieve its goals. There are stronger ways in which it is so. In the presence of an array of arguments on an issue, for example, we might draw conclusions about the certainty of some particular conclusion. Since so many people are arguing about something, the conclusion must be unknown, difficult to know, or at some point along the middle between the debating extremes. This is a kind of meta-arguing that comes in the wake of other arguments. Call these retrospective meta-arguments. We intend to argue here, however, that one might also meta-argue prospectively. One could, for example, meta-argue by making bad arguments, or, relatedly, by making no arguments at all.

Our purpose in this paper is to investigate some cases of meta-arguments for making bad arguments, which we will, after Cohen (2001), MAMBA. We will draw a distinction between weak and strong MAMBA. The paper proceeds as follows. We first go over some of the basics of meta-argument in the literature, to get the appropriate concepts on the table, in particular retrospective and prospective meta-arguments, which are our focus. We then introduce and discuss MAMBA, drawing a distinction between strong and weak versions. We conclude with some brief remarks on the normative question for such arguments.

2. Meta-argument

Considering the centrality of meta-arguing to the concept of argument, there is a notable dearth of literature on the general concept. Exceptions to this are Finocchiaro, Cohen, Breakey, and Aikin and Casey. These accounts of meta-argument in this literature vary with the shifting meanings of the prefix meta. The term meta is a relational preposition (notoriously and hilariously ambiguous at times). Aristotle's Metaphysics was so named because it dealt with matters above physical things, because it handled questions in the wake of physical things, and, more prosaically, it was a series of books placed on the shelf after the books on physical things (the Physics).

For Finocchiaro ((2013), see also (2005) and (2007) for extended discussions) to meta-argue is just to argue *about* arguing. This means to argue about argument as a *topic* of argument (Finocchiaro, 2013, p. 1). There's clearly more to meta-arguing than this, for, as in the examples just given, we argue about arguing in order to clarify something in the first order (e.g., as above, to settle who has the burden of proof or whether a fallacy was committed). To argue about arguing involves more than just arguing about arguing as a subject. That's just arguing (see Aikin and Casey (2022b). Meta-arguing, should it have any bearing *on* other arguments, ought to be something *relational*.

The relational sense of meta-argument can be found in Cohen (2001) Breakey (2021) and again Aikin and Casey (2022c, 2022a, 2022b, 2023). Severally, they maintain that meta-arguments are particular kinds of arguments, namely those that come *in the wake* of prior arguments and, most importantly, have a unique kind of *bearing on them*. This bearing is, in some sense we'll define, *evaluative*. A useful comparison is meta-theater. This is more than just theater *about* theater, plays about actors, directors, and such; it's theater that makes a commentary on theater. "Breaking the fourth wall," where actors in a play address the audience and comment on the action, performs a similar role.

For Breakey (2021), a meta-argument consists in critiquing a ground-level argument not by questioning the truth of premises or their logical connection to the conclusion, as is the practice taught in many informal logic textbooks, but rather by directing "attention to other properties of the target argument, such as its effects in a given context." Certain kinds of arguments, for example, may cause offense to a particular audience. To critique an argument on these grounds, irrespective of whether the argument qua argument is any good, is to bring meta-argumentative considerations to bear on its evaluation.

It is sometimes said that arguing is what we do when the way forward is blocked by disagreement; you see it one way, we see it another, so we work it out (Godden 2019). In the course of working it out, we'll disagree about what the argument shows, so we turn to meta-argument. Cohen (2001) maintains that the shift to meta-argument may also occur when something seems off about our arguments. While we turn to arguments to satisfy the principles of rationality, in argument we must satisfy a principle of meta-rationality, which is to say that to reason rationally is to reason *about* rationality (Cohen, 2001, p. 78). What goes for reasoning goes for arguments: to argue rationally is to argue *about arguing* rationally. Cohen describes four symmetrical meta-argumentative scenarios, all centered on the general sense that something is either right (or wrong) with our arguments. Here they are:

> (MARGA): Meta-Argument for *Resisting* Good Arguments:
> (1) This argument seems cogent but has an unreasonable conclusion;
> (2) Cogent arguments do not lead to unreasonable conclusions;
> So, (3) this argument must, in some way, be fallacious, I.e., it must fail in some way--even if I do not yet see how or why.
> (MAABA): A Meta-Argument for *Accepting* Bad Arguments:
> (1a) The argument as it stands seems fallacious, but it has a reasonable conclusion;
> (2a) All reasonable conclusions can be supported by cogent arguments;
> So, (3a) the argument can be made cogent--even if I do not yet see how.

The following two are meta-arguments to counter meta-arguments (like the above):

> (MAAGA): A Meta-Argument for *Accepting* Good Arguments:
> (1 b) This argument seems cogent and the conclusion is reasonable;
> (2b) Apparently cogent arguments with reasonable conclusions usually are genuinely cogent;
> So, (3b) it is unlikely that flaws will be found: accept the conclusion.
> (MARBA): Meta-Argument for *Rejecting* Bad Arguments:
> (1c) This argument seems fallacious and the conclusion is unreasonable;
> (2c) Arguments for unreasonable conclusions usually are really fallacious;
> So, (3c) it is unlikely that the argument can be fixed: do not accept the conclusion.

Cohen's picture of meta-argument (with all of its varieties) is a rich one and it captures several essential features of meta-argument. In the first place, Cohen's meta-arguments are retrospective, coming *after* or on the heels of first-order arguments. Second, though they come in the wake of first-order arguments, these meta-arguments *regard* them qua arguments. That is, they don't simply have them as contents, but they are about these arguments as items of reflection and evaluation according to norms of argument. Third, they further (or frustrate) the purposes of the first-order argument. They're not merely not merely evaluative, but prescriptive, too. One identifies what one should do in light of the evaluation of the argument and one resolves to do it.

Let's take MARGA as a test case. A classic example, cited by Cohen (p. 79), is the argument that $1 = 0$ that gets passed around social media every now and then (with the warning: *only for geniuses!*). In the videos, the steps seem individually logical, but then one ends up at the surprising conclusion. Regardless of the quality of arguments, $1 = 0$ can't possibly be true and so one rejects it. Just as first-order arguments can be fallacious or not, so can second-order arguments. But they're fallacious in their own peculiar way. Fallacious versions of MARGA can come in a variety of forms. One form we elsewhere (2022a) call *bothsiderism*. As the name suggests (shouldn't they always?) bothsiderism happens when, upon hearing multiple arguments on a disputed issue, an addressee invokes a deficient principle of meta-rationality to claim that the existence of different perspectives entails splitting the difference (or epistemic moderacy or skepticism). Bothsiderism a meta-argumentative move because it comes in the wake of arguing, and it is a relational evaluative judgment *about* the first order arguments without assessing their relative quality. One judges an argument to be deficient merely because there's another argument that says something different. *Whataboutism* is a similar retrospective meta-argumentative fallacy (Aikin & Casey, 2022c). Again, one judges that an argument is not effective not because of something on the first order, but because there are *different arguments* about *more important issues* outside of the argument that merit consideration first. So, to give a quick example, in response to an argument about deaths of African Americans at the hands of law enforcement in the United States, a whataboutist might ask *what about Black-on-Black crime*? The thought of the whataboutist is that the fact the one offers arguments at all on some topic is meta-argumentatively salient.

3. MAMBA

The retrospective feature of meta-arguing is central for good reason: that's what meta seems to be. But, as we've noted, the term meta is not only retrospective, as are all of Cohen's principles mentioned just now: here is argument I don't want to handle on the first order, so I move to the second order. The focus is on addressees. We think that the principles at work in here should also work forwards, in other words, that there might be meta-argumentative principles for making certain kinds of arguments, rather than merely in accepting, rejecting, or interpreting them. This is a lot. So in what follows we'll first sketch out how this might work by (briefly) returning to the broad notion of meta just discussed. Second, we'll pose some examples of meta-argumentation that parallel one of Cohen's categories (MARGA). We'll call this MAMBA: a meta-argument for making a bad argument. This is not, we should caution, another way of tailoring one's argument to an audience—i.e., making an argument one knows to have failed a standard because the target audience can't tell the difference and will buy it on that account. What's important in the cases of MAMBA we'll discuss is that the arguer intend the argument to be bad and for it to be recognized as such. MAMBA reasoning comes in various versions, but we identify two: strong and weak. Strong MAMBA licenses one to make what we will call para-argumentative moves. These are non-argumentative communicative acts relevant to the argument. Weak MAMBA is a principle allowing one to make merely bad arguments. Examples of para-argumentative communication include sarcastic abuse, ridicule, ad baculum threats; examples of weak MAMBA include merely fallacious (e.g., straw men) or weak arguments directed at certain addressees, again, for other than rhetorical effect. In addition to identifying the forms, we assess the costs and opportunities of employing MAMBA reasoning. As we conclude, the various forms of meta-argument make sense, again, around the distinction between the first and the second order. We close with some thoughts on this, namely that meta-argumentative moves are legitimate to the extent they let us make progress in the first order.

Now for some cases. One type of prospective meta-argumentative case are straw man arguments. As we've argued elsewhere (2022c), one thing that makes the straw man peculiar as arguments and fallacies is that on one level there's nothing wrong with them; critiquing bad arguments is good, isn't it? This is especially the case when the argument being critiqued is a real one that actually bad, as is the case in weak man (selectional straw man) arguments (Talisse & Aikin, 2006). Interestingly, as there are cases where critiquing bad arguments from real arguers is bad, there are cases where making bad arguments on purpose is good. What makes straw man arguments pernicious, when they are, is how they are deployed meta-argumentatively. On our (2022c) account of their meta-argumentative deployment, they are fallacious when they prematurely or illegitimately close discussion; or, as is the case in the iron man, they are fallacious when they delay the closure of an issue that ought to be closed. Socrates, in his famous exchange with Thrasymachus in book I of the Republic (340d), exemplifies the meta-argumentative deployment of the straw man (1992, 18). Plato even draws attention to it, having Socrates twice accused of being a "false witness" in argument (for a discussion of this exchange, see Aikin and Anderson 2006).

Once we have a grasp on the meta-argumentative stakes involved in making bad arguments on purpose, other cases of MAMBA come into view. Our first case comes from conservative American political commentator Ben Shapiro, who is known for his aggressive argumentative style and frequent use of meta-argument (see, for instance, his employment of the free speech fallacy, discussed in Aikin & Casey, 2023). In a 2020 discussion of a New York law banning the sale of Nazi or Confederate flags on state property, he stressed that he is not a defender of the Confederate flag, conceding that it "in fact was a symbol of slavery." He was quick to point out, however, that some people see it as central to their vision of ther southern heritage, and so he held that laws restricting the sale or display of the confederate flag are unjustified. He then made the following curious argument:

"It is an aspect of free speech in America that, (1) you can fly the Confederate flag and (2) It's just an aspect of a free speech culture that if somebody's flying a Confederate flag you probably shouldn't make the assumption that they're a baseline rote racist."[1]

It's worth noting that Donald Trump offered a similar kind of reply when pressed about his views on the Confederate flag.[2] It's not all that uncommon of a view on controversial symbols such as this. Now it might seem, however, that there already is an account of arguments such as this, the red herring. They're red herrings because they seem to be distractions from the original question, which was whether someone found the symbol, the Confederate flag, acceptable. Rather than reply straightforwardly that not, in responding one layers on another argument: I don't fly the Confederate flag because I endorse its racist agenda, I fly the Confederate flag because I favor free speech. Since this is speech that is challenged, my argument, layered on top of this very bad argument, is all the more forceful. I embody in my own person and in my own argument the thought that I may not like someone's speech but I'll defend their right to do it till the end.

The structure of the Confederate flag argument is complex: one advances, self-consciously, a bad argument or deeply implausible view (that one does not in fact hold) as a means of making another, more plausible argument, dependent on the uptake of the first argument. In the current case, the bad argument is the one embodied in the Confederate flag, which, as Shapiro himself notes, is an unquestionably racist symbol. Under this banner, one thereby advances another argument, like for instance that attempts to restrict this speech—if only by moral disapprobation—are the real villains. And embracing the symbol of disapprobation or making a manifestly bad argument is the means of making that argument is just what makes this a meta-argument. The argument rides on top of the other. Some might recognize this argument pattern from recent American politics as the "own the libs" strategy, popular with former president Trump and his followers, though it's a

---

1 The video can be found here:
https://twitter.com/jasonscampbell/status/1339301383619702785

2 https://www.youtube.com/watch?v=aZfMWRDIrBw. There are a host of other examples.

strategy popular with some liberals.³ "To own the libs" is to make some purposely wrong argument to provoke a negative reaction (and then, presumably, trade on that reaction). Relatedly, in the run-up to the 2016 United States Presidential election, journalists struggled to understand then candidate Trump's uniquely bombastic argumentative style. He would stake positions that were manifestly false only then to claim them true in some other sense. This led some journalists to claim that Trump ought to be taken "seriously" but not "literally" (Zito, 2016). Hugely exaggerated statistics, like the 3000 people killed in Chicago (one of us lives here, it's far from true), convey the seriousness of the problem.

Again, the main thought is that meta-arguments are fundamentally relational. Some meta-arguments handle normative questions in ground-level arguments. Meta-arguments are instances in which a bad or insincere argument is offered for some longer-term argumentative purpose. The question now is what this purpose is. On one view, owning the libs seems to be an immature strategy, meant merely to rile up or "trigger" some target audience. The aim of the argument, on this view, is that you get a rise of out of them, you know, for the lulz (where lulz is shadenfreude at the frustration you have caused). This is not much of an argumentative purpose: no claims are being advanced or defended, no clarity is provided on the first or even the second-order. It's pure immaturity.

On the other hand, owning the libs resembles other strategies that utilize irony meta-argumentatively. Embracing stereotypes for some marginalized groups is not only a way of taking away their sting, it can also be also done meta-argumentatively, as a way to comment on the correctness of such notions (see, for example Pauwels, 2021). The bad argument, the stereotype, functions meta-argumentatively to clarify some notion on the first order

Prospective meta-argumentation has affinities with the phenomenon of signaling. Standardly, signaling refers to the messages certain behaviors are meant to send. In many cases, these behaviors are argumentative, or at least argumentative-like. One may take a stand on some issue in order to signal one's adherence to a certain set of norms or beliefs. Bergamaschi Ganapini (2021) argues that the widely reviled practice of sharing fake stories on social media is best explained by its potential for signaling group participation. Interestingly, sharing what one knows or suspects to be fake news with others who have a similar view is a meta-argumentative deployment of a bad view. The sharing demonstrates solidarity both to compatriots and opponents. Like the own the libs strategy or the

---

3 https://www.chicagotribune.com/sns-201904161202--tms--jgoldbrgctnjg-a20190417-20190417-column.html A slightly different take: https://www.nbcnews.com/think/opinion/own-libs-was-snide-way-mock-conservatives-now-fun-way-ncna897636 fails somewhat to see the meta-argumentative possibilities.

Confederate flag case, it may demonstrate to opponents that arguing will be difficult, costly, or impossible. Relatedly, Goodwin and Innocenti (2019) have discussed cases where one offers arguments even though one won't be successful in convincing a target audience. Arguments, in such cases, may function meta-argumentatively to demonstrate the arguer's capacity for rational engagement.

So far, we've outlined a case for what we call weak MAMBA. We call it "weak," because even though one is making a bad argument, one is still operating within the purview of argument; the reasons are bad, but they're still *reasons*. Once it clear that argument might not be subject to the constraint that you can't do it badly on purpose, then it could be the case that not even offering reasons at all could be seen to be a form of arguing. Call this *para argument*.

4. Para argument

Paralegals are legal workers who work alongside lawyers, but are not lawyers themselves. They're good up to a point, in other words, for legal purposes. Similarly, para-argument moves are communicative acts that occur alongside arguments, for argumentative effect, though themselves not arguments (Aikin & Casey, Forthcoming). To some extent, the concept of para-argument should be familiar to those familiar with ad baculum arguments. Standardly, in the ad baculum argument one threatens an interlocutor in order to secure assent to some proposition. In such cases, no one would be daft enough to confuse the para-argumentative threats for reasons to assent to some proposition; they're merely incentives to agree or, what is more likely, not disagree (see Casey 2022 for a discussion of why anyone would do this). We can make the same case for argumentative bribes, where one uses some non-argumentative inducement to adopt a standpoint.

Ad baculum type appeals are typically considered fallacious because one tries to accomplish an argumentative objective by force (or bribery). It's clear, however, that not all cases of ad baculum are fallacious: threatening to cut someone's microphone should they continue to interrupt could well be a case of justified pressure on a disruptive interlocutor. The same could be said for laughing at someone's lapse in logic, acting bored during their talk, or shouting at them afterwards. Naturally, laughing, acting bored, shouting, and other such acts are not reasons, and so recourse to them, like recourse to ad baculums, is superficially a kind of bad arguing. It's at least true that it's often considered as such. Yet, it seems that such para-argumentative moves have their place in a well-functioning argumentative space. This is especially true when they're deployed to enforce the norms of a well-run exchange. Consider the following simple case. Duns Scotus, in his *de Metaphysica* holds that para-argument is the only response available when confronted with those who deny (apparently) self-evident logical principles, such as the principle of non-contradiction. Such people, he argues, ought to "be beaten or exposed to fire until they concede that to burn and not to burn, or to be beaten and not to be beaten, are not identical" (Duns Scotus & Wolter, 1987). Along similar lines, Erik Krabbe has argued that people who abuse meta-discussions—those who filibuster with endless points of order, who make specious accusations of fallacy, or who otherwise use argument to stand in the way of argument—ought to be assessed a penalty (Krabbe, 2003, p. 89). Following Krabbe, Innocenti (2022) describes a case where para-argumentative sanction would seem to be permissible. The point is that the para-argumentative sanction, though not really arguing, conveys something about the norms of arguing. In Innocenti's case, it's where the norms are violated.

The discussion so far might suggest that para-argument is hyper adversarial, along the lines of Govier's conception of ancillary adversariality (Govier, 2000, 2021), that is, all of the adversarial non-arguing (gesticulating, yelling, finger-pointing, derision, and so on) that at times accompanies argument. Certainly, para-argument comes about often in those cases where argument alone seems to have failed to do its job, as in argumentative abuses of argument, and so something a little more aggressive is on order. Argument, however, need not always concern disagreements, and so it would seem that para-argument need not serve an exclusively punitive function. Call these cases of para-argumentative reward: where clapping, nodding, snapping of fingers, shouting, or gesticulations serve as rewards for conclusions well drawn.

5. Conclusion

As we've noted, it's common to think of meta-argument as something that only comes in the wake of some first-order argument. As Cohen observed, when a conclusion from an apparently blameless argument is paradoxical, we may pause to consider whether those reasons really are as good as they seem (and thus MARGA). In this sense, meta-arguments are retrospective. We've argued here, however, that one can employ meta-arguments prospectively. This is easiest to see in cases where one offers a manifestly obnoxious (as in the Confederate flag case), exaggerated (as in the Trump case), or otherwise bad (as in the Socratic case) argument in order to direct attention to another point. In this sense, prospective meta-arguments are layered structures. The manifest badness of the argument calls attention to itself and therefore to the norms the arguer means to bring into relief. Along similar lines, para-argument is often an adversarial move that draws attention to failures of parties to abide by argumentative norms. In this sense, it too is meta-argumentative, even if it may seem to be non-argument. But its failure to rise to the level of argument underscores its function as norm-enforcement.

The looming question, so far only indirectly addressed here, is what the norms of meta-argument might be. For, it might have seemed something of a truism that one ought to defend their standpoint, and the meta-arguments seem to violate that, in some respect. We haven't the space for a full discussion here. We can say, however, that since meta-argument and para-argument often concern arguments *about* arguments, and that they arise as a consequence, it is likely that the norms of first order argument are not up to the task.

References

Aikin, S. F., & Anderson, M. (2006). Argumentative Norms in Republic I. Philosophy in the Contemporary World, 13(2), 18–23. https://doi.org/10.5840/pcw200613213

Aikin, S. F., & Casey, J. (2022a). Bothsiderism. Argumentation. https://doi.org/10.1007/s10503-021-09563-1

Aikin, S. F., & Casey, J. (2022b). Fallacies of Meta-argumentation. Philosophy and Rhetoric, 55(4), Article 4. https://doi.org/10.5325/philrhet.55.4.0360

Aikin, S. F., & Casey, J. (2022c). Straw Man Arguments: A Study in Fallacy Theory. Bloomsbury.

Aikin, S. F., & Casey, J. (2022d). Argumentation and the problem of agreement. Synthese, 200(2), 134. https://doi.org/10.1007/s11229-022-03680-4

Aikin, S. F., & Casey, J. (2023). Free Speech Fallacies as Meta-Argumentative Errors. Argumentation. https://doi.org/10.1007/s10503-023-09601-0

Aikin, S. F., & Casey, J. (Forthcoming). On Halting Meta-Argument with Para-Argument. Argumentation. https://doi.org/DOI :10.1007/s10503-023-09602-z

Bergamaschi Ganapini, M. (2021). The signaling function of sharing fake stories. Mind & Language, mila.12373. https://doi.org/10.1111/mila.12373

Breakey, H. (2021). "That's Unhelpful, Harmful and Offensive!" Epistemic and Ethical Concerns with Meta-argument Allegations. Argumentation, 35(3), 389–408. https://doi.org/10.1007/s10503-020-09538-8

Casey, J. (2022). Beliefs, Commitments, and Ad Baculum Arguments. Languages, 7(2), 107. https://doi.org/10.3390/languages7020107

Cohen, D. H. (2001). Evaluating arguments and making meta-arguments. Informal Logic, 21. https://doi.org/10.22329/il.v21i2.2238

Duns Scotus, J., & Wolter, A. B. (1987). Philosophical writings: A selection. Hackett Pub. Co.

Finocchiaro, M. A. (2005). Arguments about Arguments: Systematic, Critical, and Historical Essays In Logical Theory (1st ed.). Cambridge University Press. https://doi.org/10.1017/CBO9780511527517

Finocchiaro, M. A. (2007). Arguments, Meta-arguments, and Metadialogues: A Reconstruction of Krabbe, Govier, and Woods. Argumentation, 21(3), 253–268. https://doi.org/10.1007/s10503-007-9055-x

Finocchiaro, M. A. (2013). Meta-argumentation: An approach to logic and argumentation theory. College Publications.

Godden, D. (2019). On the Rational Resolvability of Deep Disagreement Through Meta-argumentation: A Resource Audit. Topoi, 38(4), 725–750. https://doi.org/10.1007/s11245-019-09682-1

Goodwin, J., & Innocenti, B. (2019). The Pragmatic Force of Making an Argument. Topoi, 38(4), 669–680. https://doi.org/10.1007/s11245-019-09643-8

Govier, T. (2000). The philosophy of argument. Vale Press.

Govier, T. (2021). Reflections on Minimal Adversariality. Informal Logic, 41(4), Article 4. https://doi.org/10.22329/il.v41i4.6876

Innocenti, B. (2022). Demanding a halt to metadiscussions. Argumentation, 36(3), 345–364. https://doi.org/10.1007/s10503-022-09569-3

Krabbe, E. C. W. (2003). Metadialogues. In F. H. Van Eemeren, J. A. Blair, C. A. Willard, & A. F. Snoeck Henkemans (Eds.), Anyone Who Has a View (Vol. 8, pp. 83–90). Springer Netherlands. https://doi.org/10.1007/978-94-007-1078-8_7

Pauwels, M. (2021). Anti-racist Critique Through Racial Stereotype Humour: What Could Go Wrong? Theoria, 68(169), 85–113. https://doi.org/10.3167/th.2021.6816904

Plato, Grube, G. M. A., & Reeve, C. D. C. (1992). Republic. Hackett Pub. Co.

Talisse, R., & Aikin, S. F. (2006). Two Forms of the Straw Man. Argumentation, 20(3), 345–352. **Error! Hyperlink reference not valid.**

Zito, S. (2016, September 23). Taking Trump Seriously, Not Literally. The Atlantic. https://www.theatlantic.com/politics/archive/2016/09/trump-makes-his-case-in-pittsburgh/501335/

INCONCLUSIVE ARGUMENT APPRAISAL: DIFFICULTIES IN CONCLUDING ARGUMENT EVALUATIONS AND EVALUATING ARGUMENTS WITHOUT CONCLUSIONS

DANIEL COHEN
Colby College
dhcohen@colby.edu

Abstract

Although we typically assess arguments with reference to their closure – e.g., whether the inferences successfully lead to the conclusion, whether interlocutors were successfully persuaded by rational means, or whether the arguers successfully reached resolution – inconclusive arguments are common parts of our cognitive lives and they are not all failures. They can be quite positive. Episodic arguments can make progress towards conclusions they never reach; conclusions will emerge and constantly evolve in on-going arguments; unfinished arguments improve and strengthen our standpoints; and conclusive or not, argumentation provides its participants with opportunities to gain deeper understanding and greater appreciation of others' positions. Further, since arguments can always be re-opened, even apparently conclusive arguments are only defeasibly closed. This includes argument appraisals, which are themselves arguments. In short, inconclusive arguments, like other arguments, are subject to description as logically flawed, cognitively productive, socially destructive, personally satisfying, ethically problematic, and so on.

Nevertheless, inconclusive arguments present a challenge to many cogency tests because in order to render a judgment as to whether the premises are relevant and sufficient *for the conclusion*, there has to be a conclusion. Not surprisingly, process and procedural approaches to argumentation fare better with inconclusive arguments than product-oriented approaches do. Inconclusive arguments can be judged as having been argued well or poorly *as far as they go*. Still, many of the features that can make them praiseworthy or censurable remain elusive to these approaches because these features accrue to the arguers themselves. In particular, the question as to how well the arguers *engaged* looms large. Virtue theories, I argue, are well-situated to fill this gap in argumentation theory.

1. Introduction

Athena, the Greek goddess of both war and wisdom, may be an appropriate patron for argumentation theorists, but the myth of her birth as fully formed from the brow of Zeus is not a good model for arguments. I suppose some arguments may have emerged that way from the brow of Anselm or Gödel, but the arguments most of us encounter do not come as complete packages. Our lives include fragments of arguments, arguments that are not yet completed, on-going arguments that no one expects, or even wants, to conclude, and argumentative efforts that fail to reach closure by logical, rhetorical, or dialectical measure. It is a curious and heterogeneous lot that is worthy of more attention than it has received from argumentation theorists.

Part of the reason for the lack of theoretical attention is that while inconclusive arguments may be common, they resist easy evaluation. Consider, for example, the "RSA" test, Johnson and Blair's (1977) early and influential contribution to informal logic: it is a very good and very useful heuristic for beginning to think about arguments, and where it tells us to begin is with the *premises*. In a cogent argument, the premises must be relevant, sufficient, and acceptable. Notice that despite their focus on the premises, these are all relational terms that implicate other parts of argumentation: the premises should be acceptable to the *arguers*; they must be sufficient for the *inferences*; and the need to be relevant to the *conclusion*. The RSA and similar tests target arguments that come complete with arguers, inferences, and conclusions. What about arguments that are not quite so complete?

It is important to keep the notion of an incomplete argument in mind because some common uses of the word "argument" refer to things that typically lack some of these elements. Some argumentation theorists, for example, speak of arguments as abstract inferential structures of propositions – "premise-conclusion complexes" – which are arguments without arguers, so would not be thought of as incomplete on account of that lack. (With no arguers in sight, acceptability becomes moot.) On the other hand, ordinary usage allows us to refer to any adversarial dispute as an "argument" regardless of whether it contains any premises, exchange of reasons, or any inference at all, and regardless of whether it reaches a conclusion or any other kind of closure.

The RSA criteria were not meant to apply to such un-argued, un-reasoned, and inconclusive "arguments". For the most part, argumentation theorists should be willing to cede theorizing about abstract inferences to logicians and to leave the management of personal disputes and feuds to ethicists, psychologists, anthropologists, and others. However, even if we restrict our purview to arguments with agents who are reasoning in pursuit of some argumentative goal – i.e., *verbal and social engagements in which reasons are offered in order to resolve a difference, convince and audience, demonstrate a truth, or simply commend, justify, or critique a standpoint*, to use an early Pragma-Dialectical characterization (van Eemeren, Grootendoorst, and Henkemans 1996, p. 2) – we still have plenty of incomplete arguments to consider.

The focus here is *inconclusive* arguments. They are part of our cognitive lives. Normally, the lack of a conclusion would be negative, a failure of some sort, but that is not universally the case. Inconclusive arguments can also be positive, productive, and even important. They need to be evaluated on their own merits: an inconclusive argument can be simply very good – not just very good *so far*. They cannot all be lumped into the same category. Their reasoning can be good or bad, they can be argued well or poorly, they can contribute to our cognitive fields, and they can be personally or socially constructive or destructive. There can even be *failed* arguments that still ought to count as terrific argumentation. Would it really be that odd to hear someone say, "*We had a very good and productive argument even though we never did come to and resolution*"? Granted, this might sound a bit like saying that the operation was a success but the patient died, but there phenomena here that argumentation theory should be able to explain, that informal logic should be able to assess, and about which we should have something intelligent to say.

After a brief survey of the variety of inconclusive arguments and the ways they can be valuable despite their inconclusiveness, I will consider frameworks from which those judgments about arguments can be made. Along the way, three tentative hypotheses about argument evaluation start to crystallize: first, that argument evaluations are themselves arguments; second, that argument evaluations are themselves often inconclusive; and third, that the inconclusiveness of argument evaluations as arguments is a feature, not a glitch. Unfortunately, considerations of space mean that the arguments for those hypotheses here will have to be inconclusive.

2. The variety of inconclusive arguments

I doubt that inconclusive arguments form a "natural kind" but I think the concept can still be a useful heuristic device for theorizing. As a first pass, we can divide inconclusive arguments into three rough-and-ready overlapping categories: (1) *incomplete* arguments that are cut short before finishing, (2) *failed* arguments which the arguers do not or cannot complete, and (3) arguments for which conclusions are undesirable, impossible, or *unnecessary* in some way.

2.1. Incomplete arguments

Arguments in the relevant sense are events in our cognitive lives. They involve arguers whose actions and interactions occur in some context. They have beginnings, they take place over time, and they proceed towards some resolution. Or they would, if we "had but world enough and time", but we don't. Circumstances contrive to end our arguments prematurely. Our resources are limited and might not be up to the task at hand. Rather than abandon the argument completely, we may decide to suspend the argumentation with hopes for completion, or at least continuation, in future installments.

This poses a problem for evaluating incomplete arguments because our appraisals can vary depending on how much of the argument is available. This phenomenon comes into clearer focus in negotiations, where it is often wise to withhold judgment until they are complete. What seems like a productive session at the time might, in retrospect, be seen as the turning point that led to a negative outcome. Alternatively, a contentious session that leaves a sour taste in everyone's mouth might be a cathartic moment that breaks a logjam in the negotiations. Something similar can be said about other kinds of argumentation, but that does not mean that no judgments are possible before the end. We can recognize that some episodes in ongoing arguments make more progress towards resolution than others. We can judge when argumentation has been product or counterproductive even if no resolution has been reached. We can praise an inconclusive argumentative engagement when it has proceeded in accordance with the imperatives of good argumentation and the arguers have conducted themselves well. Conversely, we should criticize it when the reasoning is shoddy, the engagement was perfunctory, or the arguers were vicious.

As in the case of negotiations, some of these interim judgments may have to be revised in the light of later developments. For example, the criticism that an arguer has missed an opportunity, whether for pursuing a promising line of reasoning or raising an objection (see Cohen 2015), would be rendered obsolete were that lacuna later filled. Similarly, what appears at first to be virtuous arguing may be unmasked as hidden motives are revealed. Other kinds of interim judgments may be less subject to retrospective revisions. It is unlikely, for example, that later developments can ameliorate blatantly fallacious reasoning or vicious conduct. We do not have to wait until the end of an argument before condemning the contributions of a troll whose only goal is thwarting argumentative progress.

Another interesting and noteworthy way for arguments to be incomplete is by the phenomenon of "mission creep" (Kvernbekk, Bøe-Hansen, Heintz, and Cohen, 2020). So-called after its military counterpart, mission creep occurs when the goals of the argument change along the way. An arguer may enter an argument with the relatively modest goal of challenging a standpoint but end up in the more ambitious project of defending an alternative. The heat of competition can turn compromise-seeking negotiators into victory-thirsty adversaries. How are we to evaluate an argument in which we start out arguing for P, soon find ourselves arguing for P & Q and then become totally focused on establishing Q? The goalposts moved! Would it be more helpful to describe this as two – or even three – overlapping arguments? On one hand, there are reasons for applauding the evolution of standpoints and goals during argumentation. After all, improving our positions is one of the reasons we submit them to trial by argumentation in the first place. On the other hand, it highlights the fuzziness of arguments' boundaries – and underscores an arbitrariness in declaring an argument done. It also shows why it is better to ask along the way, "How is the argument going?" rather than waiting until the end to ask, "How did the argument go?" or "How good was the argument?"

2.2. Failed arguments

Arguments can fail in many ways, including failing to establish their conclusions, failing to persuade others, failing to resolve differences by rational means, and failing even to engage the relevant interlocutors. The responsibility for these failures lies with the arguers: the reasoner was unable to reach the conclusion, the arguers were unable to settle things, or the proponent was unable to engage and convince her audience – unless, of course, it was a failure on the part of the argument's audience to be engaged, to recognize good reasons, or to be sufficiently reasons-responsive. These arguments would all be better if they had better arguers. This needs some qualification because when there is agent-responsibility, rather than circumstantial reasons, for the failure, it need not be assigned exclusively to the principal arguers. Incompetent judges, rogue juries, rowdy or inattentive audiences, malicious trolls, and even well-meaning kibitzers can sabotage an argument. Sometimes, what's really needed is arguers who can reason better, but other failures could be mitigated or avoided altogether by arguers who are more willing to engage, arguers who are better at the give-and-take-compromise of negotiations, or arguers – in the broad sense to include all the participants from proponents and opponents, to judges, juries, and spectators – who listen better.

There are two claims being made here. First, that the culpability for failed arguments lies with the arguers, including perhaps, secondary participants, and second, that better arguers can make failed arguments better *even though the arguments are still failures*. These claims get right

to the heart of the matter, bringing uncomfortable theoretical questions to the fore:
- How can some failed arguments be *better* than other failed arguments even though they are all failures?

This then points to a more practical question:
- What improvements in *arguers* enable those improvements in *arguments*?

That second question points us back to an obvious and important, but still often overlooked truth: *Better arguers make for better arguments.* There is no cognitive dissonance upon hearing two arguers saying to each other as they disengage from an argument, "*Well, neither one of us actually managed to succeed, but that was still a heck of a good argument.*"

The ways in which failed arguments can still be positive are as varied as the ways they fail. An arguer might come to grips with how and why the available premises are insufficient for demonstrating the conclusion; or he might come to understand an opponent's standpoint better even though she sees no change at all in its acceptability; or arguers might achieve a greater tolerance of one another's positions without in any way lessening their opposition to those positions. Those are all positive outcomes that redound to the credit of those arguments, yet none of them requires that any of the arguers succeed in their goals, whether demonstration, persuasion, or resolution of difference. It is neither helpful, accurate, nor illuminating to characterize the incompleteness of such arguments as failures and dismiss them on that account.

2.3. Conclusion-free arguments

There is a third category of arguments without conclusions that are neither incomplete nor failed. They are fine just the way they are. Their inconclusiveness is due neither to external circumstances nor agent failures but to internal factors. First, some arguments simply *cannot* have conclusions. Deep disagreements would be an example. If opposing standpoints are indeed incommensurable, then resolution would not be possible but not through any fault of the arguers–unless it is an argumentative misstep even to engage in that kind of argumentation in the first place. But that need not be the case. There can be a point to such arguments and there are benefits to be had. Thomas Nagel (1986, p. 4) wrote, "Certain forms of perplexity– for example, about freedom, knowledge, and the meaning of life–seem to me to embody more insight than any of the supposed solutions to those problems." I would add that it is argumentation that gives those insights body: the rationality that arguments manifest is found more in the acts of arguing than in the conclusions.

There are also arguments that are fine without conclusions because although they *can* have conclusions, they *do not need* them. Arguments serve many purposes, not all of which are conclusion-related. Sometimes the point of objecting and registering a complaint is just to go on record as having objected or having registered a complaint. It is part of the responsibility of political representatives, union officials, and defense attorneys to raise objections to actions and policies deemed detrimental to their clients. Doing so fulfills their argumentative obligations regardless of whether it has any effect. We could call these "protest arguments" to parallel protest votes because their efficacy is almost beside the point. When the act of arguing is what matters, the act is the proper subject for evaluation.

The limit case for when the act of arguing matters more than the conclusion is when the importance of just engaging in argument is all that matters. Reaching a conclusion would not make the argument any better. So, naturally, the contrarians among us are tempted to ask whether reaching a conclusion could ever *detract* from an argument. I suppose this could happen when an argument is a vehicle for some non-argumentative purpose. Recreational argument provides a frivolous example: if two philosophers are arguing just for entertainment, then reaching a conclusion would end the fun. It might be objected that this is not genuine argumentation, and that reaching a conclusion might detract from the engagement *as entertainment*, but not *as an argument*. A fair point, but we can still evaluate the exchange as an argument, and the example points to other possibilities. At the opposite end of the frivolity scale, we find Elie Wiesel who carried on what he describes as a decades-long argument with God. After his time in the Auschwitz and Buchenwald concentration camps, he thought the only alternative to arguing with God would be to cut off all communication completely (Wiesel 1978. p. 6). It is not an argument that he wanted to end. In this case, the attempted communication does not just accidentally take the form of an argument: argumentation is manifest *communicative* rationality, and that was his goal.

3. On evaluating inconclusive arguments

If all we can do is note that these arguments lack conclusions, then our evaluative vocabulary is impoverished. We would miss the great variety of evaluative possibilities, both positive and negative. On the negative side, some failures exacerbate bad situations and reduce the chances for future argumentative success. The failure to make a compelling case can raise doubts where there were none; a failure to settle differences can increase polarization by distorting opposing standpoints, magnifying differences, and heightening adversariality, making agreement even harder to achieve. Good arguers do not let these things happen, so again these failures are on the arguers.

The consequences of inconclusive arguments are not inevitably negative. Positive outcomes are possible, and better arguers make them more possible. Inconclusive arguments, at least in theory, keep the lines of communication open. At their best, they are capable of contributing to progress on many of argumentation's fronts: our understanding, acknowledgement, appreciation, tolerance, and commitment to standpoints can all be enhanced. Admittedly, those positive outcomes do not obviate the fact that an argument that fails to reach any acceptable closure is still a failure on that account. Therefore, it falls short *as an argument*. Inconclusive arguments, like closed, completed, and successful arguments, are subject to description as logically flawed, cognitively productive, socially destructive, ethically problematic, and so forth. They should be commended or criticized accordingly. We need the resources to say these things about the incomplete products of argumentation.

Argumentation theories are not all product-oriented, of course, and we do have resources for theorizing here. Logical approaches can evaluate the validity of the inferences that have been made, but that is a minimum condition at best. The fact that there have been no false steps *so far* has little bearing in whether those steps are heading in a direction that is at all relevant to the goals of the argument. Procedural approaches fare better, but again, the mere absence of transgressions in itself is neither a measure of progress nor much of a positive.

A more promising approach is suggested in Paglieri and Castelfranchi's "Why argue? Towards a cost-benefit analysis of argumentation" (2010). Because there may be personal, emotional, and social costs to arguing, arguing is not always the best strategy to pursue persuasion or to resolve differences. The question of whether to argue at all is a strategic one that involves weighing the costs and benefits and assessing the risks. The originating context for Paglieri and Castelfranchi's discussion is computational argumentation, and they readily acknowledge that some translation is needed before it can either be applied to ordinary, natural argumentation or generalized to decisions within arguments once they are underway. For example, the emotional toll of arguing on human arguers has no obvious analog for computational arguers. In the other direction, the task of weighing "computational costs" has to be put on hold for human arguers until the ideas of epistemic and argumentative costs are better developed.

Because our concern is argument evaluation rather than decision-making in argumentation, we need to adapt Paglieri and Castelfranchi's framework to evaluation. Our questions are, *"How well is the argument going?"* and *"How well did it go?"* rather than *"What should I do?"* Fortunately, they point us in the right direction. They never lose sight of the goals leading to the original decision to argue, what they call the "extra-dialogic" goals. These are distinct from tactical goals that may arise during the argument, the "dialogical goals." We can think of these, roughly, as the *arguers'* goals and the *argument's* goals, respectively.

Success *in* the argument does not always align with success *with* the argument. How much of a success is winning an argument but losing a friend? How much of a failure is losing to a student who learns a great deal from the argument – and you do, too? Argument evaluation should not ignore the larger picture. Sometimes the arguer's extra-dialogical goals have to take precedence over the argument's dialogical goals.

A cost-benefit approach shifts the focus back onto the arguers: evaluating inconclusive arguments has to consider the costs and benefits *to the arguers*. We are the ones who bear the costs and reap the benefits. Once we focus on the arguers, two good things happen. First, the problem of evaluating inconclusive arguments becomes more tractable because we don't need to have a conclusion in hand in order to assess how well the arguers are benefitting from the argument. Second, the problem of re-evaluation also becomes less daunting. However, because the benefits and costs might not kick in until after the arguers have disengaged, we may need to adjust where we draw the boundaries of an argument. It might seem discordant at first to say that an argument *was* good but not longer *is,* or that a failed argument was actually pretty good. The discordance disappears when we remember that this is a cost-benefit analysis of an argument, not a logical or epistemological assessment of inferences. In that sense, an argument might be compared to an investment: as the initial costs become more distant in time and benefits continue to accrue, we have to re-assess our initial or short-term judgment. It's not that those early verdicts were wrong. They might have been the right judgment at the time, but things change. Given how hard it can be to admit defeat or make concessions in the heat of an argument, the benefits are often delayed until after the dialog has concluded and the arguers have had the time and space to think it over.

Even so, it takes a special arguer to reap the benefits from what could otherwise be described as a lost argument. So once again we reach the conclusion that better arguers make better arguments. Now we can add that better arguers also make arguments better, and they can continue to do so even after the arguing has ended – whether or not it has concluded.

References

Cohen, D. (2022). You cannot judge an argument by its closure. Informal Logic, 42(4), 669-684.

Cohen, D. (2015). Missed opportunities in argument evaluation. In B. Garssen, D. Godden, G. Mitchell, & A. F. Snoeck Henkemans (Eds.), Proceedings of ISSA 2014: Eighth conference of the International Society for the Study of Argumentation (pp. 257-265). Amsterdam: Sic Sat.

Johnson, R. & Blair, J. A. (1977/2006). Logical Self-Defense. Toronto: McGraw-Hill.

Kvernbekk, T., Bøe-Hansen, O., Heintz, O., & Cohen, D (2020). The problem of mission creep: Argumentation theory meets military history. OSSA Conference Archive 5. https://scholar.uwindsor.ca/ossaarchive/OSSA12/Saturday/5

Nagel, T. (1986). The View from Nowhere. New York: Oxford University Press.

Paglieri, F. & Castelfranchi, C. (2009). Why argue? Towards a cost-benefit analysis of argumentation. Argument & Computation, 1(1), 79-91.

Van Eemeren, F. H., Grootendoorst, R., and Henkemans, F. (1996). Fundamentals of Argumentation Theory. New York: Routledge.

Wiesel, E. (1978). A Jew Today, Wiesel, M. (tr.). New York: Random House.

THE ROLE OF IMPLICITNESS IN PERSUASIVE ARGUMENTATION: LOOKING AT DIFFERENT TEXT GENRES[1]

CLAUDIA COPPOLA
Università Roma Tre, La Sapienza Università di Roma
claudia.coppola@uniroma3.it

DORIANA CIMMINO
Independent researcher

FEDERICA COMINETTI
Università dell'Aquila

GIULIA GIUNTA
University of Neuchâtel

GIORGIA MANNAIOLI
Università Roma Tre, La Sapienza Università di Roma

VIVIANA MASIA
Università Roma Tre

EDOARDO LOMBARDI VALLAURI
Università Roma Tre

Abstract

This paper argues for the manipulative potential of linguistic implicit strategies in a variety of text genres, including spoken political discourse, political tweets, spoken and written advertising, and a collection of journalistic texts. Building on the acknowledgment that linguistic implicit communication is an effective means of persuasion, the research relies on the theoretical distinction between

---

[1] Claudia Coppola has written §§1 and 3; Doriana Cimmino has written §§5.2.1, 5.2.2, 5.2.4 and 7; Federica Cominetti has written §§5.1 and 6; Giulia Giunta and Viviana Masia have written §2; Giorgia Mannaioli has written §§4, 5.2.3 and 8.

implicitation of content and implicitation of responsibility (Lombardi Vallauri 2019). Such taxonomy is presented and used to show that the text genres' features (in terms of compositional structure, thematic content and style) have an impact on the quantity and quality of implicit manipulative strategies used.

1. Introduction

Argumentation often occurs at the implicit level. When tendentious or questionable contents are conveyed through implicit linguistic strategies, manipulation can take place. This can happen not only in prototypically argumentative genres (such as political speeches or advertisements), but also in less prototypically argumentatively loaded ones (such as newspaper articles). In this paper we propose a preliminary step for the analysis of argumentation through implicit linguistic categories (presuppositions, implicatures, vagueness and topicalizations) by investigating how these are used in different text genres in the Italian context.

The paper is structured as follows: we provide a brief literature review on manipulative implicit strategies (§2) and on the relationship between argumentation and different text genres (§3); in §4 the research questions are defined and the methodology is illustrated; in §5 the corpora are described; §§6-7 include the analysis and §8 some concluding remarks.

2. Implicit communication and its manipulative uses

Many studies suggest that conveying some information through implicit linguistic strategies allows persuading the audience more effectively than expressing it on the text surface (Givón 1982; Rigotti 1988; Sbisà 2007; Lombardi Vallauri 2016a, b, 2019a; Mercier 2009; Reboul 2011; Saussure 2013; Lombardi Vallauri & Masia 2014, 2016a,b). Implicit communication rests on manifold linguistic devices, including presuppositions, implicatures, vague expressions, topicalizations. Drawing on the mainstream literature, we take presupposition to refer to any content that is presented as to be taken for granted in the conversation (Stalnaker 2002). Following Grice (1975), implicatures represent speakers' intentional meanings conveyed implicitly through the flouting or mere observance of conversational maxims. Vagueness is a property of expressions which are semantically imprecise and whose semantic boundaries are fuzzy (Russell 1923; Channell 1994; Cutting 2007). Topicalizations refer to those strategies of utterances' informational structuring which allow presenting some content as already active in the receiver's mind (Lombardi Vallauri 2009). Examples are provided below:

(1) *The Prime Minister has to **stop** making false promises* (PRESUPPOSITION = He has been making false promises)

(2) *Our country cannot grow **without keeping electoral promises*** (CONVERSATIONAL IMPLICATURE = The government does not keep its promises).

(3) *The Minister is young **but** expert.* (CONVENTIONAL IMPLICATURE = young people are usually inexpert).

(4) ***Some politicians** bribe people in exchange of votes* (VAGUE CONTENT = which politicians?)

(5) ***If politicians bribe citizens in exchange of votes**, democracy will fail* (TOPICALIZED CONTENT = politicians bribe citizens in exchange of votes)

The change of state verb *stop* in (1) presents as to be taken for granted that someone used to make false promises. Other presupposition triggers are definite descriptions, iterative adverbs like *again*, focus-sensitive operators like *also*, adverbial clauses, factives a.o. (Levinson 1983 a.o.). The utterance in (2) contains a conversational implicature, arising from the exploitation of a conversational maxim, as opposed to conventional implicatures, typically associated with the conventional meaning of a linguistic expression (e.g. *but* in (3)). In (4), it is not specified which politicians bribe people in exchange of votes. In (5) the hypothetical clause presents as already introduced in prior discourse that politicians bribe people in exchange of votes.

Implicitating strategies are very effective persuasion devices. Specifically, they divert receivers' attention and critical vigilance away from some information, as they present it either as shared (presupposition and topicalization) or as something that is constructed directly by the addressee (implicature and vagueness) (Sbisà 2007; Reboul 2011; Lombardi Vallauri 2019a a.o.).

However, not all uses of implicit communication should be seen as manipulative. Indeed, in everyday interactions some contents can be left implicit because they are actually part of an authentically shared common ground. But, when some contents are not yet known by the addressee and concern critical issues which the receiver should be aware of (as in 1-5), their conveyance by means of implicit discourse devices can prove manipulative.

Within the purview of argumentation studies, attention was devoted to disentangling the cognitive mechanisms behind implicit processing and how implicitation affects the mental representation of the meanings conveyed in a text (Maillat & Oswald 2009; Masia et al. 2017; Domaneschi et al. 2018). In this paper we are showing that a non-negligible use of implicit contents can be recognized in text genres having prototypically different degrees of argumentative features. This finding is in line with the most recent studies on text genres, as we show in the next section.

3. Argumentation in different text genres

Over the last years argumentation has been studied, within different frameworks (linguistic pragmatics, Lombardi Vallauri et al. 2018; Critical Discourse Analysis, Fairclough 1989; Pragma-dialectics, Van Eemeren 2018), in a wide spectrum of genres, ranging between more and less "prototypically argumentative" ones. Among the former, political discourse genres (e.g. meetings, Wodak 2013; interviews, Fetzer & Bull 2013; speeches, Forchtner 2013) and advertising (e.g. commercial advertisements, Kjeldsen 2012; print advertisements, Pollaroli 2015) have received considerable attention. Among allegedly less argumentatively loaded genres, media discourse has recently sparked scholars' interest (e.g. magazines' covers, Tseronis 2017; newspaper articles, Greco Morasso 2012; newspaper front page, Serafis 2022). Recent research is also showing that in some genres argumentation can be achieved multimodally (Lombardi Vallauri 2019b and Mackay 2013 for political genres; Pollaroli & Rocci 2015 for advertising; Serafis et al. 2020 for news).

The above-cited literature seems to corroborate the now well-established assumption in text genre studies that genres should be seen as social practices, i.e. as goal-oriented activities – or *activity types*, in Levinson's (1979) terms – that arise in response to recurrent situations within specific speech communities (e.g. a research article, a report)[2]. Moreover, as cognitive approaches have also recently shown[3], text genres emerge from concrete usage and gradually become recognizable on the societal level, which in turn constrain future instantiations. As Rocci (2008: 15) puts it: "genre conventions are often the main criteria that allow the reader or viewer to properly locate the communicative interaction proposed by the text in the proper spot of the relevant interaction field".

Within this framework, genres are thus to be seen as profoundly contextually and culturally situated. In this sense, they differ from the

---

[2] See a.o. the New Rhetoric school (Miller 1984), the "Sydney School" (Halliday 1985) and Swales' (1993) school of English for Specific Purposes.
[3] A more thorough review in Vergaro (2018) and Nikiforidou (2018).

notion of text types (e.g. narrative, procedural, argumentative, etc.), where the latter refer to "designations of (macro)speech acts" (Mortara Garavelli 1988: 159, our translation). Text types are categories of analysis defined on the basis of pragmatic abstract features, theoretically applicable to any empirical instantiation of discourse, i.e. to any genre. Based on this crucial distinction, argumentation should thus be intended as belonging to the text type dimension and, as such, as "cut[ting] across genres" (Labinaz & Sbisà 2018: 605). As already pointed out, there might be a prototypical correlation between a high presence of argumentative features and specific discourse genres, to be ascribed to the genre's communicative goal and social function. However, it would be inaccurate to think that "the (more or less explicit) presence of argumentation, or even the dominance of the argumentative text type, can by themselves determine which genre a certain instance of discourse belongs to" (ibid: 603).

In the light of all that, in the next sections we will analyse manipulative implicit linguistic strategies in text genres having prototypically different degrees of argumentative features.

4. Research questions and methodology

We have considered five corpora, four of which built specifically for this research. Two corpora represent political discourse, two commercial advertising (radio and visual ads), and the last one news articles. The selected genres, characterized by great impact on the public sphere, can be functionally characterized as predominantly persuasive and predominantly non-persuasive. These two groups of texts are characterized by significantly different macro-functions or macro speech acts. Indeed, while political and advertising discourse are meant to call the addressees to some action (conative function, e.g., voting or supporting a party, buying a product) news articles mostly perform an informative function, i.e., they supposedly report facts. From the argumentative point of view, they can be characterized in terms of "prototypically argumentative" text types vs. "non-prototypically argumentative"[4].

We will inquire: whether manipulative implicit strategies vary across different text genres (political speeches, tweets, advertisements, news articles) and types (prototypically argumentative vs non-prototypically argumentative); if yes, which quali-quantitative patterns can be recognized; and how implicit strategies interact with and enact the different features that characterize each genre (in terms of compositional structure, thematic content and style, following Bakhtin, 1986).

---

[4] It is worth pointing out that, given the considered corpora's sizes (apart from the political speeches' one) the present analysis does not intend to be representative of these text genres, but rather explorative of the cross-genre and cross-type variation.

Each corpus has been tagged with respect to the implicit categories presented in §1, and quantitative comparative analyses have been carried out.

5. The corpora

5.1. The IMPAQTS corpus

Many of the research questions tackled in this paper spark from a large research project currently in progress at Università Roma Tre: the IMPAQTS project. One of the objectives of the project is the building and pragmatic annotation of the IMPAQTS corpus of Italian political discourse. When completed, the corpus will include 1500 speeches held by Italian politicians since 1946 to present (around 250 hours of transcribed spoken language, around 2.5 million tokens).

The corpus is balanced according to sociolinguistic parameters, including the different communicative events in which the political speech takes place (cf. Table I):

Table I. Communicative events in the IMPAQTS corpus

Communicative event	Code	Number of speeches in the corpus
Official Assembly	A	600
Meeting, rally	C	300
Party Assembly	P	150
Declaration (broadcast)	T	450
Declaration (in presence)	D	
Declaration (new media)	N	

5.2. The *ad hoc* corpora

5.2.1 The political tweets corpus

The corpus contains 100 texts for a total amount of 2661 words, collected in two different spans of time: 2015-2016 and 2021-2022. They were partly automatically retrieved (Garassino et al. 2019: 196) and partly manually selected, considering both the influence of the politicians on Twitter (active presence) and the necessity for an overall balance among political colours (left, right and centrist). A gender-based balance was not contemplated.

5.2.2 The visual-ads corpus

The corpus collects 23 texts, counting 466 words in total, ranging from a minimum of 5 to a maximum of 82 words. Texts are taken from advertising billboards, broadcasted spots, and press. All texts were manually transcribed, if broadcasted, deriving punctuation from prosody, or simply reproducing the slogan written in the ad.

All ads were retrieved from a large collection of manipulative examples of advertising analysed for the OPPP! Website (www.oppp.it). Therefore, all contain at least one implicit strategy exploited for manipulative purposes and the corpus cannot be considered representative of the use of implicit strategies in advertising.

5.2.3 The radio-ads corpus

This corpus gathers 49 short radio-ads (2-10 sentences each). It counts 2044 words. The ads were broadcasted by Radio Globo, a Roman radio station, in 2020-2021. The audio files of the recordings were provided by the radio broadcasting station and were then transcribed and proofread.

The ads represent different merchandise categories: food, beauty products, cars, pharmaceutical products, services.

5.2.4 The news corpus

The news corpus contains four news articles, totalling 2247 words, extracted from *Oggi*, *Il Giornale*, *Il Tempo*, and *La Repubblica*, papers characterized by high circulation. The selected articles all deal with sensitive topics: a femicide, a case of child neglect, a case of young immigrant workers charged with sexual assault, and the case of a famous influencer charged with exploitation by his employees. All these topics can be expected to involve the presence of argumentative features, despite the text is supposed to be of a descriptive/informative type.

6. Quantitative analysis

As for the IMPAQTS corpus, we have analysed its currently available section (846 speeches uttered by 138 speakers, 1,813,709 tokens). In this section, numerous implicit contents were retrieved. In Table II we present the frequency of each implicit macro-category in the sections of the corpus corresponding to the different communicative events and in the entire corpus:

Table II. Implicit strategies/100k tokens per communicative event in IMPAQTS

	Implicatures	Presuppositions	Topics	Vagueness
A	7.80	8.11	1.74	3.65
C	8.58	7.67	1.54	5.24
P	6.43	7.40	1.41	5.25
D	5.03	5.60	1.23	3.25
Corpus	7.46	7.67	1.62	4.11

All the implicit macro-categories appear to be more frequent in the sub-corpus "A" (official assemblies), with the exception of vagueness, which is more frequent in speeches of kind "C" (meetings and rallies) and "P" (party assemblies) and implicatures, which are more frequent in "C" speeches. Accordingly, we can suppose that the communicative event in which a political speech takes place has a role in the kind and frequency of implicit strategies used in that speech. We can also suppose that such differences are related to the textual features typical of each communicative event or, more generally, of each text genre.

A different frequency of the implicit strategies – possibly revealing of some specific textual features – is in fact also observed in the *ad hoc* corpora. Of course, such small corpora cannot but suggest explorative tendencies. Nonetheless, the data presented in the following tables open some possibly fruitful lines of inquiry, discussed in the next section:

Table III. Frequency of occurrence of implicit strategies in the four *ad hoc* corpora

	I_con	I_cvr	P_tot	P_cs	P_def	P_adv	P_fact	Top	V_sem	V_syn
Tweet	0	66	57	12	34	2	9	16	21	10
Ad(V)	0	17	19	8	6	4	1	3	14	1
Ad(R)	0	16	13	2	8	0	0	7	3	3
News	4	8	20	0	11	2	0	1	0	0

Table IV. Implicit strategies/100k tokens in the four *ad hoc* corpora

	I_con	I_cvr	P_tot	P_cs	P_def	P_adv	P_fact	Top	V_sem	V_syn
Tweet	0.0	2450.5	2142.1	451.0	1277.7	75.2	338.2	601.3	789.2	375.8
Ad(V)	0.0	3648.1	4077.3	1716.7	1287.6	858.4	214.6	643.8	3004.3	214.6
Ad(R)	0.0	782.8	636.0	97.8	391.4	0.0	0.0	342.5	146.8	146.8
News	178.0	356.0	890.1	0.0	489.5	89.0	0.0	44.5	0.0	0.0

7. Qualitative analysis

The quantitative analysis seems to confirm the hypothesis that implicit strategies vary across different communicative settings and text genres, and that text genre-related features and text types-macro functions have an impact on the use of implicit strategies (§4).

A general tentative explanation that accounts for the variations lies in the predominant macro-function of the text genre or communicative setting considered. In particular, we observed that texts mainly performing a conative macro-function (more prototypically argumentative) exploit implicit manipulative strategies more often than less prototypically argumentative ones. In political tweets and advertising, the use of implicit strategies is higher than in news articles. This tendency does not exclude the use of implicit strategies in texts where informative type features are prevalent. In fact, overall, our data confirms that argumentation cuts across genres (see §3).

In what follows, we provide further details for each text genre considered in the four *ad hoc* corpora showing the relevance of the text's macro-function (i.e., text type) and textual features (i.e., text genre) in the variation of implicit strategies.

7.1. Tweets

The brevity of tweets makes the use of implicit strategies particularly convenient. As mentioned in §2, implicit strategies are indeed means where the economic transmission of linguistic content can overlap with a manipulative purpose. Consider example (6):

(6) Italia Viva non solo ha salvato il Paese <u>dall'esperienza tragicomica e populista del governo Conte</u>, ma ha creduto in un gruppo di ragazze e ragazzi. Questa è la cosa più bella.
'Not only has Italia Viva saved the Country <u>from the tragicomic and populist experience of the Conte Government</u>, but it [also] believed in a team of girls and boys. This is the most wonderful thing.'
(Tweet, Italia Viva)

With a *not only…but [also]* correlative coordinate structure, this tweet takes for granted that Italia Viva saved the country, and, at the same time, they believed in the young forces of their party. The tweet also contains an attack on the past government, implicitly described as tragicomic and populist by means of a definite description. In this tweet, as in many other examples, the brevity typical of the genre often results in questionable contents conveyed as presuppositions.

7.2. Advertising

In advertising as well, implicitness is often employed to smuggle manipulative contents while granting brevity. Let us consider (7), where the brand's competitors are negatively connotated through a brief and semantically vague slogan:

(7) Arriviamo <u>dove gli altri non arrivano</u>.
'We arrive <u>where others don't</u>.'
(visual advertising, BCC Bank)

Leaving the profile of commercial competitors vague (they do not arrive to what?) spares critiques or legal actions to the company, while making the vague statement unfalsifiable, and argumentatively unchallengeable[5]. Despite denotatively poor, the slogan's connotation is clearly positive, presenting the advertised company as more desirable than its competition - although the reason for this supremacy remains unspecified.

A similar mechanism can be observed in radio advertising. Buyers are not persuaded through substantiated arguments or reliable proofs, but through implicit strategies. In (8), the paratactic construction, typical of this text genre, leads the listener to imply that the advertised product is what they need. Moreover, it is presupposed that they wish they had been born with another hair colour.

(8) Nuovo Prodigy di L'Oréal Paris con tecnologia microoil. Il colore prodigioso con cui vorresti essere nata. Perché tu vali.

---

[5] See Russell (1923:90).

'New Prodigy by L'Oréal Paris with microoil technology. The prodigious color with which you wish you were born. Because you are worth it.'
(L'Oréal, radio advertising)

In conclusion, it is worth noting that both visual and audio advertising are characterized by multimodality. Images and sounds paired with the ad texts can implicitly and immediately convey information about the brand or positively connotate it. The slogan in (9) is associated with the background sounds of an award ceremony.

(9) Per la capacità di emozionare ad ogni apparizione, Peugeot 207 Coupé Cabriolet!
"For its ability to thrill you at every appearance, Peugeot 207 Coupé Cabriolet!"
(Peugeot, radio advertising)

The topicalized and presupposed argument "For its ability to thrill you at every appearance" presents the information as if it were the label of an award category, already active in the discourse and part of the knowledge of the buyers, requiring no further critical evaluation. At the same time, the background music implies that Peugeot has already beaten its competitors.

7.3. News

As anticipated, our analyses confirm that argumentation cuts across genres. We retrieved manipulative implicit strategies also in news articles, when they deal with particularly sensitive topics. Overall, the nature of the implicit strategies found in news articles differs from those in tweets and advertising, consistently with their specific textual features. For example, we did not retrieve any occurrence of vagueness used manipulatively, since vagueness would probably contrast too dramatically with the informative function of the text. On the contrary, conventional implicatures triggered by conjunctions are comparatively very frequent, probably because they are less common in short and paratactic texts such as tweets and visual and radio advertising.

Interestingly, the headlines show aspects of continuity with tweets and advertising. In (10), two presuppositions and an implicature convey disputable content.

(10) Il gigante buono e quell'amore non corrisposto.
'The good giant and that unreturned love'.
(Il Giornale, news)

In the headline, the murderer protagonist of the news is introduced as 'the good giant'. The definite description presupposes that he was good, somehow smoothing his responsibility in the killing. Moreover, mentioning 'that unreturned love' in a paratactic construction, the journalist conversationally implies that love was the reason of killing. The content implicitly conveyed is the result of an interpretation of the facts by the journalist, which makes the headline catchier for the potential reader. The fictionalization of the news, typical of the headlines (and of some articles) is thus reminiscent of the conative function predominant in tweets and advertising. This is exploited to convey an emotionally engaging opinion, likely to increase the newspaper's sales.

8. Conclusion and future perspectives

In this work we have inquired into the use of potentially manipulative implicit strategies (i.e., implicitly conveying questionable contents) across different text types and genres. In order to do so, we have analysed five original Italian corpora: monological political discourse; political discourse via tweets; visual-ads; radio-ads; (sensitive) news articles.

As for the different text types, our analysis has shown a greater use of potentially manipulative implicit strategies in prototypically argumentative texts. As for genres, the corpora in which most implicit strategies are used to convey questionable contents are the visual advertising and the tweet corpus. The use of implicit strategies reflects both the constraints characterizing the texts - e.g., the channel imposing a limited space, in the case of tweets - and the genre's typical thematic content and macro-function - e.g., attacking and discrediting adversaries in the case of political discourse.

Across the different corpora, while presuppositions are often arguably employed in the interests of brevity, it is often the case that questionable information is concealed behind implicatures or vague expressions, whose contents are for the recipients to infer. This rhetorical move is often employed in discrediting opponents and competitors - a traditionally face-threatening act, which is better performed indirectly. Interestingly, several occurrences of non-*bona fide* true implicit contents, notably implicatures and presuppositions, were also found in news articles concerned with particularly sensitive topics. Conversely, vagueness and topicalizations are less plausible for this text genre: given their function of providing new information, it would be semantically and pragmatically awkward to present it in vaguely, or as if it is already active in the receivers' mind.

In order to provide a representative analysis, in future work larger corpora of each text genre should be considered. Furthermore, the analysis should be extended to other text genres and types. Once this task is completed, our findings will allow a more fine-grained analysis of the role played by manipulative implicit contents in argumentation across different genres, possibly by crossing the methodology with other analytic tools offered by argumentation theory (e.g. Argumentum Model of Topics, Rigotti & Greco 2019).

References

Bakhtin, M. (1986). The problem of speech genres. In C. Emerson and M. Holquist (Eds.), *Speech Genres and Other Late Essays*, eds). Austin: University of Texas Press, 60-102.
Chafe, W. (1976). Givenness, contrastiveness, definiteness, subjects, topics, and point of view. In *Subject and topic*, edited by Charles N. Li, New York: Academic Press, 25-55.
Chandler, P. & Sweller, J. (1991). Cognitive Load theory and the Format of Instruction. *Cognition and Instruction*, 8(4), 293-332.
Channell, J. (1994). Vague language. Oxford: Oxford University Press.
Cutting, J. (2007). Introduction to vague language explored. In *Vague language explored*, Palgrave Macmillan, London, 3-17.
Domaneschi, F., Canal, P., Masia, V., Lombardi Vallauri, E. & Bambini, V. (2018). N400 and P600 modulation in presupposition accommodation: The effect of different trigger types. *Journal of Neurolinguistics*, 45, 13-35.
Fairclough, Norman. (1993). *Discourse and Social Change*. Cambridge: Polity Press.
Fetzer, A. & Bull, P. (2013). Political interviews in context. In P. Cap & U. Okulska (Eds.). *Analyzing genres in political communication: Theory and practice*. John Benjamins Publishing Company, 73-99.
Forchtner, B. (2013). Legitimizing the Iraq War through the genre of political speeches. Rhetorics of judge-penitence in the narrative reconstruction of Denmark's cooperation with Nazism. In P. Cap & U. Okulska (Eds.), *Analyzing genres in political communication: Theory and practice*. John Benjamins Publishing Company, 239-265.
Fried, M. (2015). Construction Grammar. In Artemis Alexiadou & Tibor Kiss (eds.), Handbook of syntax, vol. 2, 2nd ed., Berlin: Mouton de Gruyter, 974–1003.
Garassino, D., Masia, V. & Brocca, N. (2019). Tweet as you speak. The role of implicit strategies and pragmatic functions in political communication: Data from a diamesic comparison, *Rassegna Italiana di Linguistica Applicata*, 2/3, 187-208.
Garavelli, B. M. (1988). 246. Textsorten / Tipologia dei testi. In G. Holtus, M. Metzeltin, & C. Schmitt (Eds.), *Italienisch, Korsisch, Sardisch*. De Gruyter, 157–168.
Givón, T. (1982). Evidentiality and epistemic space. *Studies in Language*, 6(1), 23–49.
Greco Morasso, S. (2012). Contextual frames and their argumentative implications: A case study in media argumentation. *Discourse Studies, 14(2)*, 197–216.

Grice, H. P. 1975. Logic and Conversation. In Syntax and Semantics, Vol. 3, Speech Acts, ed. by Peter Cole and Jerry L. Morgan. New York: Academic Press 1975, 41–58.
Kjeldsen, J. E. (s.d.). Virtues of visual argumentation: How pictures make the importance and strength of an argument salient. 14.
La Rocca, D. et al. (2016). Brain response to information structure misalignments in linguistic contexts. *Neurocomputing*, 199, 1-15.
Labinaz, P., & Sbisà, M. 2018. Argumentation as a dimension of discourse: The case of news articles. *Pragmatics & Cognition*, 25(3), 602–630.
Levinson, S. C. (1979). Activity type and language. *Linguistics*, 17(5/6), 356–399.
Levinson, S. C. (1983). Pragmatics. Cambridge: Cambridge University Press.
Lombardi Vallauri, E. & Masia, V. (2014) Implicitness impact: measuring texts. Journal of Pragmatics, 61:161–184
Lombardi Vallauri, E. & Masia, V. (2016a). Misurare l'informazione implicita nella propaganda politica italiana. In R. Librandi & R. Piro (Eds.), L'italiano della politica e la politica per l'italiano, Firenze, Cesati, 539-557.
Lombardi Vallauri, E & Masia, V. (2016b). Facilitating Automation in Sentence Processing: The Emergence of Topic and Presupposition in Human Communication. *Topoi*, 37(2), 343-354.
Lombardi Vallauri, E. (2009). La struttura informativa. Forma e funzione negli enunciati linguistici. Roma: Carocci.
Lombardi Vallauri, E. (2016a). Implicits as Evolved Persuaders. In K. Allan, A. Capone & I. Kecskes (Eds.), *Pragmemes and Theories,* Cham: Springer, 725-748.
Lombardi Vallauri, E. (2016b). The "exaptation" of linguistic implicit strategies. *SpringerPlus*, 5(1), 1106.
Lombardi Vallauri, E. (2019a). La lingua disonesta. Bologna: Il Mulino.
Lombardi Vallauri, E. (2019b). "Sfruttamento di "immagini" implicite nella pubblicità e nella propaganda politica italiana". In E. M. Mirković, T. Habrle (Eds.). *Sguardo sull'immaginario italiano. Aspetti linguistici, letterari e culturali*, Edizioni dell'Università degli Studi Juraj Dobrila di Pola, 267-294.
Lombardi Vallauri, E., Baranzini, L., Cimmino, D., Cominetti, F., Coppola, C., & Mannaioli, G. (2020). Implicit argumentation and persuasion: A measuring model. *Journal of Argumentation in Context*, 9(1), 95–123.
Mackay, R. R. (2013). Multimodal legitimation Looking at and listening to Obama's ads. In P. Cap & U. Okulska (Eds.). *Analyzing genres in political communication: Theory and practice.* John Benjamins, 345-377.
Maillat, D. & Oswald, S. (2009). Defining Manipulative Discourse: The Pragmatics of Cognitive Illusions. *International Review of Pragmatics*, 1(2), 348-370.
Martin, J. R. (1984). Language, register and genre. F. Christie. Language Studies: Children's Writing: Reader. Geelong, Victoria: Deakin University Press, 21-9.
Masia, V. et al. (2017). Presupposition processing as a pragmatic garden path: evidence from event-related brain potentials. *Journal of Neurolinguistics*, 42, 31-48.
Mercier, H. (2009). *La Théorie Argumentative du Raisonnement.* PhD Thesis, E.H.E.S.S. Paris.
Miller, C. R. (1984). Genre as social action. *Quarterly Journal of Speech*, 70, 151–67.
Nikiforidou, K. (2018). Genre and constructional analysis. *Pragmatics & Cognition*, 25(3), 543–575.

Pollaroli, C. (2015). T(r)opical patterns in advertising: The argumentative relevance of multimodal metaphor in print advertisements. Unpublished PhD thesis.
Pollaroli, C., & Rocci, A. (2015). The argumentative relevance of pictorial and multimodal metaphor in advertising. *Journal of Argumentation in Context, 4(2)*, 158–199.
Reboul, A. (2011). A relevance-theoretic account of the evolution of implicit communication. *Studies in Pragmatics*, 13, 1-19.
Rigotti, E. (1988). Significato e senso. In E. Rigotti & C. Cipolli (Eds.), *Ricerche di semantica testuale* (pp. 71-120). Brescia: La Scuola.
Rigotti, E., & Greco, S. (2019). Inference in Argumentation: A Topics-Based Approach to Argument Schemes. Springer International Publishing.
Rocci, A. (2008). Analysing and evaluating persuasive media discourse in context. In M. Burger (Ed.), *L'analyse linguistique des discours des médias: Théories, méthodes et enjeux*. Québec : Nota Bene. (pp. 247–284).
Russell, B. (1923). Vagueness. The Australasian Journal of Psychology and Philosophy, 1(2), 84-92.
Saussure, L. (2005). Manipulation and Cognitive Pragmatics: Preliminary Hypotheses. In de Saussure Louis & Peter Schulz (Eds.), *Manipulation and Ideologies in the Twentieth Century: Discourse, Language, Mind, Amsterdam-Philadelphia*, John Benjamins (pp. 113-146).
Saussure, L. de (2013). Background relevance. Journal of Pragmatics, 59, 178–189.
Sbisà, M. (2007). Detto non detto. Le forme della comunicazione implicita. Roma-Bari, Laterza.
Serafis, D. (2022). Unveiling the rationale of soft hate speech in multimodal artefacts: A critical framework. *Journal of Language and Discrimination 6 (2)*, 321–346.
Serafis, D., Greco, S., Pollaroli, C., & Jermini-Martinez Soria, C. (2020). Towards an integrated argumentative approach to multimodal critical discourse analysis: Evidence from the portrayal of refugees and immigrants in Greek newspapers. *Critical Discourse Studies, 17(5)*, 545-565.
Stalnaker, R.C. (2002). Common Ground. Linguistics and Philosophy, 25, 701-721.
Swales, J. M. (1993). Genre and Engagement. *Revue Belge de Philologie et d'histoire*, 71(3), 687–698.
Tseronis, A. (2017). Analysing multimodal argumentation within the pragma-dialectical framework: Strategic manoeuvring in the front covers of *The Economist*. In F. H. van Eemeren & W. Peng (Eds.), *Argumentation in Context* (12), 335-359.
Van Eemeren, F. H. (2018). Argumentation Theory : A Pragma- Dialectical Perspective. Cham: Springer.
Vergaro, C. (2018). A cognitive framework for understanding genre: The Entrenchment-and-Conventionalization Model. *Pragmatics & Cognition, 25(3)*, 430-458.
Wodak, R. (2013). Analyzing meetings in political and business contexts. Different genres – similar strategies?. In P. Cap & U. Okulska (Eds.). *Analyzing genres in political communication: Theory and practice.* John Benjamins Publishing Company, 187-221.

NOT TO PLAY DEVIL'S ADVOCATE, BUT...

MARÍA INÉS CORBALÁN
ArgLab-IFILNOVA, NOVA Universidade de Lisboa
inescorbalan@fcsh.unl.pt

GIULIA TERZIAN
ArgLab-IFILNOVA, NOVA Universidade de Lisboa
giuliaterzian@fcsh.unl.pt

Abstract

Devil's advocacy has long been heralded as a paradigm of virtuous argumentation. Outside academic circles, however, it has acquired a notorious reputation. This paper takes a first step towards uncovering the dark side of devil's advocacy. In particular, it builds on Scott Jacobs's speech act analysis of devil's advocacy to suggest that observed vicious uses of devil's advocacy may be usefully understood to exemplify what Jennifer Saul has termed force fanleaves.

1. Introduction

Is it always permissible to voice our disagreement with another's assertion, if we know that it expresses a false or unwarranted belief? Are we sometimes obligated to speak up if we believe our interlocutor to be mistaken in his belief that p, and if so why - on what grounds? Recent work in social epistemology has lent support to the idea that sometimes, the normative balance in such situations will be tipped by distinctively epistemic considerations: positive, other-directed, collective and *pro tanto* obligations to promote epistemic goods or goals (Johnson, 2018b; Lackey, 2020; Terzian & Corbalán, 2021).

Voicing our sincere disagreement with an interlocutor, then, can be an important way of discharging such interpersonal epistemic duties, by preventing others from incurring epistemic harms (such as entertaining false beliefs). But it is not the only such way: promoting beneficial epistemic outcomes may sometimes go by way of giving voice to objections we don't ourselves subscribe to. This paper focuses on a discursive move that fits this description and is familiar from our everyday conversational lives: playing devil's advocate.

In so doing, we seek to join a somewhat scattered cross-disciplinary conversation over the normativity of playing devil's advocate: Are we ever obliged to play devil's advocate? What and who benefits from devil's advocacy? What does it take to do it well? And so on. Discussions of these and related questions can be found in various corners of different literatures, both theoretical (e.g. Mill, 1859; Aikin & Clanton, 2010; Johnson, 2018a; Stevens & Cohen, 2021) and empirical (e.g. Duran & Fusaroli, 2017; Nemeth, Brown, et al., 2001). By and large, all of these analyses tend to focus on devil's advocacy as a discursive move that is virtuous by design *and* is undertaken by virtuous interlocutors; a key presupposition is that devil's advocates pick their moments well and act in good faith. Much like any discursive practice, however, devil's advocacy is liable to be both misused and abused. Indeed, this seems to have become a frequent occurrence in certain real-world conversational contexts. The following passage, hereafter labelled [Affirmative Action], is representative of this recently observed trend:

> The man had waited in line for 10 minutes so he could tell me that his son had been denied admission to my law school alma mater because, unlike me, he was white and so couldn't benefit from affirmative action. [...] I recounted this infuriating story to a white friend of mine [...]. Instead of rolling his eyes along with me, my friend forced me to debate him –on behalf of the man from the panel [and] the devil– on whether maybe I really hadn't deserved admission to my law school. [...] [C]ross racial conversations about race have become [...] more common, and thankfully so. Unfortunately, this has invited a dangerous tendency for white people to engage in these discussions with people of color by summoning the devil himself and treating racism as a political disagreement around which two opposing viewpoints can reasonably form. [...][1]

We take it to be uncontroversial that something has gone seriously awry in the exchange reported above. Thus, we agree with the narrator's negative assessment: their friend's conversational contribution, under the guise of a well-meaning devil's advocate, was inappropriate. Moreover, as the narrator notes, the significance of this particular episode seems to stretch far beyond the merely anecdotal: conversations following this pattern seem to have become paradigmatic rather than outliers.

Now, a superficially tempting hypothesis is that these cases can simply be explained away as mistakes, merely reflecting a widespread misunderstanding of devil's advocacy. But the examples examined do not support this hypothesis: only rarely do the speakers appear to be confused or ill-informed about the use or purpose of the argumentative machinery

---

[1] https://slate.com/news-and-politics/2017/10/playing-devils-advocate-in-conversations-about-race-is-dangerous-and-counterproductive.html All webpages accessed in February 2023.

distinctive of this discursive move. A much more plausible hypothesis is that we are witnessing an abuse of the original purpose for which the move was designed.

We think, then, that [Affirmative Action] exemplifies a formally correct but vicious use of devil's advocacy. Our goal in this paper is to better understand what exactly makes it so – what exactly goes on, and goes wrong, in cases such as [Affirmative Action].

We proceed as follows. In Section **Error! Reference source not found.**-3 we offer some more detail on the profile of the ideal devil's advocate and of its real-world incarnations. Section 4 presents a speech act analysis of devil's advocacy, building on work by Scott Jacobs. Section 5 introduces what we see as a useful conceptual resource for the purpose of diagnosing at least some of the problematic uses of devil's advocacy in contemporary discourse. Section 6 briefly concludes.

2. Ideal devil's advocacy

As is known, devil's advocacy is a well-established means of introducing or manufacturing disagreement in a conversational setting. Its deployment is moreover typically (though not invariably) announced or flagged by highly recognisable preambles (sometimes added post hoc): Let me play devil's advocate here..., Let's just say, for the sake of argument..., I was just playing devil's advocate there, and so on.

In its default or textbook form, devil's advocacy has well known positive connotations. It received an illustrious endorsement from J. S. Mill as a means of staving off the danger of holding beliefs as "dead dogma", rather than "living truth":

> [The truth] is [n]ever really known but to those who have attended equally and impartially to both sides and endeavored to see the reasons of both in the strongest light. So essential is this discipline to a real understanding of moral and human subjects that, if opponents of all-important truths do not exist, it is indispensable to imagine them and supply them with the strongest arguments which the most skilful devil's advocate can conjure up. (Mill, 1859, pp. 35-36)

Following Mill, a number of scholars have sought to tease out the ways in which devil's advocacy can help further desirable epistemic outcomes and safeguard against cognitive and argumentative pitfalls. Katharina Stevens and Daniel Cohen give an especially clear-eyed and detailed articulation of this thought:

> We want an arguer who *opposes* us to *help* us: an *advocatus diaboli*, a Devil's Advocate. A devil's advocate is not merely a useful interlocutor: [...] she is the ideal other who embodies what is best and most important about argumentation. She is the opponent we need because her overall goal is to enhance the prospects of successful argumentation, i.e., getting it right [...]. [Devil's advocates] help us transcend our limits by criticizing our argument in order to strengthen it, not to defeat it. (Stevens & Cohen, 2019, p. 170)

Echoing this characterization, with a marked emphasis on the *epistemic* benefits of devil's advocacy, Casey R. Johnson notes that:

> Paradigmatically, an agent who plays devil's advocate announces her intention to defend a position she doesn't hold, and then defends that position in order to make progress on the issue at hand. [...] Paradigmatically, the term devil's advocate describes someone who, given a certain point of view, takes a position she does not necessarily agree with for the sake of debate or to explore the thought further. (Johnson, 2018a, pp. 97,99)

In turn, Scott Aikin and Caleb Clanton have stressed the value of devil's advocacy for our *collective* epistemic and deliberative practices: as long as certain background conditions are in place – most prominently, "accurate knowledge of what exactly the other side believes, what reasons they proffer (or would proffer), and what inclinations they themselves have (or would have)" – they argue that "there is a virtue to being *overtly* and *openly* uncooperative", as when one plays devil's advocate (Aikin & Clanton, 2010, p. 419).

The devil's advocate profile that emerges from the theoretical literature is, then, unequivocally positive. *Qua* argumentative or discursive practice, devil's advocacy is unanimously recognised as virtuous insofar as it is conducive to the fulfilment of desirable epistemic, rhetorical and argumentative goals, on both an individual and a collective level. In particular, devil's advocacy can be *beneficial to the arguer(s)*, by prompting them to critically reflect on the epistemic standing of their belief (Johnson, 2018a), helping them spot and repair weaknesses in their arguments (Stevens & Cohen, 2019), or equip them to face a genuine opponent (Aikin & Clanton, 2010). And it can be *beneficial to the argument(s)*, either in and of themselves (by uncovering hidden weaknesses), or as touchstones of the state of a particular debate (by broadening the space of conceivable options), or as benchmarks of deliberative quality (by safeguarding against groupthink and intellectual stagnation).[2]

---

[2] As mentioned earlier, devil's advocacy has also been the object of a smattering of empirical studies, whose findings are at least consistent with the above-described

3. The dark side of devil's advocacy

Ideal devil's advocates display recognisable clusters of traits, falling under four stable headings: argumentative attitude (cooperative), doxastic stance (private and public commitments typically but not necessarily divergent), intended beneficiary (the arguer(s), the argument), primary intellectual benefit (epistemic, cognitive, rhetorical). Obviously, these profile sketches significantly under-describe the contextual richness of real-world exchanges. Here we sketch a further profile, of a devil's advocate who has become not only recognisable but notorious in recent years.

A first observation is that vicious uses of devil's advocacy appear to strongly correlate with specific *topics of conversation*. As noted by the narrator in [Affirmative action], and echoed in the following excerpt, conversations about race- and gender-based discrimination, and corresponding corrective policies and practices, are among the most prominent examples:

> [Playing] devil's advocate when the topic under discussion is, say, whether or not we should pursue immortality or how best to end our dependence on non-renewable energy sources will probably be productive and enlightening. On the other hand, when the topic is whether or not it should be legal to shoot unarmed Black teenagers or how best to respond to sexual assault, [devil's advocacy] is a minefield of potential faux pas, triggers, and discussions that end in yelling and/or blocking each other online.[3]

Unsurprisingly, the assignation of conversational roles inevitably tracks the very same social divides made salient by these conversational topics:

> You post that article about the wage gap on Facebook, and all of a sudden, all of these cis, white, straight dudes come out of the woodwork to remind you that the statistics are faulty, that women take more time off of work, that women just don't like STEM fields —all under the guise of "playing devil's advocate".[4]

Notably, moreover, the vast majority of the observed contributions are made by devil's advocates who appear to be impervious to the

---

theoretical analyses. See for instance (Brohinsky et al., 2021; Duran & Fusaroli, 2017; Nemeth, Brown, et al., 2001; Nemeth, Connell, et al., 2001).
3 https://the-orbit.net/brutereason/2013/08/10/how-to-be-a-responsible-devils-advocate/
4 https://everydayfeminism.com/2015/09/playing-devils-advocate/

conversational score rules in play. As a result, the primary effects of these contributions are unequivocally negative, leading in the first instance to the disruption of on-topic discussions or the silencing of relevant testimony; and foreseeably hindering progress towards desirable social and epistemic goals. This is exemplified in the following episode, labelled [Street harassment] for later reference:

> I was asked to come on [a radio show] to talk about street harassment. [...] After being asked a few broad-sweeping questions that repeatedly prompted me to address the oft-claimed defense that street harassment is "just a compliment" [...], the host specifically asked for other tattooed women to call in and discuss their experiences with their body art and street harassment. So I really wasn't surprised when the first call answered was from a dude. [...] His argument was that not every situation can be tied back to structural oppression, and that my point about how so-called "harmless compliments" are actually indicative of just how much women's bodies are not respected in public spaces was absurd. [...] He just wanted to [...] give us "the other side of the story," since my stance was "one-sided" and "slanted." He just wanted to "intelligently, rationally debate" this topic [...] under the guise of "playing devil's advocate" —as if [I had] never heard these arguments before.[5]

Notice the very last quoted remark: a widely observed characteristic of vicious devil's advocacy is the tireless reiteration of objections that do not, by any stretch of the imagination, qualify as outside-the-box thinking or intellectually stimulating or beneficial:

> Some might challenge that I am shutting myself off to new ideas and censoring important opportunities for growth. But these ideas you are forcing me to consider are not new. They stem from centuries of inequality and your desperate desire to keep them relevant is based in the fact that you benefit from their existence. Let it go. You did NOT come up with these racist, misogynistic theories. We've heard them before and we are f*cking tired of being asked to consider them, just one.more.time.[6]

Importantly, as noted here, the objections being reiterated under the guise of devil's advocacy are not merely unimaginative: they coincide with still-dominant narratives in race- and gender related discourse. This speaks especially damningly to the poor quality of the proffered

---

5 https://everydayfeminism.com/2015/09/playing-devils-advocate/
6 https://feministing.com/2014/05/30/an-open-letter-to-privileged-people-who-play-devils-advocate/

contributions, and consequently their expected benefits (or lack thereof). For, a paradigmatic presupposition of devil's advocacy is (i) that the disputant(s) may have unduly neglected a particular hypothesis (and the possible reasons supporting it); and (ii) that taking into consideration an idea previously left on the margins may be beneficial to the disputant(s) and quality of their deliberation. But there is, of course, no plausible sense in which the dominant standpoints, in race- and gender-related discourse, have ever been relegated to the margins. Far from enhancing deliberative quality, the observed contributions are effectively advocating for the *status quo*.

In light of the foregoing, it is hardly surprising that several observers have highlighted the characteristically adversarial rather than cooperative attitude of self-proclaimed devil's advocates:

> It is especially harmful [...] when a man plays the Devil's Advocate to a woman trying to discuss feminist issues; it becomes another way to silence women and disregard our experiences. Disagreeing with someone who is promoting equal rights just "for the sake of argument" ultimately trivialises the oppression that marginalised people face daily.[7]

> ...if you're playing devil's advocate in order to try and help someone else, find out if that person actually wants or needs your help. Unsolicited advice is frankly annoying in almost any case, but especially when it involves a long, drawn-out debate with someone you believe to be in need of convincing [...][8]

Extant theoretical analyses, focusing as they do on the ideal form or paradigm of devil's advocacy, are –unsurprisingly – silent about almost every one of these contextual features. As a result, they are ill-equipped to understand exactly what goes wrong in these exchanges; and to substantiate the strong intuition, not only that these examples cannot be dismissed as mistakes borne of linguistic incompetence, but that they exemplify bad faith, or vicious, linguistic behaviour. Following up on this intuition we next turn to a speech act analysis of devil's advocacy.

4. A speech act analysis of devil's advocacy

Nearly 40 years ago, van Eemeren & Grootendorst (1984) proposed a systematic reconstruction of arguments as complex speech acts – that is,

---

[7] https://ashamedmagazine.co.uk/opeds/why-do-cis-men-love-to-play-devils-advocatenbsp
[8] https://the-orbit.net/brutereason/2013/08/10/how-to-be-a-responsible-devils-advocate/

combinations or constellations of at least two elementary assertive speech acts. On their enormously influential account, the paradigmatic argumentative practices of pro-argumentation and contra-argumentation are construed as illocutionary act complexes, which stand in "a justifying or refuting relation to an expressed opinion (which consists of statements acting as a claim or conclusion)" (van Eemeren & Grootendorst, 1984, p. 39). Not all argument types are easily accommodated by this binary classification, however. In particular, as argued by Scott Jacobs, this is true of "two 'nonstandard' ways of putting forward arguments that have conventional folk language descriptors: hypothetical argument and devil's advocacy" (Jacobs, 1989, p. 353). Jacobs's key thesis here is that while devil's advocacy might superficially appear to be a form of contra-argumentation, it does not carry the latter's characteristic illocutionary force –namely, of attempting to convince the hearer of the unacceptability of some standpoint. Instead, devil's advocacy carries a distinctive illocutionary force, which Jacobs characterises as 'idea-testing': its function "is not so much to try to convince one's interlocutor of the unacceptability of [the conclusion] O, as to test for the acceptability or unacceptability of O by seeing whether one's own arguments are acceptable or unacceptable to the listener" (ibid., emphases added). In fact, as we'll see shortly, the differences between contra-argumentation and devil's advocacy extend beyond just their illocutionary force. To see this, it is helpful to compare the respective sets of felicity conditions, displayed in Figs. 1 and 2 below. The first set reproduces Jacobs's formulation of the felicity conditions for contra-argumentation given in (van Eemeren and Grootendorst, 1984). The second set is our own reconstruction from Jacobs's informal discussion.

Recognition conditions
1. *Hearer* puts forward an expressed opinion O,
2. *Speaker* puts forward a series of assertions, $S_1, S_2, ..., S_n$, in which propositions are expressed.

[**Propositional content conditions**]

3. Advancing $S_1, S_2, ..., S_n$ counts as an attempt by the speaker to *convince* the hearer of the unacceptability of O.

[**Essential condition**]

Correctness conditions
4. The speaker believes that:
a) The hearer accepts O,
b) The hearer does (or will) accept $S_1, S_2, ..., S_n$,
c) The hearer will accept $S_1, S_2, ..., S_n$ as a refutation of O.

[**Preparatory conditions**]

5. The speaker believes that:
a) O is unacceptable,
b) $S_1, S_2, ..., S_n$ are acceptable,
c) $S_1, S_2, ..., S_n$ *refute* O.

Figure 1. Felicity conditions for contra-argumentation

Recognition conditions
1. *Hearer* puts forward an expressed opinion O,
2. *Speaker* puts forward a series of assertions, $S_1, S_2, ..., S_n$, in which propositions are ex-pressed.
[**Propositional content conditions**]
3. Advancing $S_1, S_2, ..., S_n$ counts as an attempt by the speaker to *test*
a) *the epistemic standing* of O, or
b) *the hearer's epistemic standing* with respect to O.
[**Essential condition**]
Correctness conditions
4. The speaker believes that:
a) The hearer accepts O.
5. The speaker *is not committed* to believing that:
a) The hearer does (or will) accept $S_1, S_2, ..., S_n$,
b) The hearer will accept $S_1, S_2, ..., S_n$ as a refutation of O.
[**Preparatory conditions**]
6. The speaker *is not committed* to believing that:
a) O is unacceptable,
b) $S_1, S_2, ..., S_n$ are acceptable,
c) $S_1, S_2, ..., S_n$ refute O.
[**Sincerity conditions**]
Figure 2. Felicity conditions for devil's advocacy (ideal profile)

Pairwise comparison of the two sets of felicity conditions reveals a number of differences between contra-argumentation and devil's advocacy. A first prominent divergence is found at the level of the respective essential conditions, which in turn track illocutionary force. This is of course as expected: whereas contra-argumentation (and, for that matter, pro-argumentation) is generally conceived of as serving a *persuasive* function, and accordingly "naturally imagined in the adversarial setting of a debate", the paradigmatic function or functions of devil's advocacy are of a distinctively exploratory nature, ideally suited to "a cooperative, joint problem-solving discussion" (Jacobs, 1989, p. 353).

Just as interestingly, the correctness conditions for contra-argumentation and devil's advocacy differ on all counts but one (item 4.a) in both lists):

> In these sorts of arguments, the speaker is not committed to believing that O is unacceptable. Nor is the speaker committed to believing that [the premises] $S_1, S_2, ..., S_n$ are acceptable or that the hearer believes this. Nor is the speaker committed to believing that $S_1, S_2, ...S_n$, refute O or that the hearer will believe this. In fact, part of the point of devil's advocacy is to avoid the characteristic

commitments of contra-argumentation. The speaker is only committed to believing that *someone might think* these things [...]. (Jacobs, 1989, pp. 353s)

Once the Jacobsian framework is in place, some of the observed vicious instantiations of devil's advocacy are easily recognisable as violations of one or more of the displayed sincerity conditions. This seems to be an apt diagnosis in the case of [Street harassment], for instance: the most plausible reading of the reported exchange is that the speaker (male caller) *is committed to believing* that the hearer's expressed opinion ('tatcalling' constitutes harassment) is unacceptable, and that his own proffered reasons (structural oppression is not as pervasive as it is made out to be, 'tatcalling' is a harmless compliment) are acceptable and they refute the main claim.

By the lights of Jacobs's analysis, then, cases such as [Street harassment] will receive one of two possible diagnoses. Depending on whether one judges that the essential condition of devil's advocacy (either version) has been fulfilled, we can conclude that the speaker has performed a *misleading* speech act, or that the speech act itself has *misfired*.

The diagnostic capabilities of the Jacobsian theoretical framework are encouraging, we think. On its own, however, this textbook speech act analysis is patently unequipped to capture or shed light on the full range of traits distinctive of real-world devil's advocacy identified in Section 3. That being said, we think it serves as a springboard for making some progress in this direction. As we'll now see, this requires paying specific attention to the work being done by the characteristic preamble(s) announcing a devil's advocate's contribution (hereafter, DA).

Notice, first, that the two initial recognition conditions of devil's advocacy (identifying the propositional content of the speech act) coincide with those of contra-argumentation: in both cases, the hearer's utterance of O is followed by the speaker's utterance of reasons $S_1, S_2, ..., S_n$. The two sets of felicity conditions diverge immediately thereafter: this divergence is tracked by the displayed essential conditions (and illocutionary force). It is a known general fact, however, that propositional content is not entirely inert when it comes to identifying the force of a speech act. For example, uttering "I promise I was at the bar last night" does not accomplish the speech act of promising; to count as a promise, the speaker's locution must refer to an act that she will perform in the future.

Following this line of thought, we might ask: What makes it the case that the very same propositional content can alternately acquire the force of refuting a claim, and of testing that same claim? One can of course point to the remaining felicity conditions for devil's advocacy, as we have done above. But these conditions refer to the speaker's doxastic state: unlike the propositional content conditions they are, on their own, opaque. What reveals, or signals, that *these* conditions are in place, and not others, is the

DA preamble. Importantly, and more fully: *the DA preamble signals* (is designed to signal), *to the hearer and/or the audience*, which set of felicity conditions will be operative in the speaker's contribution to the exchange. It thus simultaneously brings to salience one set of felicity conditions (idea-testing), and relegates another to the back seat (refutation).

It bears noting that the role of the DA preamble can also be usefully described from a canonical Gricean perspective. For, as observed earlier, one of the paradigmatic and minimal presuppositions triggered by devil's advocacy is that the speaker will go on to defend (present reasons in support of) a position that does not reflect his actual doxastic commitments (cf. Johnson, 2018a); that is, the speaker's contribution will violate the Gricean maxim of Quality. Crucially, again, this presupposition is triggered *for the audience's benefit*: the DA preamble signals to the audience that the speaker is aware of the Gricean maxims and is intent on making a cooperative contribution – even while doing so entails being insincere, by taking on the role of an imagined opponent (cf. Stevens & Cohen, 2019). In this sense, the preamble signals overall cooperativeness by framing the accompanying contribution as "*overtly* and *openly* uncooperative" (cf. Aikin & Clanton, 2010).

Against this backdrop, we turn now to introduce what we see as a useful conceptual resource, drawn from recent work in philosophy of language, that helps bridge the gap between ideal and real-world uses of devil's advocacy.

5. Devil's advocacy and force figleaves

Recall that in [Street harassment], the male radio caller presented a number of considerations intended to bear upon the narrator's claim that 'tatcalls', like catcalls, are a form of public harassment – harassment which, insofar as it is recognisable as an expression of female objectification, is also sexist. The male caller crucially prefaced these considerations – e.g.: 'It could be that catcalls and 'tatcalls' should be viewed as harmless compliments, rather than as symptoms of the objectification of female bodies' – by explicitly announcing that he was "just playing devil's advocate".

We take it as a given that a claim such as

"Catcalls should be viewed as harmless compliments",

if uttered as a pure assertive, is immediately recognisable as symptomatic of sexist attitudes on the utterer's part.

What (if anything) changes when C is attached to a DA preamble? By the lights of the extant analyses of devil's advocacy, the answer is: everything. In particular, given the sincerity-suspending effects of the DA preamble, C must be reclassified as an insincere assertion; and the speaker, as performing a speech act whose illocutionary force is *idea-testing*. Therefore, uttering C in the context of the DA preamble can no longer be treated as indicative of the speaker's sexist attitude. This is an unsatisfactory result, however. It is at odds with the intuitively correct assessment of [Street harassment] as an example of an inappropriate use of devil's advocacy. As such, it lays bare the explanatory limitations of extant analyses. We believe these limitations can be overcome, and in what follows we sketch one way of doing so. We submit that at least in some of its vicious occurrences, the DA preamble is recognisable as a *force figleaf* (Saul, 2021).

At its most general, the term 'figleaf' picks out a class or family of utterances that "provide a small bit of cover for something that is unacceptable to display in public" (Saul, 2017, p. 98). Expressions of overt sexism and racism are among the most prominent examples of publicly unacceptable discourse, across a wide range of cultural and geographical contexts. And yet, Saul maintains, in those same contexts one finds recurring usage patterns of expressions that tap into pockets of racial resentment or gender-based hostility while simultaneously allowing the speakers to publicly dissociate themselves from the attribution of racist or sexist attitudes. Saul terms these expressions racial and gender figleaves, respectively: utterances "which (for some portion of the audience) [block] the conclusion that (a) some other utterance, R, is racist [or sexist]; or (b) the person who uttered R is racist [or sexist]" (Saul, 2021, p. 161).

Alongside a variety of such utterances operating at a direct propositional level (e.g. "I'm not a racist, but...", "I have great respect for women, but..."), Saul identifies a further type of figleaf – the force figleaf – that we think helps diagnose at least some vicious uses of devil's advocacy. As before, the hallmark functional trait of a force figleaf is that of blocking the inference that either the speaker, or the utterance, are racist or sexist. But here, the screening-off effect is achieved by changing the way that the audience understands the speech act being performed by the speaker. That is, force figleaves –when successful– modify the on-record illocutionary force of the relevant speech act.

One of the examples discussed by Saul is drawn from the vast repertoire of Donald Trump's controversial public statements: "When the revelation of [the Access Hollywood tapes] led to an uproar, just before the election, Trump responded by saying that the comments were merely "locker-room talk" (Saul, 2021, p. 170).

As several commentators have noted, by (re-)framing his previous utterances as locker-room talk Trump intended to engineer the way in which those utterances would be received by his audience. To this end Trump introduced a presupposition to the effect that uttering overtly sexist expressions, while unacceptable in ordinary circumstances, *is however permissible when embedded in a locker-room conversational context*. For, in such contexts, specific felicity conditions are in place that rule out the face-value interpretation of those utterances. Thus, the intended effect of Trump's remark was to retroactively modify the intended illocutionary force of his previous speech act, thereby allowing a portion of his audience to avert the otherwise inevitable conclusion that Trump's assertion, and Trump himself, were sexist.

It should be clear from the foregoing that force figleaves share many of the recurring traits underlying real-world uses of devil's advocacy in problematic scenarios such as [Street harassment] and [Affirmative action]. Here, too, the resident devil's advocates introduced a crucial presupposition into the common ground: that those utterances, while unacceptable in ordinary circumstances, *are however permissible when embedded in the context of a DA preamble*. This is an extremely robust presupposition; much more so than its locker-room analogue. The latter merely sanctions permissibility; the former sanctions permissibility *and claims virtuousness*. The sincerity-suspending effects of the DA preamble make it so that the speaker cannot be held responsible for the bigoted views embedded in the preamble; the inference to the conclusion that the speaker himself is bigoted cannot go through. And the default illocutionary force of the devil's advocate's speech act – epistemically beneficial, cooperative idea-testing – makes it so that no assertion, in no context, is out of bounds: anything goes for the sake of open inquiry.

6. Conclusion

Increasingly, in contemporary informal conversational contexts, devil's advocacy seems to have become publicly associated with abusive, ill-willed, or otherwise vicious communicative intentions and behaviour. We think this assessment has merit, but that it is not adequately captured by extant analyses. We take [Street harassment] and [Affirmative action], among others, to exemplify a use of devil's advocacy that is both competent and inappropriate, or abusive. It is competent in the sense that the speakers' on-record utterances are fully compatible with a correct understanding of the default illocutionary force, as well as the characteristic correctness conditions, associated with devil's advocacy. And it is inappropriate, or abusive, insofar as the speakers are *exploiting* these default associations – and the adjacent presupposition that devil's advocacy is unimpeachable by design – in such a way as to "provide a small bit of cover for something that is unacceptable to display in public".

Acknowledgements We thank the ECA2022 organising committee and, in particular, the Roma Tre University students for their support on the ground. Thanks to Martin Hinton for several on-point comments and questions during and after the talk. This work was partially supported by the COST Action CA17132 *European Network for Argumentation and Public Policy Analysis*, and by the project *Reflexive verbs and conversational implicature in the semantic web* (PI: Ervas), funded by Fondazione di Sardegna.

References

Aikin, S. F., & Clanton, J. C. (2010). Developing Group-Deliberative Virtues. *Journal of Applied Philosophy, 27*(4), 409–424.
Brohinsky, J., Sonnert, G., & Sadler, P. (2021). The Devil's Advocate Dynamics of Dissent in Science Education. *Science & Education.*
Duran, N. D., & Fusaroli, R. (2017). Conversing with a devil's advocate: Interpersonal coordination in deception and disagreement. *PloS One, 12*(6), e0178140.
Jacobs, S. (1989). Speech acts and arguments. *Argumentation, 3*(4), 345–365.
Johnson, C. R. (2018a). For the Sake of Argument: The Nature and Extent of Our Obligation to Voice Disagreement. In *Voicing Dissent* (pp. 97–108). Routledge.
Johnson, C. R. (2018b). Just Say No: Obligations to Voice Disagreement. *Royal Institute of Philosophy Supplement, 84,* 117–138. https://doi.org/10.1017/S1358246118000577
Lackey, J. (2020). The duty to object. *Philosophy and Phenomenological Research, 101*(1), 35–60.
Mill, J. S. (1859). *On liberty.* John W. Parker & son.
Nemeth, C. J., Brown, K. S., & Rogers, J. D. (2001). Devil's advocate versus authentic dissent: Stimulating quantity and quality. *European Journal of Social Psychology, 31*(6), 707–720.
Nemeth, C. J., Connell, J. B., Rogers, J. D., & Brown, K. S. (2001). Improving decision making by means of dissent. *Journal of Applied Social Psychology, 31*(1), 48–58. https://doi.org/10.1111/j.1559-1816.2001.tb02481.x
Saul, J. M. (2017). Racial figleaves, the shifting boundaries of the permissible, and the rise of Donald Trump. *Philosophical Topics, 45*(2), 97–116.
Saul, J. M. (2021). Racist and Sexist Figleaves. In J. Khoo & R. K. Sterken (Eds.), *The Routledge Handbook of Social and Political Philosophy of Language* (pp. 161–178). Routledge.
Stevens, K., & Cohen, D. (2019). Devil's advocates are the angels of argumentation. *Reason to Dissent: Proceedings of the 3rd European Conference on Argumentation, Groningen, 2,* 161–174.
Stevens, K., & Cohen, D. H. (2021). Angelic devil's advocates and the forms of adversariality. *Topoi, 40*(5), 899–912.
Terzian, G., & Corbalán, M. I. (2021). Our epistemic duties in scenarios of vaccine mistrust. *International Journal of Philosophical Studies, 29*(4), 613–640.
van Eemeren, F. H., & Grootendorst, R. (1984). Speech acts in argumentative discussions: A theoretical model for the analysis of discussions directed towards solving conflicts of opinion. Fortis Publications.

VISUAL ARGUMENTATION IN THE FRAMEWORK OF THE DUAL-INFERENCE SYSTEM

HÉDI VIRÁG CSORDÁS[1]
Assistant Lecturer at Budapest University of Technology and Economics
csordas.hedi@gtk.bme.hu

ALEXANDRA KARAKAS [2]
Assistant Lecturer at Budapest University of Technology and Economics
karakas.alexandra@gtk.bme.hu

Abstract

In visual argumentation, it is widely accepted that we can analyze different images from an argumentative point of view. However, images can be many things: still and moving, drawing, painting, etc. Even though viewers seemingly detect the conclusions in some cases, this process does not mean that an image is necessarily reasoning. The paper argues that the dual-inference model in cognitive psychology developed by Kahneman, and Sperber-Mercier can be applied in the methodology of visual argumentation to enrich the framework of visual argumentation. The paper aims to address the following claims. We argue that (1) the understanding and reconstruction of visual argumentation can be enhanced by applying the dual-inference model. (2) if we accept the premises of the dual-inference model, then it can help to reconstruct the premise-conclusion structure of specific images in fine art. (3) the dual-inference model is based on the differentiation of the System 1 process and the System 2 process of thinking. The discussion of (1), (2), and (3) together aims to broaden the scope of visual argumentation as a discipline while pointing out some of the limitations of its methods and definitions at the same time.

---

1 "Back to Science!" Scholarship – BUTE 2022.
2 During my research, I was supported by the MTA Lendület Values in Science Research Group.

1. Introduction

Visual argumentation is a newer field of argumentation theory compared to previous methods that focus on argumentation in texts. This results in many theoretical and practical issues in visual argumentation that are yet to be solved. In this paper, we aim to focus on some of these issues, e.g., the problem of defining the scope of visual argumentation and the specific types of images to which visual argumentation can be applied. For instance, traditionally, fine art is not a domain of argumentation analysis; however, we argue that in some cases, it is possible to examine contemporary art with the tools of argumentation, and the so-called dual-inference model developed by Kahneman and Sperber-Mercier (Evans J., 2003; Evans J., 2008; Kahneman, 2011; Kahneman & Frederick, 2002; Metcalfe & Mischel, 1999; Smith & DeCoster, 2000; Strack & Deutsch, 2004).

The paper aims to address the following claims. We argue that (1) the understanding and reconstruction of visual argumentation can be enhanced by applying the dual-inference model. (2) if we accept the premises of the dual-inference model, then it can help to reconstruct the premise-conclusion structure of specific images in fine art. (3) the dual-inference model is based on the differentiation of the System 1 process and the System 2 process of thinking. These methods are similar to specific forms of visual argumentation. Specifically the Key-Component Table and the Waltonian scheme of reconstruction.

The discussion of (1), (2), and (3) together aims to broaden the scope of visual argumentation as a discipline while pointing out some of the limitations of its methods and definitions at the same time. The paper has three sections. First, we briefly introduce visual argumentation and some key issues we will build upon in the paper. Second, the dual-inference method and within that System 1 and System 2 are shown as possible methods for premise conclusion reconstruction. Lastly, in the third section, we analyze a painting based on the findings of the first and the second sections, which broaden the scope of visual argumentation. The paper points out the parallel between the dual-inference model and the reconstruction methods of visual argumentation. On the hand, the Key-Component method resembles System 1: its method analyses images without verbal translation. On the other hand, the Waltonian scheme method requires the verbal translation of the visual premises and conclusion of images to understand the argument of the images: this resembles the framework of System 2 of the dual-inference model.

2. Bases of visual arguments

Informal logicians since the middle of the '90s have claimed that visual arguments exist and that they can be analyzed roughly the same way as verbal arguments (Blair, 1996; Birdsell & Groarke, 1996; Roque, 2015). From the perspective of formal logic, the idea of a visual argument looks odd, to say the least. Traditionally, premises and conclusions are sentences, but O'Keefe has suggested a broader conception of argument which is more viable for visual arguments. O'Keefe claims that arguments involve "a linguistically explicable claim and one or more linguistically explicable reasons" (O'Keefe, 1982 p. 17). According to Blair, an argument does not necessarily have to be linguistic but to be linguistically explicable (Blair J., 2012 p. 207). Being linguistic or linguistically explicable is certainly not the same thing. The claim that the latter is sound to make an argument is crucial for visual argumentation. Thus, along these lines, if we accept the argument that it is eligible for claims to be linguistically explicable to argue, then images can argue just like sentences. We state that paintings, graphics, drawings, digital prints, and various images can be linguistically explicable. We can describe pictures based on what they show and what we know about them, and we can even speculate about their possible meaning, particularly in the case of contemporary art.

There are many different definitions of visual arguments. Groarke and Birdsell, for instance, claim that some pictures can be arguments in traditional premise and conclusion structures (Groarke & Birdsell, 2007 p. 103). Blair states that visual arguments are to be understood as propositional arguments in which the propositions and their argumentative function and roles are expressed. Thus, a 'visual argument is then an argument at least some of the essential elements (reasons or claims) of which are not expressed or communicated in the words of a natural language, but instead are expressed or communicated pictorially, by images and/or non-verbal signs or symbols' (Blair J., 2015 p. 214). A particular type of visual argument for Blair are multimodal visual arguments, in which verbal and nonverbal elements are mixed to a degree, 'but their successful expression also depends on their visual components.' (Blair J., 2015 p. 214).

Clarifying the exact role of images in argumentation and giving a fine-grained definition of how images can pursue is crucial if we lay the foundations of a more inclusive notion of argumentation. Birdsell and Groarke argue that images not only support traditional arguments but also have a much more critical role in argumentation for many reasons. 'In some cases, they more accurately and concisely present information and evidence that is relevant to an argument. In other cases, they have rhetorical advantages and are more forceful and persuasive than words' (Groarke & Birdsell, 2007 p. 103). Groarke highlighted the importance of visual argumentation as follows: 'a satisfactory attempt to understand argument must recognize the pervasive role that visual images play in everyday persuasion, argument and debate' (Groarke, 2005 p.186-187).

3. The Reconstruction of Visual Arguments

The main difference between visual and verbal arguments is the mode by which claim and reasons are formulated; while verbal arguments are linguistic, visual arguments make up linguistically explicable elements. Verbal arguments thus consist of a linguistically formulated premise or set of premises and conclusion, whereas in visual arguments, it is not necessary for all premises or the conclusion to be expressed linguistically; just one part is suitable to be expressed in linguistic form (Csordás & Forrai, Visual Argumentation in Commercials: the Tulip Test., 2017).

There exist two approaches in the mode of reconstructing visual arguments: the initial solution was the translation of visual elements to verbal sentences (Blair J., 1996; Csordás & Forrai, 2017; Csordás & Forrai, 2018). We will show how this reconstruction method works in René Magritte's famous painting titled *The Treachery of Images*. The other method wants to avoid the translation of the images; thus, it leaves the natural character of the pictures as it is (Groarke L., 2015; Groarke L. 2018). Csordás and Forrai claimed that 'the question we have to address now is how this difference shows up in the reconstruction of visual arguments. What informal logicians mean by reconstruction is a fully explicit and transparent statement of the argument, which contains all elements necessary for its evaluation. So reconstruction involves more than a lay understanding of the argument -- it is not a skill which everyone possesses but a learned art drawing on technical concepts. The reconstruction of an argument consists of the following elements:
1. Identifying the conclusion.
2. Identifying the premises.
3. Rephrasing the sentences.
4. Making implicit elements explicit.
5. Building up the structure of the argument.' (Csordás & Forrai, 2017 p. 174.)

The other method was developed by Leo Groarke and called the Key-Component Table. Its purpose is to assess multimodal arguments, and it is based on the view that some visual (or multimodal) units form a whole argument without being expressed propositionally (Blair J., 2004 p. 48-49). Groarke does not explicitly use the term 'reconstruction' but speaks instead of 'dressing up' an argument that covers the identification of premises and conclusions, the discovery of implicit premises, the trimming of purely rhetorical devices, and the clarification of the structure of the argument (Groarke L., 2015 p. 135). The Key-Component Table lets us illustrate many types of visual and multimodal arguments; the premises and conclusions in the table do not need verbal propositions. Therefore, it is a more authentic representation of what is happening in reality. The method identifies the argumentative acts and the elements of the argument with neutral concepts and then determines how the arguer expresses them. The essence of the method is that it 'eliminates several errors arising from interpretation' (Groarke L., 2015 p. 139). We argued that interpretation plays an essential role in identifying unexpressed premises in his method. This act is indispensable for visual reconstruction. In its absence, the conclusion would not follow from the premise, and thus it would be difficult to interpret the text as an argument. However, with the involvement of implication, interpretation problems may arise, which the Key-Component Table is trying to avoid.

4. The dual-inference model: System 1 and System

The methodology of visual argumentation as an independent disciplinary field can be derived from informal logic, but the paper claims that the dual-inference system of cognitive psychology can enrich this framework. The paper adapts the theories of Kahneman, Sperber and Mercier from the field of cognitive sciences to deepen the understanding of visual argumentative practices. Initially, Kahneman developed the dual-inference model (Kahneman, 2011), and Sperber-Mercier laid the foundations for argumentation theory to define inference as a social activity (Mercier & Sperber, 2015). According to the dual-inference system, the human mind can be divided into two major parts: System 1, which works fast, and System 2., which is slower. System 1 is mainly responsible for the subconscious mind, creating what is otherwise called the "gut" or "instinct". There are instances where of course the gut instinct can go wrong, but the fact that it can operate without much effort is a huge advantage. System 2, on the other hand, is responsible for analysis-driven decision-making. Unlike going by intuition, System 2 needs a lot of effort to conclude things (Evans J., 2003; 2008; Kahneman, 2011; Kahneman & Frederick, 2002; Metcalfe & Mischel, 1999; Smith & DeCoster, 2000; Strack & Deutsch, 2004).

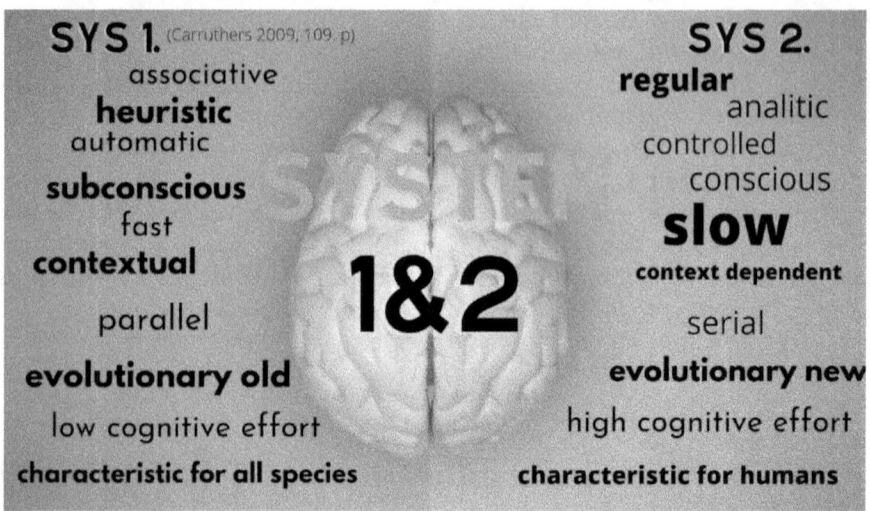

Figure 1. System 1. and System 2.

Kahneman defined the task of System 1 in such a way that, through its operation, we can create quick, intuitive conclusions and respond to impressions of events that System 2 is not aware of (Kahneman, 2013 p. 82). The processes of System 2 require effort and self-control, as without them, we cannot overcome the impressions and impulses of System 1 (Kahneman, 2013 p. 40). While understanding the images' message results from an immediate and spontaneous inference process, identifying visual arguments is a more extended, contemplative operation, allowing us to create premise-conclusion structures. We argue that the message of the picture is understood through the processing mechanism of System 1, but the processes of System 2 carry out the reconstruction of the precise visual premise-conclusion structures.

The process of visual understanding is similar to the operation of the two systems since the message of visual content is generally understood associatively, automatically, and relatively quickly within the context. It is crucial that System 2 is activated if the recipient is interested in a deeper understanding of the content of the images. We claim that understanding the message of images can happen automatically without analyzing their particular elements. The analytical, systematic apparatus considers the results of the previous conclusions and uses them to create premise-conclusion structures.

5. What type of images can argue at all?

The central topic of our paper is that visuality, as such, is a far too broad category to be helpful for any kind of analysis. The reason for this is that images can be many things, as well as many types. Visual argumentation, thus, needs to have a fine-grained definition of arguing images to be able to account for visual argumentation at all. How can we establish what type of images can argue? All images or just some of them? If not all images can argue, then there must be a solid theoretical framework for differentiating between arguing and non-arguing images. However, this framework is not developed just yet. The purpose of this paper is not to solve this problem entirely but to move towards a more analytic understanding of the scope of visual argumentation. The paper is an addition to the discipline as it helps develop a differentiation between different types of images in argumentation.

There are many possible categorizations of images. For instance, we can sort out images based on their function: what purpose do they serve? Convincing us to buy more products, communicating, evoking emotions, or are they serving purely aesthetic purposes? Or, we can analyze their domain: marketing images, fine art, scientific visual materials, etc. We can approach them by observing their technique and the time they were made, or we can ignore their context and just focus on their supposed meaning. If we accept that not all images can and do argue, then at least some of these questions need to be answered.

However, visual culture is far too complex. It is possible to state that in marketing, there seems to exist more arguing images and that in science, we tend to think that images are more neutral, but with a closer look, these statements do not make much sense. There are many not arguing images in marketing, and there is a lot of arguing images in science as well. Our present paper aims to focus on one particular domain of images, that is, fine art, to extend the scope of visual argumentation. In future papers, we will investigate other domains too, and examine whether they fit the framework of visual argumentation.

The focus of this section is to test out whether paintings can argue. For this, we apply the Key-component method and the Waltonian reconstruction method. The instance will be an example of multimodal visual arguments since it contains visual elements and textual elements as well.

6. The analysis of René Magritte's *The Treachery of Images*

The first test subject is one of the world's most famous paintings. It is by the Belgian surrealist painter René Magritte (1898-1967). The painting, made in 1928, entitled *The Treachery of Images* at first gaze, is simple. There is a pipe on it, and below it is a sentence: 'Ceci n'est pas une pipe'; in English, This is not a pipe. The ambiguous line is as complicated as straightforward as it seems. The basic idea is that the pipe in the painting is not a pipe but a painted picture of a pipe. In other words, it is only a representation of an actual or a random pipe or the idea of a pipe. As Michel Foucault argues, there is no contradiction between the two units of the painting, that is, the picture of the pipe and the sentence below it, since contradiction is only possible among two statements or within one. Here, in this case, we only have one statement. What confuses the viewer here is that we connect the text with the drawing (Foucault, 1982 p. 19-20) because of the word 'this'. This word seemingly causes a contradiction, which is not present, as there is only one statement, 'and it cannot be contradictory because the subject of the proposition is a simple demonstrative. False, then, because its "referent" - obviously a pipe- does not verify it?' (Foucault, 1982 p. 19). He then continues to say that the statement that this is not a pipe 'is perfectly true since it is quite apparent that the drawing representing the pipe is not the pipe itself' (Foucault, 1982 p. 19).

Figure 2. René Magritte's: The Treachery of Images

We will see in the reconstruction below that this causes many problems if we translate the understanding of the painting to the domain of informal logic. Foucault argues that the painting is an unravelled calligram. A calligram is defined as a text arranged to visually form an image that is usually related to the context of the text. It is a typographic tool that is popular in poetry since it enhances the context of the poem and, at the same time, offers visuality too. Let's consider whether the Key-Component Method or the Waltonian scheme can be relevant in examining the painting's supposed argumentation!

The main advantage of the Key-Component Table is that the reconstruction does not deprive the image of their natural character. Instead, images are reconstructed as arguments by themselves without having to translate their visual part. In this case, we are dealing with a multimodal visual argument, wherein a textual part complements the image, and its interpretation is only possible together. Still, the determining part is always the visual part. In the analyzed painting, the Key-Component is the depicted wooden pipe, which contradicts the textual part of the picture, which says,' This is not a pipe.' From the point of view of informal logic, the premise or a set of premises must support the conclusion. In this case, if we do not deprive the image of its natural character, i.e., do not translate it, then we will be unable to resolve the contradiction between the premise and the conclusion. Consequently, we do not understand what supports that the identifiable wooden pipe in the image is not a pipe.

Table I. Key-Commponent Table method

Key-Component	Role	Explanation
	Premise	Visual
Ceci n'est pas une pipe. This is not a pipe.	Conclusion	Verbal

Although the Key-Component Table method does not require the translation of images into verbal content, verbalization is sometimes indispensable to support the conclusion. In this case, if we do not translate visual elements, such as 'it is a visual representation of a pipe', then the conclusion that 'it is not a pipe' will make no sense. The reconstruction is made more precise by incorporating an implicit premise into the structure that 'we know the difference between the real pipe and the visual representation of the pipe.'

Table II. Waltonian reconstruction method

The Waltonian scheme reconstruction of Foucault's pipe	
PR1:	It is a visual representation of a pipe.
(PR2- implicit)	We know the difference between the actual pipe and the visual representation of a pipe.
Conclusion:	This is not a pipe.

7. Summary

The paper exposed that adapting the System 1 and 2 inference schemes of cognitive psychology can enrich the framework of visual argumentation. The dual-inference system helps in reconstructing premise-conclusion structures of images, since this framework has similaritites with already developed tools of visual argumentation, such as the Key-Component Table, and the Waltonian Scheme of reconstruction. The paper also contribute in addressing a basic problem of visual argumentation, namely the problem of the scope of the theory. There is no fine-grained definition of arguing and non-arguing images, and the sufficient and necessary conditions of an arguing image. In the future we wish to continue developing such definitions. Our first step in this research was to analyze whether the domain of fine art can be a fruitful resource for argumentation. Magritte's painting fit well into the concept of multimodal visual arguments, since it contains both visual and textual elements.

References

Birdsell, D., & Groarke, L. (1996). Toward a theory of visual argument. Argumentation and Advocacy, 33:1, 1-10.

Blair, J. (2015). Probative Norms for Multimodal Visual Arguments. Argumentation 29:2, 217–233.

Blair, J. A. (1996). The Possibility and Actuality of Visual Arguments. Argumentation and Advocacy 33:1, 23-39.

Blair, J. A. (2004). Defining The Rhetoric of Visual Argument. In C. A. Helmers, Visual Rhetorics (41-61). New Jersey & London: Lawrence Erlbaum Associates: Mahwah. .

Blair, J. A. (2012). The Possibility and Actuality of Visual Arguments. In C. W. Tindale, Groundwork in the Theory of Argumentation Selected Papers of J. Anthony Blair (pp. 205-223). Springer Science+Business Media.

Csordás, H., & Forrai, G. (2017). Visual Argumentation in Commercials: the Tulip Test. OPUS ET EDUCATIO: MUNKA ÉS NEVELÉS, 4:2, 172-182.

Csordás, H., & Forrai, G. (2018). Reconstructing Multimodal Argument: The Dove vs. Nivea. In S. Oswald, & D. Maillat, Argumentation and Inference.: Proceedings of the 2nd European Conference on Argumentation (p. 165.). London: College Publications.

Evans, J. (2003). In two minds: Dual process accounts. Trends in Cognitive Sciences, 7, 454–459.

Evans, J. (2008). Dual-processing accounts of reasoning, judgment and social cognition. Annual Review of Psychology 59, 255–278.

Foucault, M. (1982). This is not a Pipe. Berkeley and Los Angeles: University of California Press.

Groarke, L. (2005). Political cartoons in a Stephen Toulmin landscape. In a. D. David Hitchcock, The uses of argument: Proceedings of a conference at McMaster University 18–21 May 2005 (pp: 186–188). Hamilton Ontario:: Media Production Services of McMaster University.

Groarke, L. (2015). Going Multimodal: What is a Mode of Arguing and Why Does it Matter? . Argumentation, 29(2), 135-155.

Groarke, L. (2018). Auditory arguments: The logic of 'sound' arguments. Informal Logic 38(3), 312–340.

Groarke, L., & Birdsell, D. (2007). Outlines of a theory of visual argument. Argumentation and Advocacy, 43, 103-113.

Groarke, L., & Birdsell, D. (2007). Outlines of a Theory of Visual Argument. Argumentation and Advocacy, 103-113.

Kahneman, D. (2011). Thinking, fast and slow. New York: NY: Farrar, Straus and Giroux.

Kahneman, D. (2013). Gyors és lassú gondolkodás. Budapest: HVG Könyvek.

Kahneman, D., & Frederick, S. (2002). Representativeness revisited: Attribute substitution in intuitive judgement. In D. Gilovich, & D. Kahneman, Heuristics and biases: The psychology of intuitive judgment (pp. 49-81). Cambridge: MA:Cambridge University Press.

Mercier, H., & Sperber, D. (2015). A következtetés mint társas készség. Magyar Tudomány 2015:2, 219-234.

Metcalfe, J., & Mischel , W. (1999). A hot/cool-system analysis of delay of gratification: Dynamics of willpower. Psychological Review, 3-19.

O'Keefe, D. (1982). The concepts of argument and arguing. In R. Cox, & C. Willard, Advances in argumentation theory and research (pp. 3-23). Southern Ilinois University Press.

Roque, G. (2015). Should Visual Arguments Be Propositional in Order to Be Arguments? Argumentation, 29:2, 177-195.

Smith, E., & DeCoster, J. (2000). Dual-process models in social and cognitive psychology: Conceptual integration and links to underlying memory systems. Personality and Social Psychology Review, 4, 108-131.

Strack, F., & Deutsch, R. (2004). Reflective and Impulsive Determinants of Social Behavior. Personality and Social Psychology Review, 220-247.

INSINUATED VS. ASSERTED AD HOMINEM: AN EXPERIMENTAL APPROACH TO THEIR RHETORICAL EFFECTIVENESS

DANIEL DE OLIVEIRA FERNANDES
University of Fribourg, Switzerland
daniel.deoliveirafernandes@unifr.ch

STEVE OSWALD
University of Fribourg, Switzerland

PASCAL GYGAX
University of Fribourg, Switzerland

Abstract

At the crossroads between the fields of pragmatics – the study of meaning in context – and argumentation, the literature often argues that pragmatic variations can affect the rhetorical effectiveness of argumentative structures. However, experimental investigations that address this relation are still scarce, and little empirical evidence is available. This paper develops an experimental inquiry into the effect of implicit meaning, specifically of insinuation, on *ad hominem* argumentative constructions in which an opponent voices disagreement with the proponent's standpoint by launching a personal attack in the refutation. We hypothesised that insinuation can be beneficial when used as a personal attack supporting a refutation, that it is a persuasive move and that it may impact how much people agree with either the author of the *ad hominem* or its target. Our results indicate that an insinuated personal attack does not appear to be more supportive of disagreement than when the attack is asserted. However, an *ad hominem* argument with an insinuated attack is perceived as more persuasive and leads to more agreement with the opponent. Finally, the implicitness of the attack does not affect the agreement given to the proponent. Not all hypotheses were verified. We thus surmise that the rhetorical effectiveness of insinuations might play out on other rhetorical levels (e.g., preserving the image speakers want to project, disrupting conversational dynamics or staining the opponent's reputation).

1. Introduction

Insinuation. "Non-overt intentional negative ascription[s], whether true or false, usually in the form of an implicature, which [are] understood as a charge or accusation against what is, for the most part, a non-present party" (Bell, 1997, p. 36)1.

Insinuations display a series of pragmatic characteristics (Bell, 1997; Fraser, 2001): (i) the negative ascription is *implicitly* conveyed and must be inferred by the audience, (ii) as they are not part of the speakers' explicit speech, insinuated ascription can be *denied* (the ease with which speakers can deny insinuated content varying according to the context and the transparency of the insinuation, see Oswald, 2022), and (iii) for an insinuation to be deniable – at least, plausibly – the utterance must be *compatible with at least two interpretations*, one being negative-ascription-free.

These pragmatic features can yield several rhetorical advantages. Since the negative ascription is implicitly conveyed, speakers can disparage someone while having the possibility to deny having meant such a negative ascription and potentially escape negative reputational consequences. In addition, speakers may shift the responsibility for the disparaging interpretation of their utterance to the audience. As implicit messages are less critically evaluated (in particular with regard to presuppositions, see Lombardi Vallauri, 2018, 2021; Lombardi Vallauri & Masia, 2014) and people tend to trust the outputs of their own inferences (Wason, 1960, 1968; Sperber et al., 1995), implicitly conveyed content is more likely to be accommodated in the audience's cognitive environment. According to Bell (1997) and Fraser (2001), even after a speaker publicly denies the intention or content of a negative ascription, the reputation of their targets remains stained by that derogatory content.

Because it is likely to be rhetorically effective, it stands to reason that insinuations can play a role in an argumentative setting. The argumentative construction closest to this definition of personal attack is the abusive *Ad hominem* (henceforth, AH).

(Abusive) *Ad hominem*. *"The abusive [or direct] type of ad hominem argument occurs where one party in a discussion criticizes or attempts to refute the other party's argument by directly attacking that second party personally"* (Walton, 1998, p. 2).

Van Eemeren et al. (2009) experimentally investigated the perceived reasonableness of different types of AH arguments (i.e., abusive,

---

1 With this definition, Bell (1997) defines what 'innuendoes' are. The difference with 'insinuations' is that the latter does not take into account the presence or absence of the party targeted by the negative ascription as long as this ascription is implicitly conveyed to an audience. This issue will not be addressed in this paper.

circumstantial, and *tu quoque*) while comparing them to sound arguments. By asking the question, "how reasonable do you find B's reply" they addressed the normative acceptability of AH. They found that the perceived reasonableness of either AH argument is lower than that of neutral arguments.

With the rhetorical advantages of insinuations raised above, it makes sense to think that an *insinuated* AH argument (henceforth, IAH) might be more rhetorically advantageous to support a disagreement than if it was *asserted* (henceforth, AAH). In the same vein as the experiment of van Eemeren et al. (2009), we might assume that an IAH is perceived as more reasonable – or perhaps less 'confrontational' – in the context of disagreement than an AAH.

Section 2 starts by presenting our experimental protocol testing some rhetorical effects of insinuations. Then, each experiment will be developed along with its method, results and a brief discussion. Section 3 concludes with a discussion of our findings and experimental manipulations that could compensate for potential limitations or develop new directions of investigation.

2. Experimental investigation

The main difference with the experiment of van Eemeren et al. (2009) is that we were not interested in the argument's perceived reasonableness but in (i) its pragmatic and argumentative construction – or its logos, (ii) its perceived persuasiveness, and (iii) agreement given to insinuators and targets of insinuations. Moreover, our experiments only focused on manipulating the implicitness of the negative ascription in an argumentative setting by using only abusive AH arguments. Finally, we made sure that the personal attack was only present in the argument supporting the disagreement, unlike some of van Eemeren et al. (2009)'s items which seemed to prime the attack already in the formulation of the disagreement (e.g., "What do you know about ethics? [+ Argument]" or "You can't judge anything about this! [+ Argument]").

These three experiments followed the same structure (see Figure 1 for sample item and structure): a context is presented and a dialogue unfolds after it. In the dialogue, a first character (A) defends a claim with which a second character (B) disagrees. B then supports the disagreement with either (i) an argument with an asserted personal attack (AAH), (ii) an argument containing an insinuated personal attack (IAH), or (iii) a neutral argument without any personal attack (Neu). This last condition was added to the first two to control our arguments' effectiveness and to replicate van Eemeren et al. (2009)'s results with our experimental setting.

Thirty-nine dialogues were created and used, each presented under one condition. Participants were distributed into three lists according to a Latin square design and were presented with the three conditions. The analyses performed are linear mixed models considering the variability of the items and the participants (both used as random intercepts).

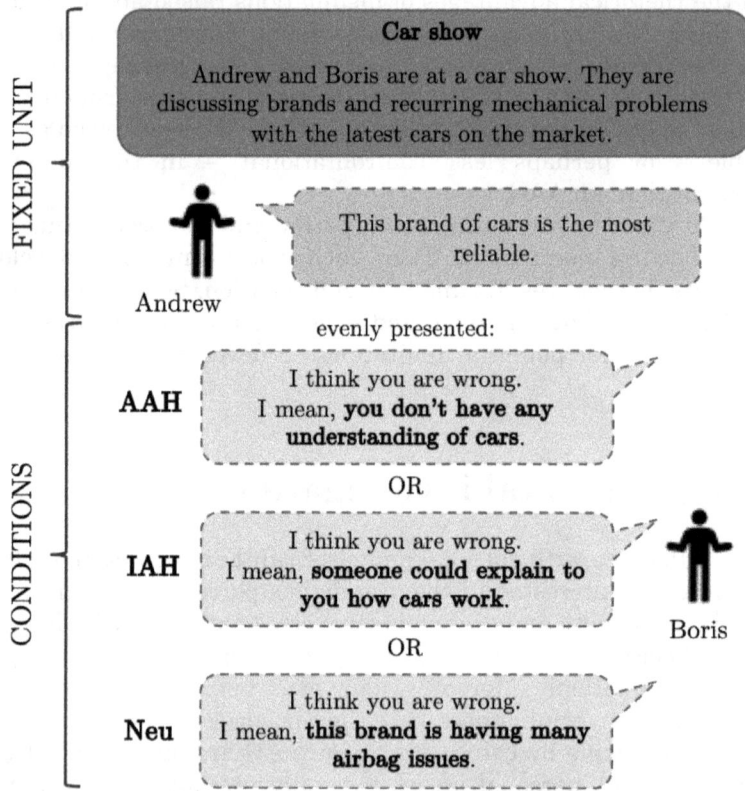

Figure 1. Sample item and structure of experiments.

2.1. Experiment 1: Logos

As a first experimental question, we wanted to know whether insinuations would make it easier to support a disagreement than asserted personal attacks would. The rhetorical effect at stake is *logos*, that is, the construction of the argumentative claim (i.e., disagreeing standpoint supported by a reason with a potential personal attack). Does an IAH give more weight to the reason given for the disagreement than an AAH?

At the bottom of each dialogue, the question "According to you, to what extent does the argument of [Character B] support his claim that [Character A] is wrong?" to which participants were asked to answer on an 11-point Likert scale ranging from 0 "Not at all" to 10 "Absolutely". One hundred and two participants were recruited on the Prolific platform (www.prolific.co) and completed the questionnaire (see Table I for sample demographics). All participants were fluent and first-language English speakers.

Table I. Sample demographics and completion time

	Experiment 1	Experiment 2	Experiment 3
Sample size	102	50	50
Age in years, m (sd)	33.12 (13.25)	34.96 (14.29)	35.66 (13.14)
Gender			
Women	48 (47%)	24 (48%)	25 (50%)
Men	52 (51%)	25 (50%)	25 (50%)
Others	2 (2%)	1 (2%)	0
Completion Time in min (sd)	22.25 (12.64)	24.08 (16.19)	24.16 (10.89)
Remuneration	£2.00 (£6.27/h)	£2.30 (£7.80/h)	£2.50 (£7.17/h)

Our results indicate an effect of the argument condition (see Fig. 2). Neutral arguments are perceived as more supportive of the disagreement than either AH. However, IAH does not support disagreement more than AAH.

One possible explanation for the non-difference between IAH and AAH might be methodological: the question did not capture the possible rhetorical effect of insinuation. By asking whether *the argument of [Character B]* supports the disagreement, participants might have been led to exclude the insinuation when they answered the question – asking someone whether an argument supports a claim amounts to instructing them to evaluate whether the content of the argument plays a justificatory role in supporting the claim. Since the content of IAH is likely almost the same as the content of AAH, it is possible that participants represented the meaning of IAH and AAH independently of both the way it was presented (i.e., insinuated or asserted) and the way they inferred the insinuated meaning. Taking the sample item in Fig. 1, the fact that someone could explain to Andrew how cars work, that is, that he does not know anything about cars, might therefore not be identified as an insinuation, but merely treated as a piece of information, however it was derived. Finally, as we asked participants to identify and judge contents, we might have abstracted the content of the insinuations from their pragmatic type and their context of occurrence.

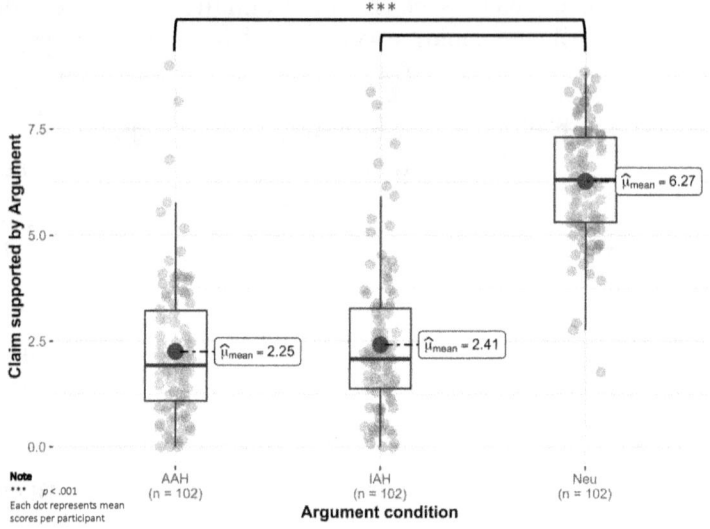

Figure 2. Results of Experiment 1

2.2. Experiment 2: Perceived persuasiveness

As a second experimental question, we wanted to know whether insinuations would be perceived as more persuasive than asserted personal attacks. We thus formulated a question which insists on considering the whole exchange and asks to judge the persuasiveness of the statement (as a whole, without any particular distinction between the disagreement claim and the argument). Is a disagreement supported by an IAH perceived as more persuasive in a conflict of opinion than when an AAH supports it?

At the bottom of each dialogue, we thus asked the question "Based on this exchange between [Character A] and [Character B], how persuasive do you find [Character B]'s statement?". Participants were asked to answer on an 11-point Likert scale ranging from 0 "Not at all" to 10 "Absolutely". Fifty participants were recruited on the Prolific platform and completed the questionnaire (see Table I for sample demographics). All participants were fluent and first-language English speakers.

Our results indicate an effect of the argument condition (see Fig. 3). Neutral arguments are perceived as more persuasive than either AH. Moreover, and as expected, IAH are perceived as more persuasive than AAH.

The measure detected a difference in persuasiveness between AAH and IAH: insinuated attacks are perceived as more persuasive than asserted ones. Nonetheless, we asked ourselves what "persuasive" would mean for participants and what explains this increase in perceived persuasiveness. Another measure that could be related to the persuasiveness of a statement is the agreement that one would have with it or with the person. In these first two experiments, we were only interested in the construction of the AH and the persuasiveness of the attacker. Neither of these experiments is concerned with the agreement with the target of those attacks. If the insinuation might work in the attacker's favour, does it also work against the target?

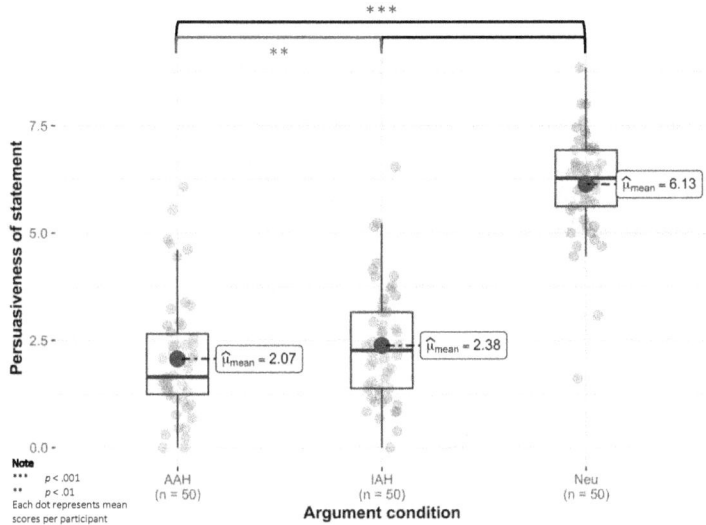

Figure 3. Results of Experiment 2

2.3. Experiment 3: Agreement

As a third experimental question, we wanted to know whether insinuations, compared to asserted attacks, would lead participants to agree with the attacker but disagree with the target. We decided to further investigate the finding of the previous experiment by examining whether it is due to higher agreement towards the attacker, lower agreement towards the target, or both.

At the bottom of each dialogue, the question "Based on this exchange between [Character A] and [Character B], to what extent do you think [Character A]/[Character B] is right?" to which participants were asked to answer on an 11-point Likert scale ranging from 0 "Not at all" to 10 "Absolutely". Participants were assigned to two lists nested within the condition lists that were displayed almost equally (i.e., 19 and 20) to the participants. Fifty participants were recruited on the Prolific platform and completed the questionnaire (see Table I for sample demographics). All participants were fluent and first-language English speakers.

Regarding agreement with the target of the attack, our results indicate an effect of the argument condition (see Fig. 4a). Neutral arguments lead to less agreement with the target than either AH. However, IAH does not lead to less agreement with the target than AAH. Regarding the agreement with the attacking character, our results indicate an effect of the argument condition (see Fig. 4b). Neutral arguments lead to more agreement with the attacking character than either AH. Moreover, as expected, IAH does lead to more agreement with the attacking character than AAH.

What could explain this non-difference between agreement with the target under IAH and AAH is that another rhetorical effect is at play concerning the target, but independently from the decrease in agreement. According to Bell (1997), insinuation is primarily used to disparage someone, to make them seem ridiculous in the eyes of an audience. The fact that the insult is insinuated and thus not frontal allows insinuators to appear more subtle and clever. Indeed, according to our results on the agreement with the attacking character, agreement seems to be preferred when the personal attack is implied rather than asserted. The latter finding is in line with the results observed in the perceived persuasiveness experiment (Exp 2).

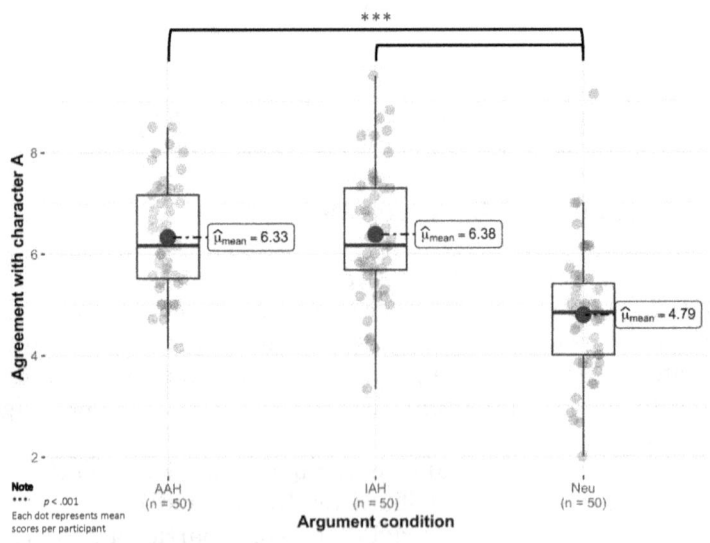

Figure 4a. Results of Experiment 3 (target of attacks; A)

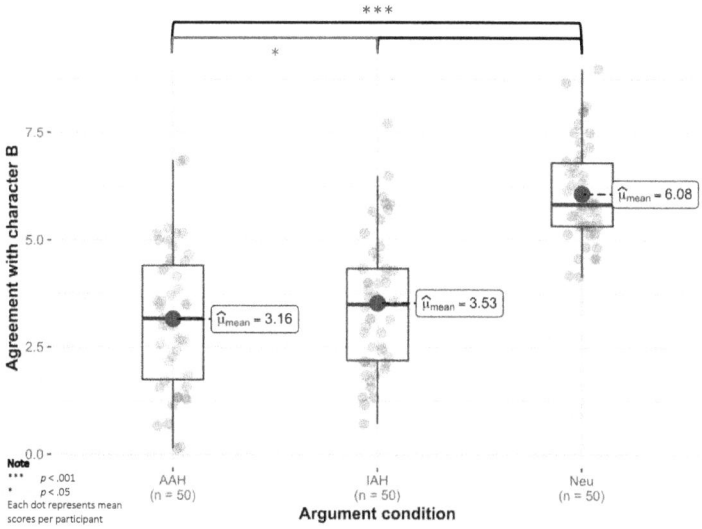

Figure 4b. Results of Experiment 3 (attacking character; B)

3. Discussion

In our experiments, we observed that pragmatic variations have consequences on argumentative effects. So far, we observed that (i) a personal attack is not considered as an acceptable argument to support a disagreement, regardless of whether the attack is insinuated or asserted, (ii) insinuations are perceived as more persuasive and (iii$_a$) lead to more agreement with insinuating speakers but (iii$_b$) not to less agreement with the target of those insinuations.

After the results we obtained on persuasion, we wondered whether an IAH is perceived as more persuasive because the AAH is considered more insulting and threatening. Indeed, a negative ascription that is asserted feels more face-threatening than a negative ascription that has to be inferred. Consequently, the asserted attack is perhaps likely to be more easily rejected, less agreed upon, and perceived as less persuasive. This issue of face-threatening raises the question of politeness. For instance, saying "you are a liar" may be perceived as more choking or blunt than saying "you have a special relationship with truth". Thus, something more elaborate and subtle might be more readily accepted. However, an opposite hypothesis suggests that implicitness may be perceived as manipulative, deceitful and mocking (Pinker, 2007). However this may be, our results support the idea that implicitness is preferred. For a follow-up experiment, we could imagine a question following the same structure as in our experiments but asking about perceived offensiveness: "Based on this exchange between [Character A] and [Character B], to what extent do you find [Character B]'s reply to [Character A] offensive?". If indeed, insinuations are perceived as less offensive, then this might explain why individuals are more accepting of them.

Besides how the attack – the argument in our argumentative setting – is expressed, the way the disagreement is expressed may have influenced our results as well. Saying "I think *you* are *wrong*" might be understood as meaning "my opinion is the right one, while yours is wrong" (Pietroiusti, 2022). In case an IAH follows a disagreement voiced in such a way, this more conflictual attitude may make the personal attack much more apparent, reducing the plausibility of an innocuous alternative interpretation. By expressing a more open attitude towards the difference of opinion, such as "I disagree", one can more easily expect an interpretation of the argument as conveying no personal attack. The principle is the same, as the second character does not share the same opinion as the first; however, with "I disagree", the conflict is about the ideas, while "I think you are wrong" is about the person. Both communicate the same – a disagreement – but highlight different attitudes. Experimental manipulation of the disagreement wording could only sometimes lead to greater persuasiveness. Still, it would leave more room for a neutral interpretation of the IAH and thus better distinguish IAH from AAH.

In conclusion, we have emphasised the rhetorical effectiveness of insinuations in terms of logos and persuasion, but we should not reduce their effectiveness to those measures. Indeed, insinuations might also be rhetorically effective in terms of the perceived image of the speakers, their effects on the audience or the dynamics of the conversation. As observed, the impact on persuasiveness and agreement is minor. This suggests that the effectiveness of IAH might be found more strikingly at another level, yet to be explored.

References

Bell, D. M. (1997). Innuendo. Journal of Pragmatics, 27(1), 35–59. https://doi.org/10.1016/S0378-2166(97)88001-0

Fraser, B. (2001). An account of innuendo. In I. Kenesei & R. M. Harnish (Eds.), Pragmatics & Beyond New Series (Vol. 90, pp. 321–336). John Benjamins.

Lombardi Vallauri, E. (2018). L'implicite comme moyen de persuasion : une approche quantitative. Corela. Cognition, représentation, langage, HS25. https://doi.org/10.4000/corela.6112

Lombardi Vallauri, E. (2021). Presupposition, attention and cognitive load. Journal of Pragmatics, 183, 15-28. https://doi.org/10.1016/j.pragma.2021.06.022

Lombardi Vallauri, E., & Masia, V. (2014). Implicitness impact: measuring texts. Journal of Pragmatics, 61, 161-184. https://doi.org/10.1016/j.pragma.2013.09.010

Oswald, S. (2022). Insinuation is committing. Journal of Pragmatics, 198, 158-170. https://doi.org/10.1016/j.pragma.2022.07.006

Pietroiusti, G. (2022). Having a disagreement: expression, persuasion and demand. Synthese, 200(1), 1-12. https://doi.org/10.1007/s11229-022-03509-0

Pinker, S. (2007). The evolutionary social psychology of off-record indirect speech acts. Intercultural Pragmatics, 4(4), 437–461. https://doi.org/10.1515/IP.2007.023

Sperber, D., Cara, F., & Girotto, V. (1995). Relevance theory explains the selection task. Cognition, 57(1), 31–95. https://doi.org/10.1016/0010-0277(95)00666-M

Van Eemeren, F. H., Garssen, B., & Meuffels, B. (2009). Fallacies and judgments of reasonableness: Empirical research concerning the pragma-dialectical discussion rules (Vol. 16). Springer Science & Business Media.

Walton, D. N. (1998). Ad hominem arguments. University of Alabama Press.

Wason, P. C. (1960). On the Failure to Eliminate Hypotheses in a Conceptual Task. Quarterly Journal of Experimental Psychology, 12(3), 129–140. https://doi.org/10.1080/17470216008416717

Wason, P. C. (1968). Reasoning about a Rule. Quarterly Journal of Experimental Psychology, 20(3), 273–281. https://doi.org/10.1080/14640746808400161

COGNITIVE MACHINE ARGUMENTATION

EMMANUELLE DIETZ
Airbus Central R&T, Germany
emmanuelle.dietz@airbus.com

ANTONIS KAKAS
Dept. Computer Science, University of Cyprus, Cyprus

ADAMOS KOUMI
Dept. Computer Science, University of Cyprus, Cyprus

1. Introduction

How can machines work with people to help them in their reasoning and decision-making? Can machines reason in a humanly cognitive compatible way that would facilitate such a human-machine interaction? How can machines argue about and debate issues with humans? Can machines change human minds?

This paper studies human reasoning and its interaction with reasoning AI machines. The ultimate aim of the work is to address questions like the ones above and contribute in forming an effective and naturally enhancing human-machine integration. Our study falls within the general approaches of Cognitive AI, (e.g. Lieto, 2021) and Explainable AI, (e.g. Adadi, 2018), with an emphasis on incorporating elements of human reasoning from studies in Cognitive Science and Philosophy.

Cognitive and Explanatory AI study how to build systems that operate in a way that is naturally connected to human thinking and behaviour. We address this goal using Argumentation and study how *Machine Argumentation* can come close to *Human Argumentation*. The choice of argumentation as a basis for Cognitive and Explainable AI systems (Dietz et al, 2022) is strongly supported by the natural and close link between human reasoning and argumentation as studied over the centuries in Philosophy and more recently in Cognitive Psychology (Mercier & Sperber, 2011). Furthermore, there is a direct connection between argumentation and explanation: the arguments supporting a conclusion naturally form a justifying explanation for the conclusion.

Cognitive Argumentation (CA) is a framework of argumentation built through a synthesis of the formal framework of Computational

Argumentation in AI with cognitive principles, drawn from the many disciplines in which human reasoning and human argumentation is studied. The task is to let these cognitive principles regulate the formulation and computation of argumentation in order to make it effective and cognitively viable within the dynamic and uncertain environment of application of AI systems. The framework of Cognitive (Machine) Argumentation has been validated by showing that it models well the empirical data from Cognitive Science in three classic human reasoning domains, that of Syllogistic Reasoning, the Suppression task and the Selection task (Dietz & Kakas, 2019; 2020; 2021).

In this paper, we concentrate on studying how machine argumentation within the framework of CA can influence, if at all, on human reasoning. In particular, we examine when and to what extent a machine's argumentative explanatory reasoning is convincing enough for human reasoners to alter their previous different decisions and to accept the machine's decision. These empirical studies are carried out with human participants interacting with COGNICA a system implemented as a realization of CA for conditional reasoning. COGNICA extends the mental models theory on human conditional reasoning presented in Johnson-Laird & Byrne (2002), from individual conditionals to sets of conditionals of different types, that together form a piece of knowledge on some subject of interest. The system operates within a subset of controlled natural language for expressing conditionals which are automatically translated by into a CA theory in the computational argumentation of Gorgias (Kakas et al, 2019). COGNICA is then able to reason argumentatively to draw conclusions accompanied with different types of verbal or visual explanations.

The next two sections overview the conceptual and formal framework of CA. Section 4, presents the COGNICA system with its cognitive faculties of reasoning and explanation. In Section 5 we present the design and our research program of the empirical studies of human-machine interaction together with the results of the first pilot study. Section 6 concludes with the currently completing and future empirical studies.

2. Cognitive Argumentation: Conceptual Overview

Cognitive Argumentation addresses the problem of formulating human reasoning in a computational and cognitively adequate way. It formalizes reasoning in terms of a general *normative condition* that defines the notion of *validity* of a (set of) argument(s). Informally, a set or a coalition of arguments, Δ is valid if and only if it renders all its counter-arguments invalid (Kakas & Mancarella, 2013). This definition of validity encompasses the obvious requirement that a valid set of arguments should not be self-attacking (i.e. that it contains a counter-argument to itself), as the set cannot be both valid and invalid at the same time. The notion of validity can be approximated by requiring that it is *not self-attacking* and that it *defends*, i.e. attacks back it forms a counter-argument, to all its counter-arguments. Such sets are called *admissible* as presented in Dung, (1995).

These notions of validity and admissibility can be formalized with precise definitions so that we can then build a computational model of reasoning based on argumentation. This model has a dialectic structure of considering the counter-arguments to a root argument that supports a desired conclusion and for each such counter-argument finding a defending argument, to add together with the root argument so that we can have a valid coalition of arguments supporting the conclusion.

CA takes the general and abstract notion of validity and turns it into *Cognitive Validity*. We consider *phenomenological or descriptive* characteristics through extensive empirical studies, of human reasoning. These descriptive elements are called *cognitive principles* of reasoning. A first such cognitive principle is the simple observation that human reasoning, in contrast with reasoning with formal logic, is not absolute. This phenomenon is linked to the recognition that we are typically reasoning with incomplete and uncertain information. This flexibility and openness of human reasoning can be naturally encapsulated via argumentation by simply defining a conclusion as one that is supported by a valid subset of arguments. In other words, and in contrast with classical formal logic, it is not necessary that all valid subsets of arguments support the conclusion or more precisely that there is no valid subset supporting a contrary conclusion. Only in some cases we would have a clear winning argument for a certain conclusion, i.e. one that would be strong enough to defeat any argument supporting a contrary conclusion. In fact, in an extreme boundary case reasoning via argumentation captures the formal logical reasoning of classical deduction (Kakas, 2019).

In effect, this first cognitive principle within CA allows this to operate under a flexible and open *dialectic rationality* where conclusions or solutions are required to be *satisficing* in the sense that they can be *justified* via a valid argument. This form of dialectic rationality allows for different conclusions to be reached based on the same set of commonly accepted premises but also based on different beliefs held by the reasoners. *Dialectic rationalism* is based in the beliefs of the reasoner, much in the same way that in Mercier, & Sperber, 2011 have accounted through argumentation for the confirmation-bias in human reasoning. From the point of view of the individual the decision can be rationally defended via subjective hypotheses, which help to justify the relative quality of the argument and conclusion of the individual.

A second general cognitive principle relates to the fact that human reasoning is immersed in a real-life practical setting. This *contextualism* heavily influences what the reasoner is hypothesizing about the current environment at the time of reasoning. The abstract and context independent dialectic model of computational argumentation can be very inefficient. It considers all possible counter-arguments on an argument chosen ad hoc to support a desired conclusion and for each counter-argument, it again chooses with no guidance a defending argument against this.

In CA different cognitive principles aim to mitigate this complexity through statements of the *relative strength*, possibly in a context-dependent way, between different types of arguments. For example, in conditional reasoning cognitive principles regulate the relative strength between arguments that are built from different types of conditional statements: statements that are based on necessary conditions, e.g. practical preconditions for an action to be carried out, give arguments that are stronger than arguments based on statements build from sufficient conditions, e.g. motivational conditions for an agent to (want to) carry out an action. The computational process of reasoning can then be focused by considering primarily arguments that are relatively strong in the current context of reasoning.

In summary, Cognitive Argumentation has a hybrid character of a *normative-descriptive* model of reasoning, combining the formal normative condition of validity of arguments within a context dependent group of currently activated arguments with descriptive cognitive principles stemming from and encapsulating empirical findings on human reasoning. Cognitive principles affect the representation of knowledge within CA and pragmatically regulate its argumentative reasoning process by guiding the selection of which arguments to use in support and defense and which counter-arguments to consider. In effect, they "calibrate" the abstract and general framework of computational argumentation from AI to give it a "cognitive" form that makes it a cognitive adequate model for human reasoning.

3. Cognitive Argumentation: Formalism

We review the basic components of the formal framework of CA. Details are found in Dietz & Kakas, (2020);(2021). An *argumentation theory or model* within CA is obtained by capturing knowledge into argument schemes together with a preference or strength relation between these schemes. Formally, it consists of a triple $A_L = (As, C, \succ)$ where As is a set of argument schemes, C is a conflict relation in the language, L, of the framework and $\succ$ is a binary strength relation on As.

Argument schemes are stereotypical reasoning patterns (Toulmin, (1958); Walton, (1995)) that are typically non-deductive. Formally, an *argument scheme, as* $\in As$, is a tuple of the form $as = (pre, pos)$ where the *premises, pre,* and *position, pos,* are (sets of) statements in the language of discourse L. Using an argument scheme $as= (pre,pos)$ we can construct an *individual argument* that *supports* the position, pos, based on the premises, *pre*. An *argument* Δ, is then a set of individual arguments that are grouped together as a coalition to support a position (e.g. a conclusion) we are interested in. The *conflict relation* C in $A_L = (As, C, \succ)$ specifies when arguments conflict with each other and is used to give the notion of attack between two arguments. Δ' *attacks* or is a *counterargument* of Δ, iff together these arguments have a conflict under C, e.g. when Δ supports "Q" and Δ' supports "not Q". The *strength relation* $\succ$ in $A_L = (As, C, \succ)$ captures the relative strength among arguments. Given two argument schemes, as and as', then as $\succ$ as' means that arguments constructed from as are *stronger* than arguments constructed from as'. This gives a notion of *defense* between conflicting arguments. Informally, an argument Δ defends against Δ' only when its individual arguments are at least as strong as those of Δ'.

Let us illustrate how we can ground the general framework of CA, into a specific framework according to some cognitive principles. Each cognitive principle results into argument schemes in As. Consider Grice's (1975) *maxim of quality* principle which states that (normally) humans communicate cooperatively so that information exchanged in conversation is assumed to hold and trusted. This is captured via a *fact scheme*: $fact(F)= (\emptyset, F) \in As$, applied for any statement F that is given as a fact in the current environment of reasoning. Similarly, the *maxim of relevance principle*, again from Grice, (1975) states that people contemplate different scenarios when they are made aware of their possibility. This principle is captured via a *hypothesis scheme*: $hyp(A) = (\emptyset, A)$ and $hyp(nA) = (\emptyset, not\ A)$, applied for any statement, A, that a reasoner is currently aware of. In other words, once we are aware of a concept we can hypothesize that it holds or that it does not hold. Humans consider at least two different types of conditions, sufficient and necessary, in relation to a consequent of interest and build different associations between them. These two different types of a condition P in relation to a consequent Q, each give two conditional argument schemes: when P *is sufficient*: $suff1= (P,Q)$ and $suff2=(not\ Q, not\ P)$, and when P is *necessary*: $necc1 = (Q,P)$ and $necc2 = (not\ P, not\ Q)$. The chosen schemes by a reasoner at any time depend on the particular interpretation of the condition, sufficient and/or necessary, by the reasoner.

The same cognitive principles give us information on the relative strength between their respective argument schemes. Hypotheses schemes are weaker than any other type of scheme *(Weakness of Hypothesis* principle) and the factual schemes are stronger than other conflicting argument schemes (*Strength of Facts* principle). Furthermore, a necessary condition is stronger than any sufficient condition giving rise to a third element in the strength relation of $necc2 = (not\ P, not\ Q) \succ suff1 = (R,Q)$, (*Strength of Necessity* principle). Similarly, we can use specific studies on argumentation, such as that of Bayesian Argumentation (e.g.

Zenkel (ed), 2013) or the evaluation of arguments from natural language in Hinton, (2020), to capture the relative strength of arguments in a CA framework.

Reasoning in argumentation is a process of analysis of alternatives, e.g. of a conclusion and its negation, by considering different arguments for and against the various competing alternatives. In CA we require that a supporting argument must (i) be *grounded* on the current state of information that the environment gives or makes the reasoner aware of, and (ii) the argument should be *cognitively valid*. Hence we can consider a *cognitive state* $S = (F,A)$ of the current reasoning, where F is a set of facts provided by the environment and A is set of propositions currently aware by the reasoner. An argument Δ is then required to be grounded on the current cognitive state, i.e. that the premises of all individual arguments in Δ are directly or recursively indirectly linked to the cognitive state. Considering only grounded arguments the cognitive validity of an argument is given by its *admissibility* in in the current CA framework, denoted by $AL(S)$, i.e. Δ is conflict-free (it is not self-attacking) and Δ defends against all the counter-arguments attacking Δ.

Reasoning under argumentation is then carried out by considering ground and admissibly valid arguments that support a concluding statement, Φ. Specifically, Φ is a *credulous conclusion* of a current CA framework, $AL(S)$, iff there exists a ground and admissible argument Δ in $AL(S)$ that supports Φ. Φ is a *skeptical conclusion* of $AL(S)$ iff Φ is a credulous conclusion and Φ is not a credulous conclusion of $AL(S)$, i.e. there is no admissible argument supporting Φ. Credulous and skeptical conclusions represent *plausible* and *definite* conclusions, respectively. If a conclusion is skeptical we can be definite and answer *YES* to the question if it holds. Similarly, if the negation of a statement is a skeptical conclusion we can answer with a definite *NO*. Otherwise, if both a statement and its negation are credulous conclusions (each one has an admissible argument supporting it) then we would give the answer *MAYBE*.

4. COGNICA: A System for Cognitive Argumentation

COGNICA is a system that simulates human reasoning based on the mental models theory of Johnson-Laird & Byrne, (2002) and using preference-based argumentation to capture and extend this theory.

Mental models are constructed by studying the human perception and comprehension of discourse. Essentially, mental models consist of various possible scenarios that are built based on what a person has in mind from which the person can then derive new information. We focus on the study of conditional reasoning, i.e. sentences of the form *If a then b*, which can

be easily translated into argumentation schemes. Conditional sentences are divided into two main categories:
1. Sentences that depend on their context and on someone's background knowledge regarding condition and consequence.
2. Sentences where the condition and consequent are semantically independent apart from their occurrence in the same conditional.

Johnson-Laird & Byrne developed ten categories for conditional sentences. The following nine categories are used in COGNICA: Tautology, Conditional, Enabling, Disabling, Biconditional, Strengthen antecedent, Relevance, Modus Tollens, Modus Ponens. We have translated the nine categories into argument schemes following the mental models theory. Also, we have extended the language with an additional conditional using a necessity modality: *"if a then always b"*. Table 1 shows these categories with their respective mental models.

Conditional Cat.	Phrase template	Mental model
Tautology	*If A then possibly B*	a b a ¬b ¬a b ¬a ¬b
Conditional	*If A then B* *If A then always B*	a b ¬a b ¬a ¬b
Enabling	*Only if A then maybe B*	a b a ¬b ¬a ¬b
Disabling	*Even if A then maybe still B*	a b a ¬b ¬a b
Biconditional	*If and only if A then B*	a b ¬a ¬b

Table 1: Mental models for Conditionals

For the first category of conditional sentences that are dependent on context we use relevant background knowledge and the result of reasoning with it to translate them into argument schemes. Regarding the conditionals of the second category, we consider the mental models in which condition is necessary or sufficient for the consequent and translate each model as an argument schemes. The cases where the mental models allow all possible outcomes, regardless of the condition, are captured via a pair of hypothetical arguments schemes supporting the consequent and its negation.

To complete the translation into the CA framework we need to specify the relative strength between the argument schemes constructed from the conditionals of different categories. This binary strength relation is defined following a policy of priority levels according to the necessity level between the condition and the conclusion. Starting from the lowest priority level this policy as follows:

Priority level 1: Hypotheses on the Consequents

Priority level 2: Sentences in which the condition is not sufficient or necessary for the consequent.

Priority level 3: Sentence in which the condition is sufficient for the consequent.

Priority level 4: Sentences in which the condition is necessary for the consequent.

Priority level 5: Sentences in which the condition is both necessary and sufficient for the consequent.

Same conditional sentence argument schemes have the same priority level. The only exception is in the conditional interpretation where a subcategory of phrases was created by adding the word *always* in the phrase: *If a then always b*. The word always indicates necessity; therefore, we give priority to the subcategory phrases over the general conditional phrases.

To illustrate the process of how the conditional sentences are translated into argumentation schemas, let us consider the following example conditional sentences in the context of swimming.

Example 1.
If the sea is hot then possibly I will swim.
If the sea is rough then I will not swim.

The first tautological sentence gives the argument scheme $arg1: (\{the\_sea\_is\_hot\}, I\_will\_swim)$ supporting the position *I will swim*. The second conditional sentence gives the argument scheme $arg2: (\{the\_sea\_is\_rough\}, I\_will\_not\_swim)$ supporting the position *I will not swim*. We also have the hypothetical arguments: $arg3:(\emptyset, I\_will\_swim)$ and $arg4:(\emptyset, I\_will\_not\_swim)$.

Arg2 and *arg1* are counter-arguments of each other, as their positions are in conflict. *Arg2* is stronger than the arg1 because the relationship between the condition and the conclusion of the tautological phrase is weaker than the relationship between them in the conditional phrase.

5. COGNICA & Empirical Studies

COGNICA has been implemented using the Gorgias argumentation framework (Kakas et al, 2018). It is available through a website at: http://cognica.cs.ucy.ac.cy/COGNICAb/login.php. After logging in to the website's main panel, users can introduce foreground and background knowledge in the system, in the form of conditional sentences as described in the previous section. Furthermore, users can also specify undisputed facts deriving from the background knowledge. A user can define two or more consequents as conflicting. The negation of a conclusion is automatically recognized as conflicting. Conflicting conclusions can also be declared through sentences of (common) background knowledge. A user can save the information entered into a file. For more details one can consult the online user manual.

After inserting some foreground and background knowledge, we can query the system for answers on whether a conclusion holds in a given scenario. We specify a scenario by entering that some conditions hold or not as facts. The system examines if the conclusion we are querying is a credulous or skeptical conclusion of the given knowledge and facts. It returns an answer, which is one of the following **Yes/No/Maybe** along with a corresponding *summary explanation* in verbal form. If the user needs more information on the system's answer then a further analytical explanation is provided. The system also offers an illustrative graphical explanation showing the arguments that support the conclusion and their relationship, i.e. attacks and strength, to arguments supporting the opposite conclusion.

Consider the phrases of the Example 1. We provide the following facts: *the sea is hot, the sea is rough*. Figure 3.1 shows COGNICA's answer and summary explanation. The system's answer is *No*. Note how the system's explanation (Figure 3.1) is both *attributive* and *contrastive*, i.e. it explains the reason supporting the answer (attributive part) but also it explains why this answer against the opposite answer (contrastive part). By pressing the *further explanation* button, we have access to the analytical verbal explanation and the graphical explanation as shown in Figure 3.2.

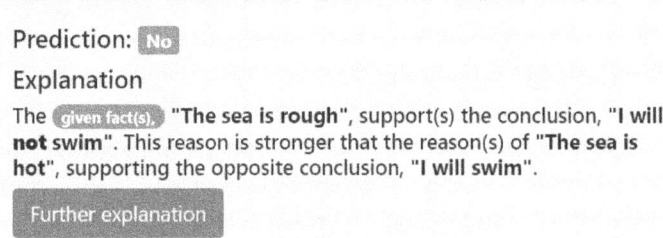

Figure 3.1 COGNICA's answer and summary explanation.

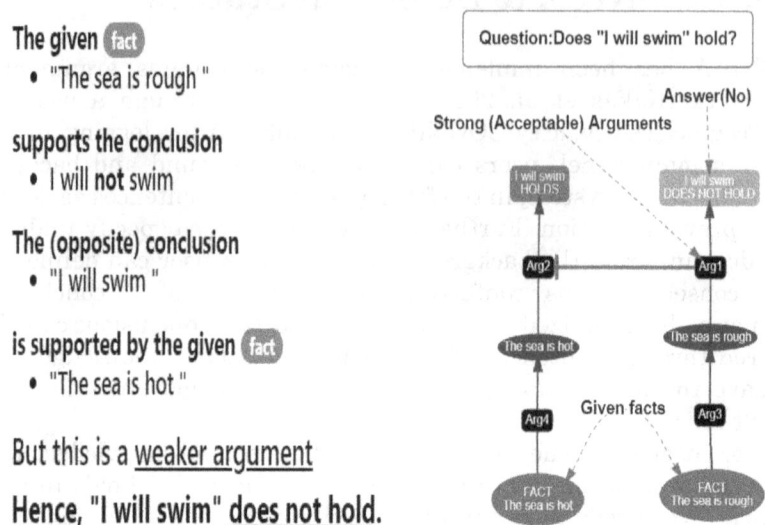

Figure 3.2 Analytical explanation and graphical explanation.

Similarly, if we are given just the fact that *The sea is hot* COGNICA will return the answer **Maybe** with a summary explanation in Figure 3.3.

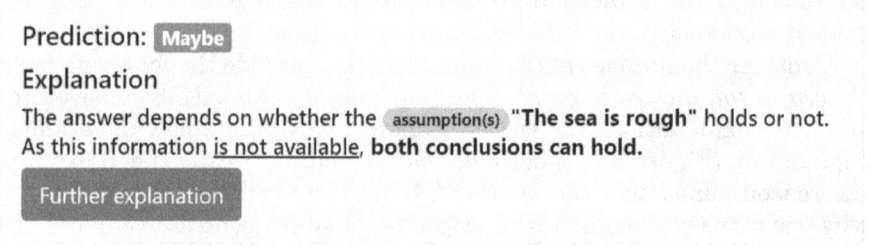

Figure 3.3 COGNICA's answer and explanation.

5.1. Empirical Studies

We are using the COGNICA system to perform empirical studies in order to evaluate whether the answers of the system are correct according to the participants, but also to evaluate whether the system has an influence on the participants to alter their answers based on its explanations.

In a first completed such study the participants were provided with short pieces of knowledge made up from 3-4 conditionals and asked to answer questions under specific scenario conditions. For each question the participant was asked to select one of the following answers: *Yes/No/Maybe*, alongside with an associated level of confidence (*High/Medium/Low*) regarding their specific answer. Then the system's answer was shown to the participants together with a verbal and a graphical explanation. After that, the participants were given the opportunity to alter their response, along with their associated level of confidence.

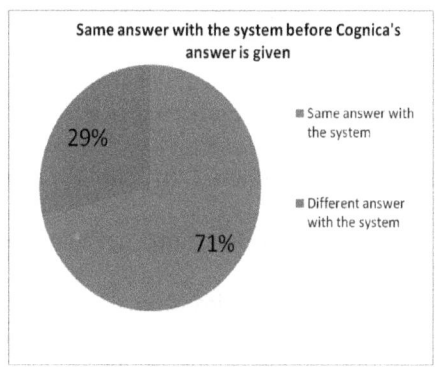 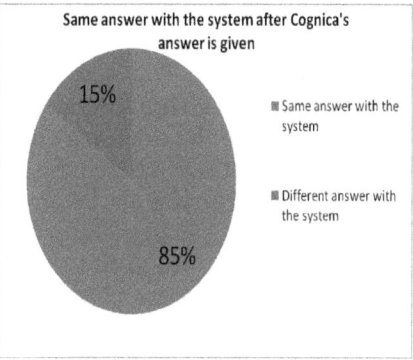

Figure 3.4: Agreement with COGNICA before & after its answer.

Fifty people participated in the experiment. Each participant needed to answer ten questions. 71% of the participants provided the same answer as COGNICA's answer before they were presented with the system's answer. This shows that the system's responses are cognitively valid.

The percentage of people who gave the same answer as the system, after seeing the system's response and explanation, increases by 14% and reaches 85% (see figure 3.4). The increase of the percentage leads to the conclusion, that many participants altered their response after studying the system's response.

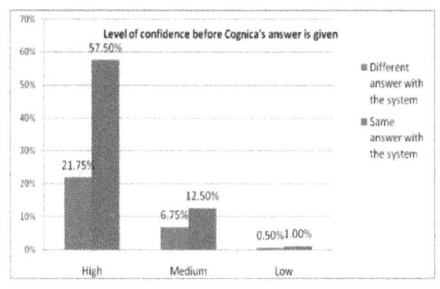 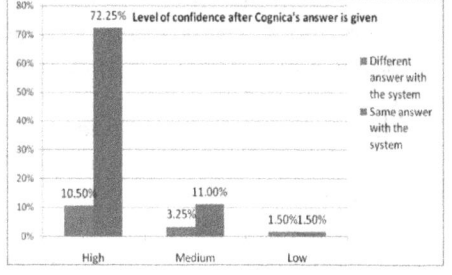

Figure 3.5: Level of confidence before & after COGNICA's answer.

Another way to observe the effect of the machine's answer and explanation on the human reasoning is to observe that in 50% of the cases where the participants' answers differ from those of COGNICA, the participants changed their answer. Furthermore, in those cases where participants adopted the same answer as the system, they had declared a high level of confidence, after reading the result and the explanation of the system. There is a low percentage (~2%) of people who declared a low level of confidence and had the same answer with the system, after reading the answer of the system. In addition, the percentage of participants who gave from the start the same answer as the system and a medium level of confidence, decreased from 13% before the answer of the system, to 11% after it was given. The percentage of participants who gave a different answer from the system and declared a high level of confidence, decreased from 22% before the system's answer to 11% after it was given. This leads us to observe that participants are positively affected by the system developing a level of trust towards the system.

The interested reader can try an example of the reasoning exercises we are using in these studies at the following link:
http://cognica.cs.ucy.ac.cy/cognica_evaluation_2022_se/index.html.

6. Conclusions

Human-machine integration requires meaningful forms of interactions between them. Interactions that make transparent the underlying reasons for their positions and decisions offering explanations at a common cognitive level of understanding. We have presented Cognitive Argumentation as a general framework under which such interactions can be realized. The framework facilitates principled empirical studies with human participants interacting with machines build under the CA framework.

Based on such a machine, called COGNICA, we have set up a long-term research program of carrying out large-scale experiments of comparison between human and machine argumentation. A first study has shown that there is a statistically significant effect of influencing the human participants to change their decision when this differs from that of the machine and the machine explains its different position. Interestingly, this interaction motivates participants to "drift" to more "careful reasoning" as they progress in the experiment, in accordance with the argumentation theory of Mercier & Sperber, (2011).

We are currently completing a new experiment where we are investigating further, how this human-machine interaction varies across the population with different cognitive and personality characteristics. Since the completion of the collection of data for this latest large-scale experiment, the Large Language Models machine of ChatGPT was released. COGNICA and ChatGPT are very different. COGNICA is model-centric, programmed explicitly to reason via argumentation, whereas ChatGPT is data-centric emerging from self-supervised learning experiences. This opens up new opportunities for comparative studies across different AI technologies as well as studies of neural-symbolic integration and interaction between the two types of arguing machines.

References

Adadi, A. and M. Berrada (2018), "Peeking Inside the Black-Box: A Survey on Explainable Artificial Intelligence (XAI)," in *IEEE Access*, 6, 52138-52160.

Dietz, D., Kakas, A., & Michael, L. (2022). Argumentation: A calculus for Human-Centric AI. Frontiers in Artificial Intelligence, 5, https://doi.org/10.3389/frai.2022.955579.

Dietz, E., & Kakas, A. (2021). "Cognitive argumentation and the selection task," in *Proceedings of the Annual Meeting of the Cognitive Science Society, 43*, 1588–1594.

Dietz, E., & Kakas, A. (2019). Cognitive argumentation for human syllogistic reasoning. *Künstliche Intell.* 33, 229–242. doi: 10.1007/s13218-019-00608-y

Dung, P.M. (1995). On the acceptability of arguments and its fundamental role in nonmonotonic reasoning, logic programming and n-person games, Artificial Intelligence, 77(2), 321-357.

Grice, H. P. (1975). Logic and conversation. In P. Cole & J. L. Morgan (Eds.), Syntax and semantics (Vol. 3). New York: Academic Press.

Hinton, M. (2020). Evaluating the Language of Argument Volume 37 of Argumentation Library, Springer.

Johnson-Laird, P. N., & Byrne, R. M. J. (2002). Conditionals: A theory of meaning, pragmatics, and inference. *Psychological Review, 109*(4), 646–678.

Kakas A. & Mancarella, P. (2013). On the semantics of abstract argumentation. *J. Log. Comput. 23(5)*: 991-1015.

Kakas, A., Moraitis, P. & Spanoudakis, N. (2019). GORGIAS: Applying argumentation. Argument Computation. 10(1), 55-81.

Kakas, A. C. (2019). Informalizing formal logic. *Informal Logic* 39, 169–204.

Lieto, A. (2021). Cognitive Design for Artificial Minds. Routledge.

Mercier, H. and Sperber, D. (2011). Why do humans reason? Arguments for an argumentative theory. *Behavioral and Brain Sciences*, 34(2):57–74.

Toulmin, S. (1958). *The uses of argument*. Cambridge University Press.

Walton, D. (1995). Argumentation schemes for presumptive reasoning. Bradford Books.

Zenker, F. (Ed.) (2012). Bayesian Argumentation: The Practical Side of Probability. (Synthese Library). Springer.

SINISTER INTEREST AS ARGUMENTATIONAL VICE. BENTHAM'S *HANDBOOK OF POLITICAL FALLACIES* AND VIRTUE ARGUMENTATION THEORY

IOVAN DREHE
Technical University of Cluj-Napoca
drehe.iovan.p@gmail.com

Abstract:
Jeremy Bentham's *Handbook of Political Fallacies* (1824) was usually overlooked or treated only summarily when compared to other figures such as Aristotle, Locke or Mill (e.g. Hamblin 1970; Secor 1989; Hansen 2020). His treatment might be indeed dismissed as irrelevant for reasons such as its psychologistic undertones. Even if from a rather "formal" perspective on fallacy theory his approach might be considered less relevant, from the angle of a newer approach to argumentation, i.e. the virtue theoretic approach (VAT) (e.g. Cohen 2005, Aberdein 2010), his work might provide some relevant insights to the field.

The purpose of my presentation is to discuss the following issue: what is the cause of fallacious argumentative behavior and what answer does Bentham provide that would be relevant to VAT in particular and argumentation theory in general? Several answers have been provided to this question from the perspective of VAT (e.g. vice, *akrasia*). The starting point of the presentation will be the character trait of "sinister interest", considered by Bentham as the main cause of fallacious behavior, and I will consider it both as a vice and as *akrasia* of purpose.

1. Jeremy Bentham's *Handbook of Political Fallacies* (1824) was usually overlooked or was treated only summarily when compared to other works by philosophers such as Aristotle, Locke or Mill (with a few exceptions: e.g. Hamblin 1970; Secor 1989; Hansen 2020) His treatment might be indeed dismissed as irrelevant for reasons such as its psychologistic undertones, therefore Bentham's approach might not elicit too much interest for standard established argumentation theories. However, the virtue theoretic approach (VAT) (e.g. Cohen 2005, Aberdein 2010) is better positioned to benefit from insights the work of Bentham can provide.

The importance of historically inspired studies in VAT cannot be stressed enough. All the important approaches in argumentation that emerged starting with the second half of the 20th century., e.g. Perelman & Olbrechts-Tyteca's new rhetoric, informal logic (Johnson, Blair), pragma-dialectics (van Eemeren, Grootendorst), formal dialectics (Krabbe) etc. "went back to the Greeks" at some point and this entailed a great benefit to their theories. Classical texts such as Aristotle's *Topics* or *Rhetoric* were re-examined from the perspective of the research programs of the respective conceptualisations. VAT should not make an exception and this paper will focus accordingly on the way a certain mental disposition might influence argument production, starting from what Bentham has to say in his book on sophistry in politics. However, I should mention here that this paper should be considered as one oriented primarily on drawing insights from the works of Jeremy Bentham and only secondarily should be looked upon as a piece of scholarship historiographical in nature, focused on shedding light on the thought of Jeremy Bentham. Also, the political orientation of Bentham will not be taken into account in this case. The premise I start from is the following: an eventual profile of an argumentative agent needs to take into consideration as many of her/his dispositions as possible and even more the potential interactions between these dispositions.

My paper aims to discuss the following issue: what is the cause of fallacious argumentative behavior (intentional or paralogistic) and what answer does Bentham provide that would be relevant to VAT in particular and argumentation theory in general? There are already several discussions regarding this question from the perspective of VAT (e.g. vice or akrasia in argumentation; e.g. Aberdein 2016; Drehe 2016a).

2. The focus of the presentation will be the character trait of "sinister interest", considered by Bentham as the main cause of fallacious behaviour, and it will be considered both as a vice and as a cause for an akratic break of purpose/aim (as in Rorty 1980). At a passing glance, sinister interest is nothing more than private interest as contrasted with the public interest in political context; and it seems there is nothing more to it from the perspective of argumentation studies than the occasion for a passing ethical treatment in, for example, the context of dialogue types (the particular interests of the agents involved in a negotiation for example) or pragma-dialectics (e.g. dialogical failure due to the interference of a participants sinister interest). Since "sinister interest" is considered a root vice/argumentational disposition of an agent, it might be interesting to ponder on how particular fallacious argumentative behaviours and the used fallacious arguments that join the fray might be influenced or determined one way or another by the root vice/disposition (cf. with medieval discussions on pride as a root vice).

Bentham enumerates the following four main causes for fallacious argumentative behaviour (Bentham 1824, 362):
1. Sinister interest (in the case of a person that is fully aware of having it in the detriment of others);
2. Interest-begotten-prejudice;
3. Authority-begotten-prejudice;
4. Self-defence ("the need of self-defence against counter-fallacies").

Among these, "sinister interest" is considered bye Bentham to be the main cause. In short, it can be considered as private interest, or group interest, as opposed to the public, or common interest:

"The mind of every public man is subject at all times to the operation of two distinct interests; a public and a private one. His public interest is that which is constituted of the share he has in the happiness and well-being of the whole community, or of the major part of it; his private interest is constituted of, or by, the share he has in the well-being of some portion of the community less than the major part: of which private interest the smallest possible portion is that which is composed of his own individual – his own personal – interest." (Bentham 1824, 362)

Usually one of these interests, private or public, are from Bentham's perspective, in opposition and one must be inevitably sacrificed (in most cases the public interest being the unlucky one). Bentham insists that persons who are exposed to what he calls "sinister interest" acts negatively upon opinions (theirs or the ones of others) in two ways: what is relevant in a certain context is ignored, or, if not ignored, the focus on what is relevant is significantly diminished (Bentham 1824, 33-34). Thus, it produces either false opinions or misrepresentations of opinions (Bentham 1824, 38) and Bentham insists that there are cases where a person is fully aware that it is affected by sinister interest and the more informed that person is, the more it should be seen with scepticism and lack of confidence (Bentham 1824, 38). This is of course a case of obvious inclination towards sophistry. For Bentham, the persons usually affected by this vice are the ones who are or want to be in positions of authority: evidently, politicians, followed by those whose careers are in the fields of institutional religion, jurisprudence or medicine (e.g. Bentham 1824, 57 sqq.). And these persons will, by habit, use their (argumentative) abilities towards furthering their own self-interest in whatever context they consider fitting, and they are obviously led by a central disposition that dictates purpose and means, negative ones in this case.

3. Thus, the question I consider here is the following: is there an interaction between the different dispositions of an argumentative agent? For this, first, I consider virtues defined as "internalized normative dispositions" (Drehe 2016a following Oackley & Cocking 2003). Also, I take dispositions to be predisposed to interact between themselves, some having the potential to become dominant. In (Drehe 2016a and Drehe

2016b) I distinguished between "teleological" virtues (or vices) and "instrumental" virtues (or vices), argumentational virtues being of the latter variety, while epistemological or ethical ones being related to purpose. It follows that there is an interaction at least between a teleological virtue/vice and an instrumental virtue/vice. Of course, teleological-teleological and instrumental-instrumental interactions are possible, but I will not delve on these here, since sketching the first kind of interaction takes precedence. For example, in certain deliberative contexts the purposes are set by ethical or civic virtues and then the agents employ whatever instrumental virtues (argumentation virtues) they have in their arsenal to achieve their goal. Something similar happens in what Bentham says about sinister interest and the way it acts upon the agent driven by it.

Naturally, we take sinister interest to be a vice, or a negative disposition, and we also accept that it interacts (following Rorty 1980 here – having in mind a disposition that can cause an akratic break of purpose or aim) with other dispositions. Also, from the perspective of virtue epistemology, we can observe its motivational component (cf. Zagzebski 1996). The inner workings of this interaction are not apparent. However, we can try to ponder on this in the light of what I would call "dispositional synergy" and thus sketch an eventual way to deal with these kinds of interactions. Of course, in common language usage, synergy means cooperation between two or more entities whose cooperation will give rise to a greater result than one of the cooperants is capable to achieve alone. Usually, the concept was used in religious, sociological, biological, or economic contexts. In this case I would like to adapt its usage to VAT.

Dispositional synergy can be of two kinds: 1. Positive, when different dispositions augment each other, and the purpose (even a morally questionable purpose) of the teleological disposition is reached following the interaction with one or more instrumental dispositions (for example, an argumentational virtues). 2. Negative: when a disposition might hinder or derail the working of another, and the case I consider here is the one where the purpose given by the teleological disposition is not achieved and this happens because of the interference of the wrong instrumental disposition or dispositions (for example, an argumentational vices). Therefore: when a disposition such as sinister interest interacts/cooperates with a disposition such as knowing certain rules and techniques of logic/argumentation/persuasion, then the result has the chance to be positive from a persuasive standpoint (even if morally wrong, as sophistry in most cases is). This way of seeing sophistry is a hypothetical development of what I already presented in (Drehe 2016b) where I tried to show that a fallacy theory specific to VAT should consider *akrasia* or incontinence as an explanatory concept appropriate to sophistic or paralogistic behaviours, or in (Drehe 2017) where I considered these issues from the perspective of the audience.

4. This present picture of interaction between virtues and virtues, virtues and vices and vices and vices (argumentational, epistemological, ethical or of other kinds, seen as dispositions) is sketchy at best. In future papers I will try to develop this venue of inquiry and I plan to draw on insights taken not just from historical authors but also from more recent work in psychology and cognitive science. Given the possible informative import from this interdisciplinary research, it is probable that the understanding of the interactions and synergies between argumentative dispositions will at least in part pass beyond a presentation that is rather phenomenological and descriptive in nature.

References

Aberdein, A. (2010). Virtue in argument. *Argumentation*, 24 (2): 165-179.
Aberdein, A. (2013). Fallacy and Argumentational vice. In Mohammed, D., & Lewiński, M. (Eds.). Virtues of Argumentation. Proceedings of the 10th International Conference of the Ontario Society for the Study of Argumentation (OSSA), 22-26 May 2013. Windsor, ON: OSSA.
Aberdein, A. (2016). The Vices of Argument. *Topoi* 35 (2): 413-422.
Bentham, J. (1824). *The Book of Fallacies: From Unfinished Papers of Jeremy Bentham*, by a Friend, London: John and H. L. Hunt.
Cohen, D. H. (2005). Arguments that backfire. In D. Hitchcock and D. Farr (eds.), *The Uses of Argument*, OSSA, 2005, pp. 58-65.
Drehe, I. (2016a). Argumentation Virtues and Incontinent Arguers. *Topoi* 35 (2): 385-394.
Drehe, I. (2016b). Fallacy as vice and/or incontinence in decision-making. In D. Mohammed and M. Lewinski (Eds.). *Argumentation and Reasoned Action, Proceedings of the First European Conference on Argumentation, Lisbon 2015*. Vol. 2. Pp. 407-416.
Drehe, I. (2017). The Virtuous Citizen: Regimes and Audiences. *Studia Universitatis Babes-Bolyai – Philosophia*. Vol. 62. Issue 2. Pp. 59-76.
Hamblin, L. C. (1970). *Fallacies*. Methuen & Co. Ltd.
Hansen, H. (2020). Fallacies. In *The Stanford Encyclopedia of Philosophy* (Summer 2020 Edition), Edward N. Zalta (ed.), URL = <https://plato.stanford.edu/archives/sum2020/entries/fallacies/>.
Oakley, J., Cocking D. (2003). *Virtue ethics and professional roles*. Cambridge: Cambridge University Press.
Rorty, A. O. (1980). Where Does the Akratic Break Take Place? *Australasian Journal of Philosophy*, Vol. 58, No. 94; December 1980, pp. 333-346.
Secor, M. J. (1989). Bentham's 'Book of Fallacies': Rhetorician in Spite of Himself'. *Philosophy & Rhetoric*. Vol. 22, No. 2 (1989), pp. 83-94.
Zagzebski, L.T. (1996). *Virtues of the mind*. Cambridge: Cambridge University Press.

TWO POINTS FOR A FEMINIST VIEW ON ADVERSARIALITY

LUCIJA DUDA
University of Manchester
lucija.duda@manchester.ac.uk

Abstract

Audrey Yap (2020) argues that overreliance on Govier's (1999) minimal adversariality renders impossible the access of marginalized philosophers to philosophy. That is why philosophy should aim for a methodological pluralism, i.e., promote collaboration as an argumentative practice. According to Yap, marginalized philosophers are unable to produce a speech act of polite disagreement because social stereotypes about politeness influence the uptake. Although her goal is to reveal how marginalized philosophers are silenced, I argue that the theoretical framework which treats uptake as the constitutive of speech act (1) reproduces the regimes of representation of 'other' in Western philosophy, (2) perpetuate white and male embodiment as normative and (3) struggles to recognize the forms of argumentative agency that marginalized philosophers have. I suggest two programmatic points that feminist argumentation theory should consider: it is necessary to (1) theoretically shift to audience, and (2) stop treating collaboration as an alternative to minimal adversariality because stereotypes persist the change of argumentative practice as they emerge from power imbalance between arguer and audience. These points will allow feminist scholars to better scrutinize argumentative situations and recognize some other important phenomenon, such as adversarial listening (Davis and Godden, 2021).

1. Introduction

In this paper I address the way in which the perception of the social location of philosophers affects practice of philosophical argumentation. More specifically, I reply to Audrey Yap's (2020) take on minimal adversariality. She argues that philosophers belonging to underprivileged groups are unable to produce a speech act of polite disagreement because the stereotypes about the social norms of politeness and agressivness affect the uptake. Therefore, the overreliance on adversary method undermines marginalized philosophers' ability to put forward an argument, thereby preventing their access to philosophy. Building on the work of Hundleby (forthcoming), Yap suggests that philosophy ought to make space for methodological pluralism, and introduce alternatives, such as explicit collaboration as argumentative method.

The theoretical framing of speech act theory that Yap uses is ill-suited for the liberatory goals of feminism. I argue that the framing used to describe *inability* of marginalized philosophers to produce speech acts of polite disagreement is what creates *unable* arguers from BIPOC philosophers.[1] It does so in three ways. First, (1) it maintains the 'regimes of representation of other' (Hall, 1997) specific for philosophy. Second, (2) it ratifies as normative the embodiment of philosopher as white and male. Third, (3) it prevents the recognition of the specific forms of argumentative agency philosophers from underprivileged groups have. Drawing upon Yap's work and feminist argumentation theory, I suggest two programmatic points beneficial for feminist metaphilosophical work on adversariality: (1) it is necessary to shift to audience as unable to give an uptake, and (2) it is necessary to stop perceiving collaboration as the alternative to adversarial method in philosophy, since it does not guarantee that social location will not affect philosophers' argumentative practices. These points allow feminist argumentation theories to better describe the dynamic of adversarial argumentation, and to recognize some important phenomena, such as adversarial listening (Davis and Godden, 2021).

2. Marginalized philosophers and the (in)ability to argue

In her landmark article 'A Paradigm of Philosophy: The Adversary Method', Janice Moulton (1983) argues that the adversarial method, a method of subjecting claims to the strongest objections, ought not to be considered the best evaluative tool and method to do philosophy. Adversariality emerges from the conceptual conflation of aggression with positive traits like success or ambition. However, the positive associations are accepted only when it comes to men. Reflecting on Moulton, Govier (1999; 2021) distinguishes between minimal adversariality and ancillary adversariality. Briefly, minimal adversariality is a matter of logical opposition between two claims. A person arguing for 'p' is simultaneously rejecting '~p'. The opposition between claims might slip into opposition between persons, known as ancillary adversariality. For Govier (1999) the latter is a strident form of communication featured by "lack of respect, rudeness, lack of empathy, name-calling, animosity, hostility, failure to listen and attend carefully, misinterpretation, inefficiency, dogmatism,

---

1 Black, Indigenous, and people of colour.

intolerance, irritability, quarrelsomeness, and so forth." (p.245) But aggression can be mitigated if arguers apply the norms of politeness and respect throughout the discourse.

Yap in her paper "Argumentation, Adversariality and Social Norms" argues against Govier's minimal adversariality, claiming that overreliance on it restricts BIPOC philosophers' access to philosophy.[2] For Yap, marginalized philosophers are *unable* to produce a speech act of polite disagreement (minimally adversarial speech act) since the stereotypes about social location and norms of politeness and agressivness affect the performative force of the speech. For instance, Asian Americans are stereotyped as passive, unsociable and disengaged interlocutors, but intellectually capable. This racialized perception affects the performative force of Asian American philosophers; hence they end up not participating in the polite adversarial debate regardless their intentions. Given that, Yap argues that in cases of power imbalance, minimal adversariality and politeness make philosophy inaccessible to non-white identities. For this reason, philosophy should make space for new methods, such as explicit collaboration.

Although Yap is correct in saying that stereotypes about social location play a role in argumentation, I argue that the framing of speech act theory that Yap is committed to creates from marginalized philosopher individual *unable* to produce an objection. Namely, the framing yields the implication that for something to count as an objection the necessary condition is the uptake of the socially privileged listener. This entails three problems: (1) it invokes the historical and oppressive relation between non-normative identities and practice of reasoning, feeding into the regimes of representation of 'Otherness' specific for Western philosophy, and (2) the fact that performative force of speech act is determined by the powerful hearer, confirms that to be a philosopher one needs to satisfy the norms of white and male appearance. Finally, (3) it ratifies the dominant ways of limiting what philosophy and argumentation theory recognize as philosophical and argumentative agency. Before I present my argument, let's first look into the tradition Yap is drawing from for her case, and then into the details of her claims.

Yap (2018) claims that 'overreliance on the adversary paradigm undermines the ability of relatively marginalized philosophers to make philosophical arguments, including arguments relevant to their own oppression' (p. 14). This *inability* is a result of something external to the

---

[2] Given the lack of space, I cannot do justice to the vast array of feminist argumentation theory on adversariality, politeness and aggression. On how the norms of politeness and aggressiveness are gendered and present double binds for women, see Hundleby (2013) and Burrow (2010). On how feminist non-adversarial models commit to gender essentialism by treating white women's politeness as universal, see Henning (2018, 2021a); on how feminist non-adversarial models struggle to recognize passive-aggressive white talk, see Henning (2021b).

speaker. Yap uses Quill Kukla's (writing as Rebecca) (2014) constitution theory of uptake (CTU), which states that uptake is constitutive of speech act.[3] This means that what speech act one ends up performing is determined by the hearer rather than by the speaker's communicative intentions. For instance, in the context of feminist philosophical spaces, Black women are unable to disagree with white women because their contribution will not count as polite academic disagreement. That is so because the stereotype "Angry Black Women" will affect the white women's uptake of the speech act of polite disagreement. So, what Black women philosopher ends up doing is an aggressive attack. Black women lack the *ability* to participate in philosophical argumentation only because her white audience interprets her wrongly.

Yap's goal is to reveal the ways in which marginalized philosophers lack access to philosophy and end up silenced by the powerful audience. However, her project is limited by the theoretical framing she is using to explain the inability to engage marginalized philosophers' experience in philosophical adversarial practice. One of the epistemic duties of a philosopher is to seek different opinions to reach the truth and epistemic betterment (Rooney, 2013, p. 322). Given the adversary paradigm, to disagree or to raise an objection is one of the fundamental tasks of any philosopher. Therefore, the ability to fulfil that task is one of the most important normative abilities we expect a philosopher to manifest. According to the CTU framework that Yap uses, the implication is that the necessary condition for a marginalized philosopher to produce an objection, a golden nugget of philosophical practice, is an appropriate uptake given by the white (or other socially privileged) audience. In other words, whether marginalized philosophers argue 'appropriately' and whether they fulfil their epistemic duty depends on the members of the white audience.

I have mentioned that the problem with this framing is threefold. Let's start with the first problem. The CTU entails that marginalized philosophers are *unable* to produce polite forms of philosophical disagreement. I argue that this maintains the regimes of representation of 'Other' in philosophy. The concept of regime of representation was introduced by the cultural theorist Stuart Hall (1997). In Western culture exists a recognizable pattern of representation of the non-white and non-male identities through their 'difference' from the identity of the elite master, taken as a universal norm. According to Hall, the whole imagery through which 'difference' is represented from text to text at some historical point is called regime of representation.

The representation practices have constructed difference as otherness located in the realm of the inferior because they are structured by the key dualisms of Western thought: mind/body, white/Other, masculine/feminine, civilized/primitive, and reason/emotion. A

---

[3] I take the name 'constitution theory of uptake' from McDonald (2021)

considerable amount of philosophical and feminist literature (see Plumwood, 2002; Jaggar, 1983; Irigaray, 1985; Derrida, 1981) has been published debating the assumed hierarchy between the relata of dualisms, and their socially attributed associations. Feminist criticisms argue that in the mainstream Western philosophical tradition the ideal of reason is constructed as masculine, white and a property of a 'master' whilst presented as universal. Meanwhile, the feminine and the 'other' are constituted through exclusion from reason, and associated with emotionality, animality, body and primitivity. The imposed hierarchical relationship and the denial of overlap between dualistic pairs form an alignment between the righthand side of each pairing, assigned lower value and depicted as inferior, whereas the lefthand side of each pairing is granted the status of universality and superiority. In philosophy the evidence for gendered and racialized dualisms comes from overtly sexist and racist philosophical writings which claim that women and slaves are less prone to rational thought (James, 2000), hence representing them as anomalous. Additionally, there is a recognizable pattern of devaluation of underprivileged groups in the philosophical usage of metaphors and examples in Western tradition (Lloyd, 1993). The metaphors and examples used, together with sexist and racist claims, have created an oppressive regime of representation which has constructed non-white and non-male subjects as different others, marking the difference with exclusive possession of qualities as emotionality, primitiveness, bestiality, being closer to nature.[4]

Let's go back to Yap's framing. I have already argued that CTU framing implies that the *ability* of marginalized philosopher to produce a polite disagreement is conditioned by the right uptake of their presumably white audience. The regimes of representation of Western philosophy have figured 'Other' and 'feminine' as excluded from reason. In combination with 'polite objection' as the standard form of philosophical reasoning, the metaphor of inability attached to marginalized philosophers' acts invokes the traditional lack of overlap between, on the one hand, reason and civility as the quality exclusively possessed by the male white master and, on the other, complementary qualities exclusively possessed by marginalized other. In Yap's framing, marginalized philosophers are marked by the gender and racial differences in their ability to produce something that socially privileged philosophers do with ease. Although she makes sure that this inability is not a cognitive one, the metaphors reflect others' reasoning capacity as depending on their gendered and racial difference.

The second problem with CTU framing is that it ratifies the norms of white and male appearance in philosophy. I think about philosophy as a

---

[4] Gendered reason is not only an embellishment of philosophical writing but constitutes the notion of reason. For more details see Lloyd (2002) and Plumwood (2002).

field of appearance highly regulated by norms. To appear in adherence to a norm is the precondition for appearing at all. Therefore, recognition can come only from the satisfaction of appearance norms (Butler, 2015). Given the cultural hegemony of white male practitioners and white male subjects in the Western philosophical tradition, and racist and sexist regimes of representation, it is not surprising that the field of appearance in philosophy is regulated by whiteness and maleness. To be recognized as philosophical agent, one needs appear as white and male. This explains the underrepresentation of marginalized philosophers. As said before, according to CTU, Black women's speech act of polite disagreement count as such only when it receives an appropriate uptake by the white audience. What follows is that for her contribution to be a polite objection there is a norm of appearance she needs to satisfy. The suggestion is that a philosopher who raises objections must be white and male, otherwise the objection is not made. This re-establishes the traditional embodiment of philosopher as white and male.

Moreover, the third problem with Yap's framing is that it re-establishes the dominant ways of limiting what philosophy and argumentation theory recognize as philosophical and argumentative agency. This linguistic problem pours into the metatheoretical. My argument here is analogous to Butler's (2015) argument about the language we use to describe those who are excluded by the politics. According to Butler (2015): 'If we claim that the destitute are outside of the sphere of politics — reduced to depoliticized forms of being — then we implicitly accept as right the dominant ways of establishing the limits of the political. (...) Such a view disregards and devalues those forms of political agency that emerge precisely in those domains deemed prepolitical or extrapolitical.' (p.78)

For Butler, the destitute are located outside the field of appearance of political theory, i.e., political theories do not recognize them as normative political agents. Butler alerts us that by treating groups historically considered apolitical agents as still belonging outside politics we risk ratifying the dominant boundary between who does and who does not count as a political agent, and between what is and what isn't political agency. An analogy can be drawn with metaphilosophy and argumentation theory. With the aim of illuminating the ways in which injustice is done by minimal adversariality, the language used to claim that the marginalized philosopher is *unable* to produce a speech act of polite disagreement – standard form of philosophical reasoning – locates them beyond the limits of the space considered philosophical. Moreover, besides ratifying the Western norms of appropriate philosophical agency, the CTU's language of *inability* causes disregard of non-traditional sites of philosophical knowledge, leaving them as not philosophical. Black feminists and philosophers of colour have been urging the rest of the philosophical community to recognize blogs, pamphlets, literature, and philosophers' critical autobiographies (Henning, 2020; Yancy, 2012; Davis and Yancy, 1998) as sources of philosophical knowledge and tools to reframe philosophy. But this is not possible when the theoretical framework maintains the image of BIPOC philosophers as without recognizable agency.

However, Yap makes it clear that marginalized philosophers do not lack the general ability (cognitive or professionally necessary skills) to argue. Her claim is that they lack a specific ability to argue here and now; or that they have what is called, a narrow ability, but given the external surroundings they do not exercise it.[5] They are still able to argue, just not capable of exercising the ability in front of a white audience. Therefore, marginalized philosophers are agents who just do not exercise their agency because they do not receive appropriate uptake. But this is not the case: CTU entails that marginalized speakers are no agents at all. Speakers' intended speech acts creates rights and duties for them, but according to CTU, argues McDonald (2021, p. 3159), hearer have the power to change the normative status of a speaker and create rights and duties in their name. This implies that speakers do not possess any recognizable autonomy. According to McDonald it is hard to see how speakers with no minimal autonomy can become agents at all.

Feminist argumentation theories that intend to expose the injustice BIPOC arguers experience should not harm BIPOC philosophers more. The suggestion that marginalized philosophers are not recognizable agents has detrimental consequences. It justifies their subordination and presents marginalized philosophers only as passive victims of white privilege. In her critique of Lacanian feminism, Fraser (2012) argues that to assume that women are only passive victims of male supremacy is a disabling assumption which totalizes male dominance and treats men as the only social agents. Additionally, the assumption erases the existence of feminist theorist and activists, thereby making impossible to understand how women contribute to culture even in conditions of subordination. Although Yap's intentions are more than welcomed in metaphilosophy and argumentation theory, my critique of the CTU framework to realize those intentions entails the message like Fraser's. CTU does not help to reframe the ethnocentric western practices of argumentation and philosophy. Its disabling assumptions erase the BIPOC's philosophical and argumentative contribution needed to reach that goal. Additionally, it erases the contributions for general epistemic betterment of our argumentative and philosophical practice. Changing the framework does not make gender and racial material reality magically disappear, but it points us in the direction where to search for this materiality to change.

---

[5] For an overview on practical abilities and agency see Small (2017).

3. First programmatic point: focus on audience

The programmatic points I propose are suggestions of the ways in which feminist argumentation theory and feminist metaphilosophy can better fulfil their liberatory goals. My first programmatic point is to shift the focus from marginalized philosophers to the *inability* of the audience. I argue that it allows us to better assess the argumentative situations. The visual identity of the speaker triggers the audience to apply to the speaker their own norms and beliefs about the cultural identity associated with the speaker's visual identity. The stereotyping happens when there is a great power dynamic between the audience and the speaker (Hall, 1997). So, it is better to say that the stereotypes affect the audience's ability to give an appropriate uptake, rather than the speakers' ability to produce the speech act they intend. Let's consider two examples of Black women sharing their experiences in philosophy classes and philosophy conferences.

Example 1: Philosophy class

"We are discussing the effects of privilege and oppression on members of dominant groups when Jackie, an African American woman, states her frustration that white women and men get sympathy for the damage they endure from occupying privileged positions, but white women simply can't hear anything about the damage she endures as a woman of color. She then proceeds to offer numerous ways this is occurring in relation to the class debate on the text we are discussing and how it happens to her in daily life. Almost immediately, Rachel, a white woman in the class interjects: "That's not true! I am really sympathetic to your problems, but it doesn't mean that I am not also damaged by racism." Jackie calmly suggests that she is not denying that Rachel may face some issues, but that it is not an excuse to ignore her or other women of color. Rachel responds to Jackie by declaring that she cares about her, but that Jackie cannot see that because she is too angry. To this, Jackie responds: "So now I am an 'angry black woman.' I'm screwed again." (Wolf 2020, p. 902)

Example 2: Epistemic diversity

"On a panel entitled 'Epistemic Diversity,' one presentation attempts to quantify intersectionality; however, the statistical model included proxies such as height, wearing of hats, and respondent's personal shirt preferences in addition to race and gender. During the Q&A, I commented that some of their examples weren't really the kinds of identities that intersectionality is concerned with. The presenters responded

"Intersectionality is for everyone and every type of identity. Crenshaw uses identities. Why wouldn't you want this important work universally used?" (Henning 2020, p.261)

According to the CTU framework, in the Philosophy Class example Jackie produced an aggressive disagreement or an attack rather than a thoroughly thought disagreement. Whereas in the Epistemic Diversity case, Henning's disagreement that 'X and Y examples are not the kinds of identities intersectionality is concerned with' is taken as a speech act of implicature expressing her reluctance for the intersectionality to do the good it can.

Firstly, the focus on audience reveals that the audience's ability to give an appropriate uptake is affected because the audience (mis)perceives only the quality in which arguer enacted their role of a philosopher. In other words, in assessing Henning's and Jackie's argument, audience only assesses how they behave in performing their role - angrily or maliciously. But the qualities, misperceived or not, cannot be reduced to the concept of social role (Downie 1971, p.133). That audience (mis)perceives the quality in which the role is enacted suggests that for them some other role is salient in the discourse. People occupy various social roles, and each of them come with a set of norms, expectations, and permissible moves in a dialogue, and which of them becomes salient for a hearer depends on the external factors (Popa – Wyatt and Wyatt, 2017).[6] Given the context, Henning and Jackie would ideally occupy the social role of a philosopher and a philosophy student, respectively. However, their visual identity as an external factor triggers their biased hearers to make the role of race and gender salient. This discursive move affects the audience's ability to give an appropriate uptake.

Focus on audience and salient roles helps to avoid the problems with CTU. Contrary to CTU, in my suggestion a marginalized philosopher performs a speech act of polite disagreement if she fulfils the duty of her role as a philosopher, the role that should be privileged given the context. According to Rooney (2013, 322), one of the epistemic duties of a philosopher is to seek a different opinion about X to reach an epistemic betterment. In Philosophy Class and Epistemic Diversity cases both Jackie and Henning proposed a different opinion. Just because for the audience the salient roles are roles of gender and race, that does not undermine the fact that the marginalized philosophers have fulfilled the duty of their professional role successfully.

Moreover, shifting to how social location affects the audience's ability to give an appropriate uptake will allow feminist argumentation scholars and metaphilosophers to better analyze cases of polite minimal

---

[6] The short-term social roles salient in a discourse are discourse roles. But for the lack of space, I only take Popa-Wyatt and Wyatt's (2017) notion of role saliency in a discourse.

adversariality. In the cases above, both the presenter and Rachel refused to exchange reasons with Henning and Jackie. While Rachel refuses to listen to Jackie by accusing her of being angry and irrational, the presenter refuses to carefully attend to Henning's disagreement by imputing that she does not want intersectionality to proliferate. With this refusal, the reciprocity condition necessary for argumentation to continue is unsatisfied, so it can be said that the audience blocked the communication. Focus on the refusal to exchange reasons allows to see the refusal as a case of Davis and Godden's (2021) adversarial listening. In both cases, there is a-rational exercise of power where the audience (Henning and Jackie after uttering a disagreement) is fully subjugated to the will of the speaker (the presenter and Rachel), who refuses to grant audience to their audience. Therefore, the shift in theorization to audience opens to feminist analysis the possibility to perceive adversarial listening as an argumentative move.

4. Second programmatic point: collaboration is not a solution

Yap, following Hundleby (forthcoming), argues that explicitly cooperative argumentative practice should be introduced as a philosophical method. I agree that methodological pluralism is beneficial. However, I argue that collaboration does not grant that the social location of the arguer will not affect the audience's uptake. That is why my second programmatic point suggests that feminist argumentation theorists should stop indicating collaboration as an alternative to minimal adversariality.

Although Yap does not state clearly what is explicit collaboration, if it is anything like the kind of collaboration Hundleby (2013) suggests in her previous work, it is hard to see how change of argumentative practice secures the right uptake. According to Hundleby, collaborative exchange of reason resembles epistemic division of labour in science. Each scientist contributes to the development of some argument from their own expertise to show that the method used is the best to do the job. The difference is that "there is no opposition to [...] claims of expertise, only inadequate understanding that can be overcome by sharing some of the expert or testimonial evidence." (Hundleby 2013, p. 254)

First, I argue that stereotypes about social location affects the hearer's uptake even if argumentative practice is changed. The condition for stereotypes to be invoked is the visual identity of an interlocutor and the power imbalance between participants. And these are stable factors external to argumentative method. For instance, in collaboration with white feminists, Black feminists testified that the mere presence of their bodies was enough for them to be ascribed as a cause of tension, and as a disruption to the solidarity of white feminists who are present (hooks,

2000:56 as cited in Ahmed, 2009). Moreover, collaboration is often associated with politeness. Henning (2018) showed that the cooperative practice feminist philosophers have in mind would interpret practices of politeness and respect of Black women who speak African American Vernacular English as adversarial. Additionally, Henning (2021b) also argued that collaboration might easily slip in unrecognized passive-aggressive argumentative tactics, i.e., 'white talk', addressed at BIPOC women in academia.

Second, socially privileged speakers may judge marginalized philosophers as lacking the innate talent to do philosophy only given their social identity, regardless their argumentative method. In philosophy, the audience is inclined to think about philosophical talent as an innate ability (Saul, 2012), and this is an ideological baggage of the hegemonic history of Western philosophy. Bodies of colour are believed to be ontologically incapacitated, not circumstantially (Yancy 2012, p.10). Therefore, no argumentative practice will change the audience's engagement with the arguer as the belief about lack of talent is not tied to minimal adversariallity, but to the history of philosophy.

Finally, it is unclear how collaboration would guarantee equal access to philosophy for all philosophers. A final thought that this paper wishes to convey is that if feminist argumentation theory aims to combat racist and sexist formations, then feminist scholars need to do the metatheoretical step of questioning the alleged non-neutrality of its theoretical frameworks. Otherwise, it is impossible to successfully reframe philosophical practice to be inclusive towards BIPOC philosophers and to legitimize BIPOC's contributions.

References

Ahmed, S. (2009). Embodying diversity: Problems and paradoxes for Black feminists. Race Ethnicity and Education, 12(1), 41-52. https://doi.org/10.1080/13613320802650931
Burrow, S. (2010). Verbal sparring and apologetic points: Politeness in gendered argumentation contexts. Informal Logic, 30(3). https://doi.org/10.22329/il.v30i3.3033
Butler, J. (2015). Notes toward a performative theory of assembly. Harvard university Press.
Davis, J., & Godden, D. (2021). Adversarial Listening in Argumentation. Topoi. 40, 925–937. https://doi.org/10.1007/s11245-020-09730-1
Derrida, J. (2016). Of Grammatology (5th ed.). (Trans: G. C. Spivak). The John Hopkins University Press. (Original work published 1976)
Downie, R. (1971). Roles and Values. An Introduction to Social Ethics. Routledge. https://doi.org/10.4324/9781003021995
Fraser, N. (2013) Fortunes of Feminism From State-Managed Capitalism to Neoliberal Crisis. Verso

Govier, T. (2021). Reflections on Minimal Adversariality. Informal Logic, 41(4), 523-537. https://doi.org/10.22329/il.v41i4.6876

Govier, T. (1999). The philosophy of argument. Vale Press.

Hall, S. (1997). The Spectacle of 'Other'. In S. Hall, J. Evans and S. Nixon (Eds.), Representation. Cultural representation and signifying practices (pp. 223–291). Sage Publications.

Henning, T. (2018). Bringing wreck. Symposion: Theoretical and Applied Inquiries in Philosophy and Social Sciences, 5(2), 197-211. https://doi.org/10.5840/symposion20185216

Henning, T. (2020). Racial Methodological Microagressions. When Good Intersectionality Goes Bad. In L. Freeman and J.W. Schroer (Eds.), In Microaggressions and Philosophy. (pp. 251–273). Routledge New York and London. https://doi.org/10.4324/9780429022470

Henning, T. (2021a). "I Said What I Said"—black women and argumentative politeness norms. Informal Logic, 41(1), 17-39. https://doi.org/10.22329/il.v41i1.6687

Henning, T. (2021b). "Don't Let Your Mouth": On Argumentative Smothering Within Academia. Topoi, 40(5), 913-924.

Hundleby, C. (forthcoming) Beyond Adversary Paradigm: Argument Repair. In R. Cook and A. Yap, Feminist Philosophy and Formal Logic. University of Minnesota Press

Hundleby, C. (2013). Aggression, Politeness, and Abstract Adversaries. Informal Logic, 33(2), 238–2. https://doi.org/10.22329/il.v33i2.3895

Irigaray, L. (1985). This sex which is not one. Cornell University Press.

Jaggar, A. (1983). Feminist Politics and Human Nature. Harvester.

James, S. (2000). Feminism in philosophy of mind. The question of personal identity. In M. Fricker and J. Hornsby (Eds.), The Cambridge Companion to Feminism in Philosophy (pp. 29–49). Cambridge University Press.

Kukla, R. (2014). Performative force, convention, and discursive injustice. Hypatia, 29(2), 440-457.

Lloyd, G. (1993). The Man Of Reason. 'Male' and 'Female' in Western Philosophy (2nd ed.). Routledge London.

Lloyd, G. (2002). Maleness, metaphor, and the" crisis" of reason. In L. Antony and W. Charlotte, A mind of one's own. Feminist Essays On Reason and Objectivity (pp. 73-89). Routledge.

McDonald, L. (2021). Your words against mine: The power of uptake. Synthese, 199, 3505–3526. https://doi.org/10.1007/s11229-020-02944-1

Moulton, J. (1983). A paradigm of philosophy: The adversary method. In S. Harding and M.B. Hintikka, Discovering reality (pp. 149-164). Springer, Dordrecht.

Plumwood, V. (2002). The Politics of Reason: Towards a Feminist Logic. In R.J. Falmagne and M. Hass, Representing reason. Feminist Theory and Formal Logic (pp. 11–45). Rowman & Littlefield.

Popa-Wyatt, M., & Wyatt, J. L. (2018). Slurs, roles and power. Philosophical Studies, 175, 2879–2906. https://doi.org/10.1007/s11098-017-0986-2

Rooney, P. (2012). When philosophical argumentation impedes social and political progress. Journal of Social Philosophy, 43(3), 317-333.

Saul, J. (2013). Implicit bias, stereotype threat, and women in philosophy. In K. Hutchison and F. Jenkins (Eds). Women in philosophy: What needs to change, (pp 39-60). Oxford University Press. https://doi.org/10.1093/acprof:oso/9780199325603.001.0001

Small, W. (2017). Agency and practical abilities. Royal Institute of Philosophy Supplements, 80, 235-264. https://doi.org/10.1017/S1358246117000133

Wolf, A. (2017). "Tell Me How That Makes You Feel": Philosophy's Reason/Emotion Divide and Epistemic Pushback in Philosophy Classrooms. Hypatia, 32(4), 893-910. https://doi.org/10.1111/hypa.12378

Yancy, G. (2012). Introduction: Inappropriate Philosophical Subjects? In G. Yancy (Ed), Reframing the practice of philosophy: Bodies of color, bodies of knowledge. (pp. 1-19). SUNY Press.

Yancy, G., & Davis, A. (1998). Angela Davis. In G. Yancy (Ed.), African-American Philosophers. 17 Conversations (pp. 13–31). Routledge.

Yap, A. (2020). Argumentation, Adversariality, and Social Norms. Metaphilosophy, 51(5), 747–765. https://doi.org/10.1111/meta.12458

MINIMAL ARGUMENTATION: A RESEARCH PROGRAM

MICHEL DUFOUR
University Sorbonne-Nouvelle
michel.dufour@sorbonne-nouvelle.fr

Abstract
This paper introduces the Minimal Argumentation Project which focuses on the basic (minimal) conditions making a discourse an argumentation. Taking the word "argument" to mean a linguistic product of the premise/conclusion type and "argumentation" to refer to a longer discourse supposed to involve arguments, the project aims at clarifying how they are related. The paper discusses various questions raised by the requirement of arguments in an argumentation. It also sketches the possibility of an argumentation without explicit arguments. The dispersion of the results of students asked to find arguments in a text shows that beyond the paradigmatic case of a manifest controversy, the boundaries of the field of argumentation require specific attention.

1. Introduction

The goal of this paper is not to discuss a theoretical point or empirical results, but to introduce an emerging research project, the Minimal Argumentation Project (MAP).

We know that the English term "argument" is equivocal. O'Keefe (1977) notoriously stressed a difference between "argument" understood as a product (typically a premises/conclusion system) and "argument" understood as a verbal process involving conflicting points of view. The same kind of equivocation may exist about the word "argumentation", as is the case in French, for instance. These double meanings, linked with double aspects which are taken as related, directly matter for the MAP, for it provisionally accepts but also questions the principle that arguments are included in an argumentation. From now on, to prevent any confusion, the word "argument" will mean argument-as-product – understood as a premises-conclusion system or, in a less technical vocabulary, reasons supporting a point of view – and "argumentation" will be used to mean a longer discourse, typically a plea or a whole verbal exchange that is supposed to be argumentative.

The MAP grants that there are paradigmatic or normative argumentations for instance the critical discussion dear to pragma-dialecticians (Van Eemeren & Grootendorst, 2004). However, the goal of the project is to explore situations that may appear marginal or border-case by comparison with them. As we are interested in verbal communication, we can say that an argument is a short discourse, whereas an argumentation is a long one in the sense of longer than the argument(s) it involves or is expected to involve. A limit case occurs when what is taken as argumentation is just the presentation of a single argument.

Yet the most interesting point for the MAP is the link between discourses of different sizes: argumentation and argument. Its leading question could be worded as follows: What makes a long discourse an argumentation, or what is the minimal necessary condition allowing to call a discourse an argumentation? The adjective "minimal" presupposes that if there are several necessary conditions, at least one appears more important than the others. This way to formulate the question presupposes strict definitory conditions. However, the MAP is open to the possibility that what makes a discourse an argumentation is a matter of degree – in other words, that a discourse could be more or less an argumentation –, or that the boundaries of the concept of "argumentation" are fuzzy. An easy answer to the previous questions comes quickly to mind: it contains argument(s). This seems reasonable and we often are satisfied with it, but this common-sense response poses serious practical problems.

In the first part of this paper, I discuss several of those problems and the challenges they raise. In the second part, I illustrate them with an important case for argumentation theory and critical thinking, namely some difficulties currently encountered by students in the recognition of the arguments of a text. It is not always easy to get through this step that is supposed to be crucial to determine whether a discourse is an argumentation or not. Finally, I broaden the topic in a conclusion where I also list a few topics especially relevant to the MAP.

2. From argument to argumentation and beyond

The MAP is mainly interested in discourses that are not obviously argumentations. Yet, a first impression may be misleading and be a consequence of the difficulty to identify arguments, for it is often because we identify manifest arguments that we say that a particular discourse is an argumentation.

2.1. The Nawa principle and its limits

The presence of arguments can be taken as a necessary and perhaps also sufficient condition, for considering that a discourse, or part of it, is an argumentation. We can summarize this in a short principle: "No argumentation without arguments" (Nawa).

Indicator words or some other typical grammatical constructs can, in turn, be seen as criteria ensuring that we face an argument. Accordingly, if this argument is included in a discourse, this discourse is an argumentation. But we know that some indicator words, or other argumentative markers, are equivocal because of uses other than indicating a logical link. So, there may be no argument where such a word suggests there is one. Hence, you may doubt that the discourse containing the alleged argument is an argumentation. In short, indicator words are not always sufficient to warrant that a discourse is an argumentation. Moreover, you can abuse the argumentative potential of indicator words to give the impression that you put forward arguments. For instance, adding a few "therefore" here and there can be a trick to give the illusion that you argue. This raises the more general problem of the border between identification and evaluation: for instance, is a bad example an example? In the same way, is a bad argument an argument?

It often happens that pragmatic considerations come into play when deciding if a discourse is an argumentation. For example, is adversariality necessary? Notice that adversariality may concern an opposition between arguments, but also between persons or parties, and sometimes both as suggested by H. Marraud (2015). One doesn't involve the other. Some authors believe that a dialectical opposition based on a difference of opinion of participants is necessary to say that a discourse is an argumentation. This is notoriously the case of pragma-dialecticians (Van Eemeren, 2003) but also of authors like Govier (1999, 2021) or Casey (2020). Making an opposition a necessary condition leads one to consider that the use of arguments is not sufficient to make sure that we deal with an argumentation. In short, this view seems to opt for the view principle that a minimal condition for having an argumentation is the conjunction of arguments and opposition. The challenge faced by people holding this view is then to qualify a discourse where arguments are used in a non-agonistic context or a situation that is indeterminate from this point of view. This is a good topic for the MAP.

Besides the risk of taking for an argument what is not an argument (for example as a consequence of a misinterpretation of indicator words) or, conversely, of not taking an argument for an argument, the implementation of the Nawa principle raises another issue.

For an argument to be a good one its premises must be relevant to the conclusion. In the same way, taking a discourse containing a single argument for an argumentation requires that the conclusion of this argument is relevant to the whole discourse. Otherwise, the argument will be interpreted as a digression or a puzzle. In a theater, for instance, during the rehearsal of a declaration of love (free of arguments) the actor suddenly argues, in the same tone: "Someone has left the backstage door open, for we freeze on this stage". Nobody moves and the rehearsal goes on. The actor's argument is irrelevant to the declaration and although strictly speaking it is included in the actor's discourse, it does not make it an argumentation. Most of us probably think that its irrelevance takes it out of the declaration of love, even if we can look for fancy ways to make it relevant.

Thus, the literal formulation of the Nawa principle is not sufficient. The relevance of arguments is required if we want to preserve the unity of the discourse, in the spirit of Grice's principle of cooperation. In the theater example, you could say that there is not a single discourse but two: one declaration of love and, between its two parts, an argumentation limited to a single argument about the backstage door. Although it seems reasonable to separate the argument from the rest of the actor's discourse, this solution may seem theoretically *adhoc*. Moreover, we can easily imagine less clear-cut cases where you may hesitate about the relevance of the argument and then refuse to put it aside. These cases of balancing what should be retained or abandoned in terms of arguments are particularly relevant to the MAP.

The Nawa principle raises a similar set of problems with the management of implicit aspects of arguments. We know that when actual arguments are reported to an ideal norm, some expected components may be missing. It may be an indicator of inference: "Lucy will not participate. She broke one leg". It may also be one or several premises: "Socrates is a man, therefore Socrates is an animal". A conclusion too may be missing: "You cheated and the law states that all cheaters must be punished." A major problem raised by implicit elements is whether you are right to think that you are facing an argument rather than logically independent statements. What is said or written is often insufficient to conclude, unless you take into account some pragmatic aspects. In his *Rhetoric,* Aristotle already insisted that during a trial you can dispense with the conclusion of your arguments because every participant knows what is at stake. Uncertainty about implicit elements remains probably more often unresolved in rhetorical exchanges –in the ancient sense of the term – than in dialectical face-to-face dialogues, because the addressee is an indefinite audience that cannot immediately reply or ask questions. The crucial point is less the rhetorical form as such than the impossibility of asking for explanations (Blair 1998, Govier 1999, chap.1). On the contrary, in a dialectical exchange an interlocutor can, in principle, immediately ask the speaker to make her argument more explicit. For the same reason, an interpretative indetermination about some possible implicit elements is likely to occur about written discourses whose readership is remote in space, time, or social status, and therefore cannot ask for clarification.

Questions about implicit elements follow a preliminary interpretative work of what is stated and are motivated by the interpreters' desire for clarification. This agrees with the fact that many current expressions stress a kind of priority of the arguer over the framing of an argument: arguments are "given", "put forward" or "presented". So, it seems that an arguer argues, but not her audience. However, each member of the audience is supposed to understand the argument (as it was designed), that is has to remake and appreciate it.

Thus, while arguers are supposed to have complete control over what they say, their arguments may seem difficult to identify, unclear or weak from the point of view of receivers who may be interlocutors, members of an audience, readers, or analysts. So, the interpretation initiative is supposed to come from them, who can or cannot request the assistance of the arguer. If this assistance is not possible, the arguer can only bet on the fact that the receiver understands, but only on the basis of what has been stated. The identity of the arguer's argument and the understood argument is a norm rather than a fact. We can also imagine that the intended and the understood argument are identical but differ from what has been stated. Think for instance of the use of a coded language, studied by a person who does not know the code. This suggests that even if the Nawa principle holds, the single argument that makes of a discourse an argumentation may be not identified as such by someone who, accordingly, would not acknowledge an argumentation.

So, the recognition of an argument by a receiver is as important as the act of the arguer, but in the case of remote communication, this balance and cooperation may not be applicable in practice. For various reasons that can range from shyness or social intimidation to physical or temporal distance, asking for a clarification of what looks like an argument or is left implicit may sometimes hardly be done, if only it can be done. This confirms that even if you grant the Nawa principle, taking into account some empirical pragmatic factors is required to allow the identification of the argument necessary to establish that a discourse is an argumentation. The minimal conditions making a discourse an argumentation spread beyond the strict Nawa principle.

2.2. Argumentativity

The concept of argument may not be the key to argumentation. When you look for a definition, there is a well-known alternative to the search for satisfactory necessary and sufficient conditions. Drop the idea of an abrupt boundary, grant the possibility of a fuzzy one, and allow a progressive or more or less continuous entrance into the conceptual field.

This kind of idea has been applied to argumentation by some linguists, apparently mainly French-speaking ones, whose training in logic does not drive them to immediately link argumentation and manifest arguments. One of their key concepts is argumentativity. The word is suggestive but a bit obscure, at least for people more used to a tradition framed by the basic concepts of logic. What is argumentativity? You can find this term in Ch. Plantin's *Dictionnaire de l'argumentation* (2016), recently translated into English (2018). "Argumentativity, degrees and forms" (2016, p 94) is not a standard entry word followed by a definition, but what Plantin calls a sub-entry word. It sends you to the full entry word "Argumentation", and more precisely to its paragraph "Argumentation, a way to manage disagreement". Unfortunately, the word "argumentativity" does not appear in its few lines. This may confirm that it is not an easy word. You find more help on a web page devoted to "Argumentativity". There, Plantin (2021) makes a distinction between "extended" and "limited" theories of argumentation. Limited ones "link argumentation with different kinds of discourses (deliberation, epidictic, judiciary, advertising, preaching …)", whereas the extended theories contend that "the language (Ducrot)or the discourse (Grize) are essentially argumentative". In both cases, this may cancel the very question of minimal conditions making a discourse an argumentation, since any discourse would be an argumentation. But it also depends on what is meant by "essentially". In the case of Ducrot, Plantin reports that this means that "what the utterance E1 (and the utterer as such) *means* is the conclusion E2 targeted by this utterance" (Plantin 2016, p 417). So, according to Ducrot, meaning is closely related to the idea of suggesting a conclusion. From this point of view, if "essentially" means "necessarily", to speak and to argue amounts to the same. If "essentially" does not mean "necessarily", the MAP will ask where the border is.

However, Plantin says that "the argumentativity of a sequence [of terms] is not a matter of all or nothing". For him, argumentativity is not an intrinsic property of a sequence but is a matter of degree, as already suggested but the (sub)entry of his dictionary. This degree can be said pragmatic, for "Concerning degree, an exchange begins to become argumentative when an opposition arises between two orientations of speech", the orientation of an utterance being (according to Ducrot who introduced this notion) the selection of "the utterances that can follow it" (Plantin 2016, p 417). Notice that Plantin's view on this topic appears less radical than Ducrot's, for Plantin does not seem to preclude the possibility of a low and perhaps even a zero degree of argumentativity, which, for him, would amount to a zero degree of opposition.

According to Plantin, the argumentativity of a discourse is a matter of degree but also of what he calls "form":

" […] we can distinguish two main forms of argumentativity in the development of discourses that develop in an argumentative situation.

— Two juxtaposed monologues, contradictory, without allusion to each other, constitute an argumentative diptych: each partner elaborates, repeats, and reaffirms his position. (See Antithesis)

— This "argumentative diptych" articulates and becomes more complex by incorporating the rebuttal of the opposite position."

The fact that there are partners, presupposes a mutual acknowledgment. So, according to Plantin (and contrary to Ducrot), argumentativity is not intrinsic to a discourse, i.e. "semantic", but bound to a pragmatic situation, at least because of this juxtaposition of partners. This reminds us of the debate about the necessity of adversariality to characterize an argumentation, or of the implicit dialectic suggested by what I have called Protagoras' principle, namely that there always are two divergent views on any topic (Dufour, 2020).

I will not delve further into theories using the concept of argumentativity, but I will highlight just a few key points related to the concept of argumentativity that seem important to the MAP.

First, the key concepts of these linguists are not the same as those of philosophers or logicians. First, linguists speak of utterances or sequences rather than of propositions and truth values as logicians do. But all share the basic idea of a linguistic expression (utterance, sequence, or proposition) that "follows from" a previous expression. For instance, Ducrot's original view on meaning is based on the claim that an utterance U1 can or cannot be followed by utterance U2.

How to interpret this "can"? It seems that it should be understood as a normative or statistically "normal" possibility rather than a factual one because nothing prevents you to utter U2 after uttering U1, even if it sounds abnormal. So, this basic equivocation about the meaning of "can" should be avoided.

Now, it could be tempting to make an analogy between the linguists' concept of "following from" and the logical consequence, since both of them distinguish right and wrong followers. But the approach of linguists seems broader because it sometimes applies to entities that are below the threshold of the proposition. The interesting point is that the argumentativity of a discourse may appear at a linguistic level more elementary than the level of the proposition, hence more elementary than the level of an argument if you grant that an argument is made of propositions.

However, the idea of consequences drawn from elementary constituents of an utterance, even from single words, is not new. It seems difficult to deny that "No job, no money" can be interpreted as an argument. Logicians will probably claim that the two parts of this sequence are only abbreviated propositions and so, the presumed inference from words finds a place in regular propositional logic. But some very empirically minded linguists could object that logicians cheat because they change what is uttered.

Even if we grant that a consequence driven from a noun is foreign to most formal systems of logic, the idea of a concept "contained" in another one, or a predicate included in a subject is a very old one. The concept of connotation was already discussed by medieval thinkers, and the inclusion of a predicate in a subject is at the root of the analytic/synthetic distinction, notoriously associated with the Introduction of Kant's *Critique of pure reason* (1999). Kant's well-known example is the inclusion of the predicate "extended" in the concept of "body". To speak like Ducrot, we could say that "body" selects "extended" as a possible follower but discard "non-extended". We could add among consequences drawn from a noun or a descriptive term, those expressly authorized or forbidden by field-dependent or topical rules of inference.

We can also suggest an analogy between, on the one hand, linguists' analysis of the argumentative value of some prepositions, conjunctions, or adverbs and, on the other, the constraints on consequences set by the "logical constants" of logicians or the syncategorematic terms of medieval thinkers. In both cases, however, one remains at a rather normative level indicating what to do or not to do.

In a way, the notion of argumentativity broadens the set of words and expressions that can count as clues of the presence of an argument. Thus, the extension of this set would go well beyond standard indicator words such as "therefore", "because", "as", and other expressions whose primary function is only to invite inference. In addition, by highlighting the potential but implicit inferences of a discourse, the notion of argumentativity blurs the line between overt argumentation and other discourses whose argumentativity only remains at a potential level. In any case, this topic is relevant for the MAP.

3. Students understanding

One of the first reasons for the MAP was the various difficulties met by many untrained students taking an argumentation course and the dispersion of their results in identifying arguments of an ordinary text, even when some are made explicit by indicator words. By ordinary text, I mean a text that does not depend on a specific field of knowledge and has not been modified to be more easily analyzed by students. It typically comes from a daily newspaper, a magazine, a book, or an essay accessible to a wide readership. Notice that this is a case of what I have called remote argumentation: no direct interaction with the author of the text is possible.

Over many years of teaching argumentation theory and practice, I noticed two common tendencies among students. If I can rely on similar testimonies of colleagues from different countries, these tendencies are not specific to an education system, a language, or even to the personality of the teacher. The first one is to find arguments, sometimes surprising ones, almost "everywhere and anywhere", granted that not all arguments are marked by indicator words or syntactical devices. The second tendency, on the contrary, is to miss part or the totality of arguments that I expected to be easily identified, for instance because of the presence of indicator words. Students, at least mine, often have a rather loose knowledge of the meaning/use of connectors.

At least two circumstantial explanations come to mind: some students have a loose involvement in the course and this may justify that they "over-find" or miss some arguments. On the contrary, others are seriously involved in the class but, at least at the beginning, their goodwill may push them to discover arguments where common sense would hesitate to see a single one. This is certainly interesting from a psychological and pedagogical point of view and both of these explanations are plausible. Yet, beyond some blatant extreme cases, the identification of arguments and sometimes the understanding of their role in the global text remain a difficult task for a significant number of students. I presume that this dispersion in the identification of arguments is not limited to students and suspect that the results of more mature adults would also show a high dispersion.

In the previous part of this paper, I have stressed several factors than can make difficult or controversial the identification of an argument. The case of over-finding arguments is particularly interesting when you ask students to explain what makes them think that they have found an argument. This is sometimes quite difficult for them, for instance when other people judge the propositions supposed to constitute the argument irrelevant to one another. But there are other interesting limit cases.

Here is the text of a first example:

"Today, hirudotherapy – from the Latin *hirudo*, meaning leech – is booming in Russia. Five centers have just been established in Moscow, with pharmacies selling "85000 animals" each month, according to Guennadi Nikonov, director of the International Leech Research Centre. A joint venture has also been created with a French company to manufacture by-products in the form of a cream." A few students saw "Today, hirudotherapy [...]is booming in Russia" as a premise for "Five centers have just been established in Moscow", whereas most of the others rather disagreed about the presence of an argument. The first ones interpreted the creation of the five centers as a consequence of the booming of hirudotherapy, whereas according to the others, the first sentences of this text were a succession of descriptive statements. Notice that nobody suggested that the first sentence was a conclusion supported by the next sentence.

Some similar cases can be seen as a form of the "Post hoc, ergo propter hoc" fallacy, not in the sense where a given argument illustrates this fallacy, but rather in the very fact of interpreting a mere temporal succession as a causal argument. For instance, "She did work hard and is now famous", is interpreted as "She is now famous because she did work hard."

Another kind of example is the interpretation of the expression of a goal as a final cause justifying a behavior. "She did work hard to become the famous person she is" is readily reconstructed as "As she wanted to be famous, she did work hard" or even as "She did work hard, hence she became famous". Here, using the concept of argumentativity, we could say that a weak argumentativity is interpreted as a strong one, even if the reconstructed argument is not deductively valid.

We have already noticed that the concept of argumentativity is especially interesting when it concerns only a possible consequence, not an explicit one. If we consider that what makes an argument explicit is that it does not require rephrasing what is actually said or written, then all the previous examples can be said implicit arguments. So, someone who would stick to the Nawa principle but doubt or deny that they are arguments could argue that it is the reconstruction that produces the (manifest) argument required by the Nawa principle. In other words, the reconstruction would unfairly increase the argumentativity of the discourse. Of course, if you grant the concept of argumentativity and agree that meaning essentially amounts to arguing, as suggested by Ducrot, then you don't need to increase the argumentativity of a discourse to make it an argumentation. So, students may not always over-find arguments but only show their sensitivity to an argumentativity that is below the threshold of standard manifest arguments.

This common and almost daily situation appears a good example of the border or hybrid cases the MAP is interested in.

4. Synthesis and conclusion

This brief overview of the MAP highlighted that it is organized around the relationship between argument and argumentation, where "argument" refers to a linguistic structure of the premises/conclusion type, and "argumentation" means a longer discourse, typically a plea or a whole verbal interaction supposed to contain arguments. The MAP tries to connect these two different levels of discourse. This project addresses multiple themes and problems, some of which have been sketched or discussed in this paper. Let us recall a few of them and briefly highlight others. This will not be a comprehensive list but is just intended to offer a broader overview of the concerns of the project.

Some problems about a discourse can be said internal insofar as they can be discussed on the only basis of what is said or written, thus without being obliged to take into account other contextual specificities, in particular the status or intentions of stakeholders, such as author, interlocutor, audience, etc. Other problems are therefore external in the sense that they require the integration of some of these pragmatic aspects.

We have seen that a recurring issue is the manifestation of arguments in the concerned global discourse. Recognizing, identifying, and circumscribing an argument is not always self-evident. This is of decisive importance if one characterizes an argumentation by the presence of arguments. We have seen that this problem is often related to another, namely the interplay between explicit and implicit aspects of the discourse. Once an argument has been identified, the question arises of its relationship to the general orientation of the discourse in which it is embedded. This may not be clear. In particular, this relationship can be obscured by ambiguous, irrelevant, or confused verbal expressions. All of these issues have their place in the MAP.

In addition to these internal problems, we have seen that determining whether a discourse is an argument or not may largely depend on the practical modalities of communication between stakeholders. Thus, the possibility or not of real interaction is crucial, and therefore the fact that the stakeholders are or are not remote from each other, in the various possible senses of this term. The use or not of some mediation, like a technology as old as writing, shows some consequences of remoteness on the understanding of what is at stake in a particular discourse, especially whether it is an argumentation or not. We have seen that the hermeneutic problems sometimes raised by the identification of arguments by students are closely related to the fact that they concern texts from which the authors are distant.

An ideal normative approach to human communication usually presupposes that the intentions of the participants are clear and acknowledged as such, their interpretations are right, and they share some background knowledge. The MAP takes the more limited perspective of argumentation and is particularly interested in situations deviating from these standards. This may occur, for instance when there are misunderstandings between stakeholders, perhaps about their intentions or goals, or when some of them have insufficient information on the conditions of production of the discourse or what is at stake. Some arguments may also not be acknowledged as arguments put forward by X, and others may wrongly be ascribed to Y. If, as some argue, an argument always presupposes antagonism or disagreement, as in the paradigmatic case of controversies, stakeholders may also underestimate, over-estimate or misestimate the nature or extent of the disagreement.

So, the MAP does not seek to bring back problems under an already existing theory, but rather to explore challenging limit cases whose theory is still to be done.

References

Aristotle. (1937). *The Art of Rhetoric,* trans. E. S. Foster. Loeb Classical Library. Cambridge, Mass.: Harvard University Press.

Blair, J. A. (1998). The limits of the dialogue model of argument. In *Argumentation and Rhetoric*, ed. H. V. Hansen, Ch. W. Tindale & A. V. Coleman. CD-Rom. St Catherines, Ontario: Ontario Society for the Study of Argumentation.

Casey, John. 2020. Adversariality and argumentation. *Informal Logic* 40(1), 77-106.

Dufour, M. (2020). Protagoras' principles, disagreement, and the possibility of error. *Reason to dissent: Proceedings of the 3rd European Conference on Argumentation (ECA). Studies in Logic 86,* Vol II, pp 231-242. London: College Publications.

Govier, T. (1999). *The Philosophy of Argument*. Newport News, VA: Vale Press.

Govier, T. (2021). Reflections on minimal adversariality. *Informal Logic,* 41(4), 523-537.

Kant, E. (1999). *Critique of Pure Reason,* trans. P. Guyer. Cambridge: Cambridge University Press.

Marraud, H. (2015). Do arguers dream of logical standards? Arguers' dialectic vs. arguments' dialectic. *Revista Iberoamericana de Argumentacion,* 10, 1-18.

O'Keefe, D. J. (1977). Two concepts of argument. *Journal of the American Forensic Association,* 13, 121-128.

Plantin, Ch. (2016). Dictionnaire de l'argumentation: Une introduction aux études d'argumentation. Paris: ENS Editions. English trans. (2018). Dictionary of Argumentation: An Introduction to Argumentation Studies. London: College Publication.

Plantin, Ch. (2021). http://icar.cnrs.fr/dicoplantin/argumentativite/

Van Eemeren, F. H., Grootendorst, R. (2004). *A systematic theory of argumentation: The pragma-dialectical approach.* Cambridge: Cambridge University Press.

FRAMES AND INFERENCES

ISABELA FAIRCLOUGH
University of Central Lancashire
ifairclough@uclan.ac.uk

Abstract
I argue that framing works via an inferential process, by 'inviting' a conclusion which an audience may accept or reject, from premises they may accept or reject (resulting in a successful or failed 'framing effect'). I take framing to be, covertly, a directive speech act, with two main mechanisms: the audience is either being 'invited' to see a situation as X (not as Y) or is being 'invited' to take a reason Z as one that ought to outweigh all other reasons in deciding what to believe or do. I will draw on the Argumentum Model of Topics to represent a few arguments where framing occurs either in the minor or major premise, which may be either accepted or rejected; in the latter case, the 'invited inference' is replaced by an alternative counter-inference. The AMT seems very well placed to account both for how a framing source intends an argument to be reconstructed and accepted by an audience, and for how the audience may reject the invited inference and draw another conclusion.

1. Introduction

Framing theorists, usually working in the fields of behavioural psychology or political and media communication, are not argumentation theorists, and a connection with argumentation is not explicitly made. It is however easy for argumentation scholars to see how one of the most widely cited definitions of framing can be translated into the language of argumentation theory, involving various argument schemes:

> Framing essentially involves selection and salience. To frame is to select some aspects of a perceived reality and make them more salient in a communicating text, in such a way as to promote a particular problem definition, causal interpretation, moral evaluation, and/or treatment recommendation for the item described. Typically frames diagnose, evaluate, and prescribe... (Entman, 1993: 52).

According to this definition, what framing does is define or categorize a situation, or suggest a causal explanation, or evaluate, or recommend what ought to be done about it. The so-called 'framing devices' being used seem therefore to be part of implicit arguments from classification or definition, from effect to cause or cause to effect, or from various premises towards a practical conclusion.

In the above-mentioned disciplines, the concept of frame is usually defined very vaguely. A recent assessment of the state-of-the-art, coming from behavioural psychology, takes it for granted that:

> there is no single way of thinking what a frame is that all those who work with the concept would agree upon. The concept of a frame is something that can be framed in many different ways, one might say. Trying to give it a watertight definition is surely a fool's errand. (Bermúdez, 2021: 11)

This agnosticism is not shared by linguists, who seem to have a fairly good idea of what frames are, taking their understanding from Fillmore's cognitive semantics (Fillmore 1982, 1985, 2006), including the evolving FrameNet dictionary at Berkeley California. Argumentation scholars should also be in a good position to explain how framing actually works, by a process of inference from premises to a conclusion, leading to either acceptance or rejection of the conclusion.

In the purely semantic interpretation, frames are syntactic-semantic structures, indicating for example which semantic (thematic) cases a verb can take, e.g. an agent as subject, or a patient as object. Semantic frames model shared linguistic meaning. In a cognitive frame, however, this minimal structure is expanded to include background experiential knowledge, which varies from person to person. For example, the MIGRATION frame, as a semantic frame, is a structure that includes a concept of agency and one of purposeful movement, usually across country borders. This kind of frame is a source of entailments (logical implications), as inferences. If it is true that this man is a migrant, then it has to be true that he has moved, or trying to move, from his usual place or country of residence to another one.

Cognitive frames, by contrast, include vast amounts of extra-linguistic information, grounded in an individual's experience, in addition to shared linguistic knowledge. A speaker's beliefs, values, political commitments, his whole life experience will add new layers of meaning, and therefore the possibility of new inferences, to the word 'migrant'. For some people, all migration is beneficial and necessary, and all migrants are welcome and contribute to economic growth. For others, the MIGRATION frame will contain beliefs about pressure on infrastructure, on the living standards and wages of local populations, about changes to national identity, culture and demographics, etc. From different cognitive frames, people will draw different inferences (no longer purely logical), namely that migration ought to be encouraged or discouraged, that it is good or bad for their country, and so on.

Fairclough & Madroane (2020) argued that frames as structures must be distinguished from framing as process. Taking that idea further here, I suggest that framing as process involves the use of a frame in a simple speech act (e.g. assertion) or a complex one (argumentation, explanation, narrative). In argumentation, framing can take place in the premises, but also in the way the disagreement itself is formulated, or in the choice to focus on one issue rather than another, thus perhaps framing out even more relevant aspects of a controversy. My scope here is fairly narrow, and involves framing that takes place at the argumentation stage of a critical discussion (van Eemeren 2010).

I will introduce the notion of 'invited inference'[1], i.e. the inference that an audience is invited to draw. In framing a situation, and apparently representing or describing it, a message framer *directs* an audience to take that representation as the most acceptable and relevant way of seeing that situation, or presents a particular reason as the most relevant and significant one that the audience should consider in deciding what to believe or do. The 'invited' inference is one the audience is being directed to draw, steered towards drawing, based on accepting to see a situation in the way it is being represented, or accepting a particular weighing of reasons, given by the framing act, as the most reasonable one.

In Sections 2 and 3, I briefly address a few important questions asked in framing research, and suggest how they may be answered from the perspective of argumentation theory. In Section 4, I suggest that argumentation scholars are already in possession of a model for the representation of arguments which does justice to how framing works, in the form of the *Argumentum Model of Topics* (Rigotti 2006; Rigotti & Greco Morasso 2010; Palmieri, 2014; Rocci, 2017; Rigotti & Greco, 2018). The AMT can explain framing as a rhetorical process, as aiming at either audiences which may be already persuaded, or at least open to persuasion, or audiences that are 'hostile' (Tindale, 2022) and might reject the framing speech act and the inference they are invited to draw.

2. Are 'framing effects' rational?

The connection between framing and argumentation/rhetoric seems straightforward. It suggests itself in the most obvious questions asked in framing research: *What is the point or purpose of framing?* The production of 'framing effects'. *What are 'framing effects'?* They are changes in an audience's attitudes, beliefs, behaviour, possibly in their decisions and action, induced by the way an issue is framed, for example by the 'selective

---

1 I borrow the concept from Elizabeth Traugott (Traugott & Dasher, 2009), who used it in a different context, to explain how semantic change happens in language.

salience' given to one or another consideration in a decision-making situation, or the salient way in which an issue is represented. This invites a further question: *Are framing effects rational?* Is it rational to be swayed from one belief to another only by the way an issue is framed?

The way an issue is framed will always 'invite' an inference to a particular conclusion. This 'invited inference' may be accepted or rejected (i.e. the intended framing effect may succeed or fail). From an argumentation perspective, a successful framing effect is either (1) the acceptance of a (new) conclusion about what one should believe or do, as a consequence of being exposed to a particular way of framing a situation; (2) a change from one previously drawn conclusion to another, as a consequence of an alternative framing; (3) a reinforcement of previously held views. Rejecting the 'invited inference' may happen when audiences disagree with the premise(s) where the framing speech act is put forward, based on beliefs they have which lead them to an alternative inference (i.e. a counter-argument with a conclusion that negates the one they were invited to draw).

An interesting exchange of views was hosted by the journal *Behavioural and Brain Sciences*, in response to an article on the rationality of framing effects by Bermúdez (2022). Almost thirty scholars took up the invitation to comment on the article (also developed as a book – Bermúdez, 2021). Taking equivalency framing as the paradigm of framing (*the glass is half empty or half full*), it is easy to conclude that framing effects are not rational. Equivalent descriptions should not lead to different choices. However, in different ways, both Bermúdez and his commentators agree that framing effects are (almost always) rational: people always have a reason to prefer one description to another because, even when alternative descriptions seem equivalent, audiences often do not see them as such (a half-full glass seems a better choice than a half-empty glass).

From the point of view of argumentation theory, I take it that framing effects are by definition rational (even in 'equivalency framing'), unless the shift in the conclusion (decision) being reached occurs in a wholly arbitrary way, i.e. when the audience has no particular reason to choose one description rather than the other, taking them to be equivalent. In 'emphasis' framing, which makes salient different aspects of a situation, framing effects cannot be but rational. Because I take framing to work via an inferential process, there will always be some reason why a framing effect is produced or not. The conclusion reached may not be reasonable, and may not withstand critical scrutiny, but at least it will have been rationally arrived at, for a reason.[2]

---

[2] The distinction between 'equivalency framing' and 'emphasis framing' originates in political communications studies (Chong & Druckman 2007).

3. What can be framed?

The question *What exactly can be framed?* has always seemed to be getting nowhere, or worse, to lead to various 'apples and oranges' pseudo-taxonomies of entities. In search of a more systematic answer, Hallahan (1999) surveyed more than two hundred framing studies, concluding they deal with the framing of seven types of entities: (1) situations (including relationships between individuals and interactions); (2) attributes (characteristics of objects and individuals); (3) choices (alternative decisions presented in different ways); (4) actions; (5) issues (problems, disputes, controversies); (6) responsibility (causes); (7) news (media reports).

To ask what kind of entities or phenomena can be framed is to ask: *What is the 'domain ontology' of framing theory?* Drawing on work in linguistic semantics (Mourelatos, 1978, on aspectual classes of verbs) and on formal ontologies developed in philosophy (e.g. Basic Formal Ontology – Arp et al., 2015), I suggest that what can be framed are, in most general terms, *situations*, themselves broken down into *states* and *occurrences*, and involving *participants* (e.g. objects, individuals) with *attributes,* as follows:[3]

(1) States (including relationships). The modern left's *hatred of the working class* is a paradoxical development. *Friendship* is a precious gift. There is a *massive left-wing bias* at the BBC.

(2) Entities (objects and individuals, instantiating attributes or participating in events, activities, processes). *These people* are not refugees but economic migrants. *Greta Thunberg* is a zealot, the new Joan of Arc. *Greta* is leading the Net Zero crusade.

Note that the objects being framed can be material artefacts or institutional/ cultural artefacts, including institutions, theories. *Universities* are businesses and *students* are customers. *Woke capitalism* is a cultural cancer. *[The theory of] Global warming* is a hoax/ a myth/ a religion.

(3) Attributes, either inherent or institutional (and nominalisations thereof). Greta's *zealotry* is ridiculous. The Ukrainians' *bravery* is astonishing. *Being woke* is being a virtue-signalling radical. *Wokeness* is a religious cult. (Also note that status functions – being Prime Minister – also belong here.)

(4) Occurrences: events – *Brexit* was an expensive divorce; processes – We shall witness a real *weather apocalypse, hell on earth. The culture wars* are dividing America; activities and actions – *Climate change activism* has become an End-of-Days cult. *Saving the planet* is narcissistic self-delusion. What is the BBC *doing*? Brainwashing Britain.

---

3 Examples are adapted from the British press.

Subsumed to (4), as part of the structure of events, processes or actions, there can be framing of the consequences of action (Deporting illegal migrants to Rwanda *will end the trade in human misery*), causal responsibility (Sinful man has brought fire and floods upon himself with *his wicked hubristic behaviour*), goals (*Net Zero* is a myth/ a dangerous trap), and so on.

4. Frames and inferences

How does framing work? I suggest it works by 'inviting' an inference from a combination of premises, one or more containing a so-called 'framing device' (a definition, classification, analogy, metaphor, etc.), in accordance with one or more argument schemes. Framing attempts to lead an audience to a conclusion for what they ought to believe or do, i.e. an essentially normative conclusion. If the audience accepts the conclusion, and thus decides to change their understanding of the situation, or decides to act in a particular way, then the framing effect has been successful.

Two main mechanisms[4] were first proposed in Fairclough & Mădroane (2020) and are restated more clearly here. Framing works either by telling the audience that (1) they ought to 'see' (or understand) an entity or situation as X (not as Y); or that (2) they ought to take a particular reason Z (and not some other one) as the most significant in deciding what to believe or do; in other words, Z ought to outweigh other possible reasons. These reasons (X, or Z) are presented to the audience as rationally acceptable, i.e. as relevant, true and sufficient in view of the conclusion the audience is invited to draw. If the audience actually finds them so, then the framing effect may succeed. Relevance for an audience means relevance to their (emotional, cognitive and evaluative) 'concerns' (Blackburn 1998; Fairclough & Fairclough 2012: 70-71). Without a 'concern', without being interested in the matter or caring about it, the audience may not be moved to draw the normative conclusion. Framing is ubiquitous, but few topics will possess that immediate relevance to what matters to us that can generate motives for decision and action.

While the simplest framing speech act is an assertion (with framing occurring either in the NP or VP position (i.e. '*The refugees* have arrived by boat from Calais' vs. 'These people *are refugees*'), the illocutionary force of such an act is in fact that of a covert *directive*. In framing a situation, a message framer (e.g. a media source) *directs* an audience to take that representation as the most rationally acceptable way of seeing the situation (e.g. you ought to see them as *refugees*, not as *economic*

---

[4] A third one described there can be subsumed to the other two. For example, narrative framing works by analogy (Fairclough & Mădroane, 2020), and the audience is being invited to *see* someone *as* a 'villain' or a 'hero'.

migrants), or presents a particular reason as the one that should carry most weight in decision-making (e.g. *humanitarian considerations* in deciding which migration policy to support). A premise that asserts the priority of one consideration over others can also function as a 'framing device'.

If this is correct, and the final conclusion is always a normative or practical one (Fairclough & Fairclough, 2012; Fairclough, 2022), about what the audience ought to believe or do, or what is recommended, then a normative-evaluative element, capable of providing a 'motive', must be already present somewhere in the premises. Certainly, the argument, as given to me, as target audience, by the framing source, does include a covert normative element (you *ought to* see this as X, or take Z as most important reason), but I would have to accept this 'invitation' in order to change my view of the situation to X, or allow Z to outweigh other reasons. Where can this normative-evaluative dimension come from, that I, the audience, already accept? What can make the framing relevant to my concerns, and capable of motivating me?

Many 'frames' (in premises the audience may accept) include an inherent evaluative dimension (i.e. 'good' or 'bad', 'desirable' or 'undesirable'). A look at FrameNet (n.d.) shows that evaluation is built into the definition of many frames, i.e. in semantic frames. For example, CRIME or ROBBERY involve a perpetrator who 'wrongs' a victim. A CATASTROPHE is an 'undesirable' event which affects the patient 'negatively', etc. If a normative-evaluative dimension is already part of denotational meaning, then framing a situation as 'X' in a minor premise will combine with an implicit major premise saying, for example, that 'X is undesirable' or 'X is wrong' (entailed by the meaning of 'X'), to yield the conclusion that 'X is not recommended'. Such a conclusion will presumably be shared by everyone, in virtue of their knowledge of language and their common experience, as part of the 'X' frame.

If the evaluative dimension is not already part of the denotational meaning, it can often be reliably provided by social consensus, e.g. by shared moral norms. If Boris Johnson is framed as a 'liar' (via a DECEPTION frame), the public can be trusted to provide it – lying is wrong, morally unacceptable – and thus predictably draw the invited conclusion. So, if evaluation is not built into semantic frames, it is often included in partially shared cognitive frames.

The following three examples involve the framing of sugar in debates about public health (Fig. 1-3).

Example 1. *Sugar is addictive. (If a substance is addictive, then it is a drug.) Therefore, sugar is a drug.*

Example 2. *Sugar is a drug. (If a substance is a drug, then it is dangerous to human health.) Sugar is dangerous to human health.*

Example 3. *Sugar consumption can undermine health. The goal of not undermining one's health outweighs other goals. Therefore, sugar consumption is to be avoided.*

In these examples, sugar has been framed as an addictive substance, as a drug, and also in terms of its impact on health, itself framed as a goal which outweighs other goals. The first two examples are telling the audience to *see* sugar in a new way, taking it for granted that they can supply an acceptable major premise (shown in brackets). The third example both tells the audience that sugar can undermine health (thus framing it in terms of salient negative consequences, in the minor premise), and puts forward an explicit major premise which audiences allegedly ought to accept as true, namely that health concerns ought to outweigh all other concerns. This last example illustrates the second mechanism I mentioned earlier, where framing attempts to provide a reason that ought to outweigh all others.

In examples (1) and (2), framing is being done in the minor premise, presented to the audience for acceptance, and it relies on a presumably widely accepted major premise. Seeing sugar as an addictive substance and as a drug might however not be universally accepted. In these cases, the 'invited inference' will not be drawn (though the major premises will not necessarily be rejected), and the framing effect will fail. In example (3), framing occurs in both premises, and either or both may be accepted or rejected. In situations such as (3), the audience is not just being told about a particular situation (and trusted to supply the missing premise), but also told how to think about that situation, in general lines, or how competing considerations ought to be weighed together (i.e. the major premise is also explicitly given).

These evaluative judgments (X is good/bad, beneficial/dangerous), grounded in particular concerns (e.g. for safety and health, or for figuring out the nature of sugar), indicate the relevance of the framing premises to the audience. An existing concern for my health will make the framing of sugar as a dangerous drug relevant to me. If I judge this description to be acceptable, the framing might persuade me; if I find it false, the framing might still be relevant but might not influence my beliefs and behaviours.

The AMT framework (Rigotti & Greco, 2018) allows for a more detailed representation of these arguments, as the confluence of two syllogisms, one containing an abstract maxim (a general abstract rule, itself derived from an abstract 'locus' or 'topos', expressing a semantic-ontological relation[5], and applicable to many situations), represented in AMT and in the figures below on the right-hand side, and the other rooted in the

---

5 These may be relations of semantic entailment (logical implication), cause and effect, analogy, indexicality, means-goal relations, etc. – those relations that underlie the argument schemes recognized in the literature.

particular case at issue – on the left-hand side. This double syllogistic structure (graphically represented in the form of a Y) shows how a given audience's set of beliefs and values (*endoxa*, as major premises) combine with the (minor, *data*) premises provided, as in (1) and (2) by the framing act, but also with a range of abstract warrants (*maxims*), to yield final conclusions that may be accepted or rejected.

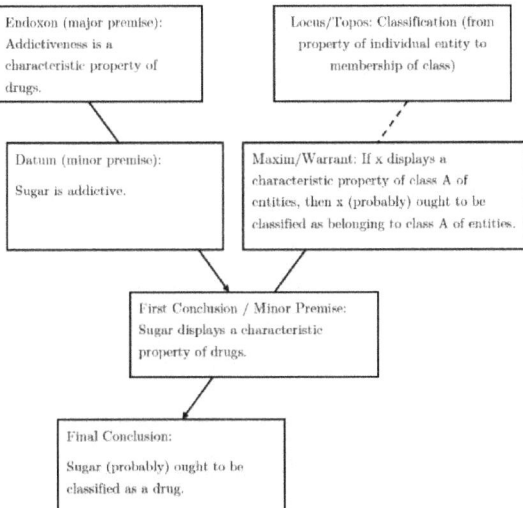

Figure 1. Sugar is addictive (salient property)

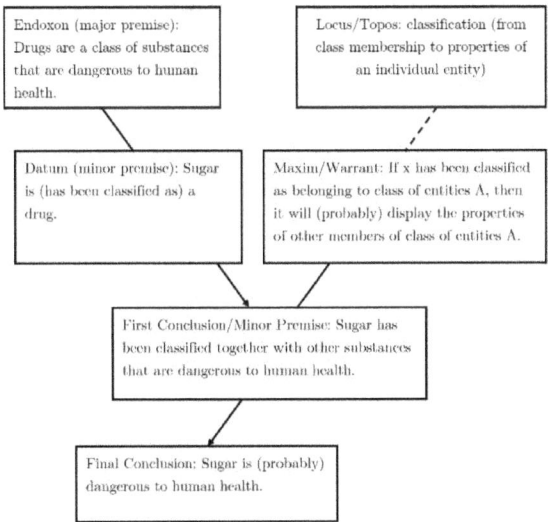

Figure 2. Sugar is a drug (salient classification)

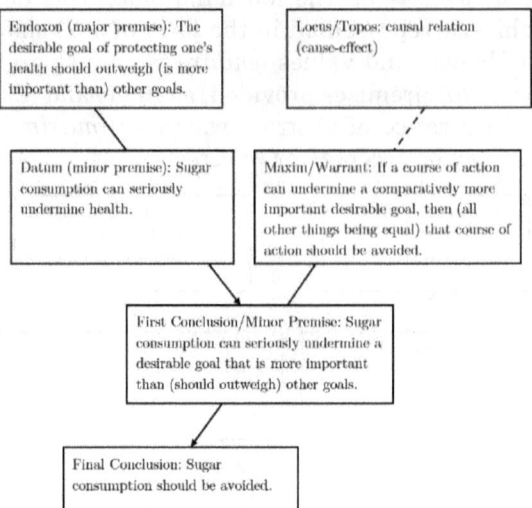

Figure 3. Salient impact on salient overriding goal

I suggest the representations in Figures 1-3 for the arguments in examples 1-3, assuming the invited inferences are actually accepted (based on accepting both the *datum* and the *endoxon* premises). While the maxim itself is intersubjectively shared (and the right-hand inference may be valid), the actual truth-value of the premises it combines with (the left-hand side) may be contested by the audience.

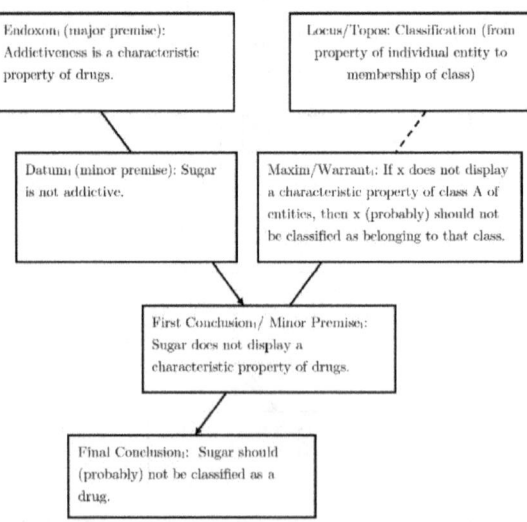

Figure 4. Sugar is not addictive, therefore not a drug.

If the *datum* premise in examples (1) or (2) is rejected as false, another inference might replace the invited inference – let us call this alternative inference a 'counter-inference', activating an alternative *endoxon* (which may replace the original one or be added to it). For example, someone might reason like this: it is false to claim that sugar is addictive ($Datum_1$), and because ($Endoxon_1$) addictiveness is indeed a characteristic property of drugs (and, if it is not addictive, a substance cannot be a drug), it is false to say that sugar has drug-like properties ($Conclusion_1$). This counter-inference converges with an alternative $Maxim_1$, saying that, if the properties that characterize a class of entities are not displayed by an entity considered to be a potential member of that class, then that entity should not be categorized as belonging to that class. The *Final Conclusion$_1$* (Fig. 4), if accepted, is the opposite of the one invited by the original argument (Fig. 3).

Figure 4 illustrates therefore a counter-inference, based on rejecting a *datum* (minor) premise: it is false that sugar is addictive, therefore sugar ought not to be classified as a drug. But the audience may also reject an (un)expressed major premise (endoxon) that suggests a desirable order of priorities (where one reason allegedly outweighs others). They may refuse to accept that everything ought to take second place to the goal of not endangering one's health. In addition, audiences may reject a general statement presented as something they ought to believe as 'common sense' (also in an expressed major premise). An example involving the framing of illegal migration in the UK press, illustrating the latter situation, is discussed below.

The MIGRATION frame introduced by a *datum* premise (e.g. 'Current UK policy has allowed the number of migrants crossing the Channel in boats in 2022 to rise to 45,728') may not contain any overt and unambiguous normative-evaluative dimension, either inherent in lexical meaning or widely agreed upon. Is 'migration' good or bad? – the word itself does not say. The task of providing the 'ought' or the evaluation ('good' or 'bad'), that can lead to the conclusion, will fall to the major premise, but the *endoxa* activated in reasoning about migration may be very different from person to person. Audiences will draw on the factual information at their disposal, their values, their political orientation, and will be deeply divided on what ought to be done about the constant inflow of migrants crossing the Channel by boat, in relation to very different concerns. (No such deep divisions seem to exist in judging lying or crime to be wrong or undesirable.)

It will therefore not be possible to ensure that the desired 'invited inference' is drawn, unless the audience is also given some evaluative-normative content to reason from, to make the issue relevant for them. An argument about migration in the left-wing press (*The Guardian*), typically favourable to a continuous rise in migration numbers, on account of alleged positive effects and humanitarian considerations (presumably intended to resonate with the practical and moral concerns of their readership) will usually contain an explicit positive framing of migration in the major premise, presented as part of uncontroversial common-sense (*endoxa*). This alleged 'common sense' will be probably contested by a *Daily Mail* or *Daily Telegraph* typical reader, who will remain unpersuaded and hostile to the positive framing. For a right-wing reader, the framing of migration in terms of its positive effects, in *The Guardian*, will activate a counter-inference, from a different *endoxon* ('accepting more migrants has a negative effect'), leading to the opposite conclusion about what the right migration policy ought to be. The positive/negative evaluation is likely to be already shared amongst typical Guardian/Telegraph readers, and part of their cognitive frames. For them, the framing effect will be a reaffirmation and strengthening of existing commitments.

There is no space here to represent arguments of this last sort, where the framing is done in a *major* premise expressing a general statement. This kind of framing happens, for example, whenever a media or political organisation is attempting to naturalise, as potential and desirable 'common sense', a highly controversial general proposition ('All migration is beneficial', 'All white people are by definition racist', or 'Transwomen are women'). A lot of ideological work is clearly being done in this way.[6]

5. Conclusion

A framing speech act (for example, an assertion where a salient expression occurs either as part of noun or verb phrase), functioning as a premise in an argument, is an attempt to shape the way the audience sees or understands a situation, or to influence a decision outcome. The illocutionary force of such a speech act is, covertly, that of a directive, to the extent that the audience is being steered towards seeing a situation as X not Y, or towards taking reason Z to outweigh other reasons in deciding what to believe or do. A 'framing effect' will be successful when the audience actually draws the 'invited inference' and accepts the invited conclusion about what they ought to believe or do, based on accepting the premises as true, sufficient and relevant to their own (cognitive, practical, moral) concerns, thus turning the normative reasons they have been given into motivating reasons. In this way, 'You ought to accept Z as the most relevant reason becomes 'I accept that Z is the most relevant reason'. Audiences hostile to the message – in relation to their personal experience, factual knowledge, moral-political values – may reject the premises where the framing occurs as false (or irrelevant, or insufficient), and supply different premises (either data or endoxa) that steer their own argument towards another conclusion.

---

6 I am drawing here on a view of ideology as 'naturalized common sense', supporting or subverting relations of power (Fairclough, 1992: 87).

References

Arp, R., Smith, B. & Spear, A. D. (2015). *Building Ontologies with Basic Formal Ontology*. Cambridge Mass.: MIT Press.

Bermúdez, J. L. (2021). *Frame it Again. New Tools for Rational Decision-Making*. Cambridge: Cambridge University Press.

Bermúdez, J. L. (2022). Rational framing effects: A multidisciplinary case. *Behavioral and Brain Sciences 45*, 1–59. DOI: 10.1017/S0140525X2200005X

Blackburn, S. (1998). *Ruling Passions. A Theory of Practical Reason*. Oxford: Clarendon Press.

Chong, D. & Druckman, J. N. (2007). Framing Theory. *Annual Review of Political Science 10*, 103-127, DOI: 10.1146/annurev.polisci.10.072805.103054

Eemeren, F. H., van (2010). *Strategic Maneuvering in Argumentative Discourse*. Amsterdam: John Benjamins.

Entman, R. M. (1993). Framing: Toward clarification of a fractured paradigm. *Journal of Communication 43 (4)*, 51-58.

Fairclough, I. & Fairclough, N. (2012). *Political Discourse Analysis*. London: Routledge.

Fairclough, I. & Mădroane, I. D. (2020). An argumentative approach to 'framing'. Framing, deliberation and action in an environmental conflict. *Co-herencia 17(32)*, 119-158. DOI: https://doi.org/10.17230/co-herencia.17.32.5

Fairclough, I. (2022). The UK Government's 'balancing act' in the pandemic: rational decision-making from an argumentative perspective. In S. Oswald, M. Lewinski, S. Greco, R. Villata (Eds.), *The Pandemic of Argumentation* (pp. 221-242). Dordrecht: Springer.

Fairclough, N. (1992). *Discourse and Social Change*. London: Polity Press.

Fillmore, C. J. (1982). Frame Semantics. In The Linguistic Society of Korea (Eds.), *Linguistics in the Morning Calm. Selected Papers from SICOL-1981* (pp. 111-137). Seoul: Hanshin.

Fillmore, C. J. (1985). Frames and the semantics of understanding. *Quaderni di Semantica 6(2)*, 222-254.

Fillmore, C. J. (2006). Chapter 10: Frame semantics. In D. Geeraerts (Ed.), *Cognitive Linguistics: Basic Readings* (pp. 373-400). New York: De Gruyter Mouton.

FrameNet (n.d.). Lexical Database, International Computer Science Institute, Berkeley California. At: https://framenet.icsi.berkeley.edu/fndrupal/frameIndex (last accessed 28 January 2023).

Hallahan, K. (1999). Seven Models of Framing: Implications for Public Relations. *Journal of Public Relations Research, 11(3)*, 205–242.

Mourelatos, A. P. D. (1978). Events, Processes, and States. *Linguistics and Philosophy, 2(3)*, 415-434.

Palmieri, R. (2014). *Corporate Argumentation in Takeover Bids*. Argumentation in Context 8. Amsterdam: John Benjamins.

Rigotti, E. (2006). Relevance of context-bound loci to topical potential in the argumentation stage. *Argumentation 20(4)*, 519–540. DOI: 10.1007/s10503-007-9034-2

Rigotti, E. & Greco Morasso, S. (2010). Comparing the Argumentum Model of Topics to Other Contemporary Approaches to Argument Schemes: The Procedural and Material Components. *Argumentation 24(4)*, 489-512. DOI: 10.1007/s10503-010-9190-7

Rigotti, E. & Greco, S. (2018). *Inference in Argumentation: A Topics-Based Approach to Argument Schemes*. Argumentation Library 34. Dordrecht: Springer.

Rocci, A. (2017). Modality in Argumentation: A Semantic Investigation of the Role of Modalities in the Structure of Arguments. Argumentation Library Book 29. Dordrecht: Springer.

Tindale, C. (2022). *How We Argue: 30 Lessons in Persuasive Communication*. London: Routledge.

Traugott, E. C. & Dasher, R. B. (2009). *Regularity in Semantic Change*. Cambridge Studies in Linguistics 97. Cambridge: Cambridge University Press.

THE REASONABLENESS OF FALLACY ACCUSATIONS: AN EXPLORATORY STUDY

JOSÉ ÁNGEL GASCÓN
Departamento de Filosofía, Universidad de Murcia
jgascon@um.es

Abstract
Empirical research is needed to understand how knowledge of the fallacies influences argumentative behaviour. In this study I take a look at how accusations of fallacies are made on Twitter. 865 accusations were analysed according to seven criteria, which incorporate aspects of what a reasonable criticism should consist in, such as whether the accusation was substantiated or whether the accuser was willing to discuss it. The findings point in the direction of further research.

1. Introduction

Since the beginnings of informal logic, the concept of fallacy has taken a central place in the study and pedagogy of argumentation. The claim that knowledge of the fallacies is a cornerstone of argumentation and critical thinking instruction is nowadays almost a triviality. From time to time there have been, however, certain dissenting voices. According to Finocchiaro (1981), "the concept of a fallacy as a type of common but logically incorrect argument is a chimera" (p. 17), since the arguments usually presented as examples of fallacies in textbooks are either not common or correct. Instead, textbooks usually misinterpret arguments, exaggerating their intended strength, to make them fit in the categories of fallacies. Along similar lines, Boudry, Paglieri and Pigliucci (2015) argued that characterisations of fallacies are not particularly useful to detect flawed arguments due to what they call the "fallacy fork": on the one hand, if those characterisations pick out common patterns of argument, then it turns out that many instances of such patterns are cogent arguments; on the other hand, if they manage to identify only flawed arguments, then they are not commonly found in real-life argumentation.

Regarding the effects of teaching fallacies, Blair (1995) has held that when properly taught, fallacies can be pedagogically useful and "are obviously parts of learning how to evaluate arguments" (p. 333). However, Hitchcock (1995) disagrees: "everything that can be said with the use of these [fallacy] labels can be said without them, and in general said more clearly" (p. 325). He argues that teaching fallacies fosters an attitude of looking for "a basis on which to convict and sentence" arguments, rather than making a genuine effort to understand them (p. 326). And, finally, Hundleby (2010) has shown that current textbooks on argumentation and critical thinking promote precisely such an adversarial attitude.

Independently of one's position in this discussion, one aspect of it that we should notice is that many of the disputed claims are *empirical*. It is an empirical question whether fallacies are commonly committed, whether knowledge of the fallacies improves argumentative skills, whether it encourages superficial and oversimplified criticism or whether it fosters an adversarial attitude. Unfortunately, however, to my knowledge very few empirical studies have been conducted specifically to test those claims. Some works purporting to provide empirical evidence, such as Jason (1987) on whether fallacies are common or Finocchiaro (1987) on what kinds of fallacies there are, merely rely on particular debates or texts, not a broad corpus. There have also been quite a few studies on the effects of critical thinking courses—with rather depressing findings (Willingham, 2008)—but none focusing specifically on fallacies. The studies conducted by van Eemeren, Garssen and Meuffels (2009) are the closest we have to empirical research on fallacies, but they focus on whether the pragma-dialectical rules have conventional validity—i.e. whether ordinary arguers endorse them—rather than on the effects of teaching fallacies.

I here present what may be called an initial exploration, a glance at how arguers who know about fallacies criticise arguments. I analysed accusations of having committed a fallacy on the social network Twitter in order to see how they are made. This study can give us an initial hint on how the concept of fallacy is used by ordinary arguers.

2. Methodology

Twitter was chosen as the medium from which to collect accusations of fallacies. 2080 "tweets" (i.e. messages) containing the word "fallacy" were retrieved on 7th of July 2022 using the on-line tool Vicinitas.[1] Tweets were then sorted into those who were accusations of having committed an argumentative fallacy and those who were not, in order to exclude the latter from the study. As expected, most of the tweets (1215) did not contain an accusation of fallacy. The reasons for exclusion were:

- Repeated tweets ("retweets"): 558.
- Falsities: 199.
- Replies to accusations: 85.

---

[1] https://www.vicinitas.io

- Automated messages (bots): 63.
- Bias: 57.
- Theoretical reflections: 48.
- Tweets in foreign languages: 19.
- Jokes: 13.
- Other reasons: 173.

The number of tweets containing an accusation that were analysed in this study were then 865. Each of those tweets was analysed (by myself) according to seven criteria, which were formulated as yes-no questions. A third value, "Dubious", was also used in those cases in which it was too difficult—even taking the words and the surrounding context of the tweet into account—to provide an answer for the criterion. The criteria were chosen because they point to what can arguably be considered as indications of a reasonable accusation of fallacy. For certain criteria, the ideal answer would be "Yes", whereas for other criteria it would be "No". The criteria are the following.

Criterion 1. *Is the kind of fallacy identified?*

When someone accuses an arguer of having committed a fallacy, at the very least the accuser is expected to specify *which* fallacy has been committed. That claim, however, although it seemed initially plausible when this study began, can be put into doubt. In fact, we will see in the next section that there are reasons to believe that accusations of *unidentified* fallacies might turn out to be more informative, since the arguer often explains the fallacy instead of naming it. But that will have to wait until we discuss the results.

Criterion 2. *Is the fallacy misidentified?*

This criterion refers both to accusations in which the accuser correctly spots a fallacy but uses the wrong label—such as accusing someone of committing a strawman when the arguer actually committed an *ad hominem*—and to those in which the arguer did not commit any fallacy at all. It also includes labels of fallacies that are not recognised in the studies of argumentation and do not fit into current conceptions of fallacies—for instance, "false assertion fallacy".

Only when it was absolutely clear that the identification was wrong was this criterion answered in the affirmative. When there was even the slightest possibility that the accuser could be (even partly) right, the answer to this criterion was "No". On the other hand, sometimes the thread was missing (deleted tweets or suspended Twitter accounts) and I could not check the allegedly fallacious tweet, and in those cases I assigned the value "Dubious" to this criterion.

Criterion 3. *Is the accusation substantiated?*

This is one of the most significant criteria used in this study, since it is easy to see whether an accusation fulfils it or not, and also because it undoubtedly marks an important characteristic of reasonable accusations of fallacies: they must be explained or justified. For an affirmative answer to this criterion, *any* kind of substantiation was considered enough—explanation of the fallacious pattern, of the context, an analogy or whatever.

Criterion 4. *Does the substantiation make reference to the context?*

The possibility of answering this criterion with a "Yes" obviously depends on an affirmative answer to the previous one. If the accusation was not substantiated in the first place, the answer to this criterion will be "No". Any reference to the context (the circumstances of the arguer or the audience, contextual information related to the content of the argument, or whatever) was considered enough for an affirmative answer to this criterion.

Criterion 5. *Does the accusation rely on the taxonomic technique?*

The taxonomic technique is the assumption that "the presence of an argument scheme that *may* be fallacious always makes an argument fallacious" (Hundleby, 2010, p. 287). That is, it is a method of identifying fallacies that relies solely on whether a particular argument instantiates a fallacious scheme.

An accusation of fallacy was considered as relying on the taxonomic technique when the accuser explained the fallacious scheme instead of the precise flaws of the particular argument. Depending on the context of the accusation, the value for this criterion could also be "Yes" when the accuser just named the fallacy without any explanation or justification.

In the absence of evidence that the accuser relied on the taxonomic technique, the value for this criterion was "No" by default. Furthermore, for certain kinds of fallacies—such as strawman, cherry-picking, red herring, or *non-sequitur*—there is no specific fallacious scheme, so in those cases the accuser cannot rely on the taxonomic technique. And, of course, if the fallacy is not identified (criterion 1), then the value for this criterion will also be "No".

Criterion 6. *Does the accusation rely on a problematic theory?*

This idea of a "problematic theory" may be problematic in itself, but the criterion was introduced in order to identify accusers' reliance on conceptions of fallacies that are manifestly at odds with (traditional or modern) scholarship. An example is the use of wrong names of fallacies, such as "fallacy of reductio *ad absurdum*". Another example is the presence of very questionable claims in the substantiation of the accusation, such as "common sense is a fallacy" or "all appeals to authority are fallacious" (this latter case also involves taxonomic technique, which is a kind of problematic theory).

The difficulty in analysing this criterion is that many accusations do not make manifest the theory on which they rely—especially when it is a bare, unsubstantiated accusation. In the absence of any clear sign of a problematic theory, the value for this criterion was "No". As we will see, there were also several kinds of fallacies that do not fit easily into current lists of fallacies in the argumentation scholarship but that cannot be said to be at odds with them; in those cases, I assigned them the value "Dubious".

Criterion 7. *Is the accuser open to discuss the accusation?*

Finally, an accusation of fallacy can be either a definitive judgment or an initial criticism of the argument that could be discussed. Therefore, the issue here is whether the accuser is willing to discuss the fallacy accusation or not. Sometimes this can be seen clearly—when, for instance, the accuser "blocks" the other user on Twitter or says that he or she will not discuss the matter further. But other clues were considered as well. An accuser was also considered unwilling to discuss the accusation when she or he made a bare accusation without substantiating it or referred to hidden interests or bias on the part of the arguer. On the other hand, an accuser was considered open to discussion when the accusation was well substantiated, when the accuser gave reasons or asked questions together with the accusation, or when the accusation was formulated without too much confidence ("I think", "there might be").

Admittedly, criteria 2, 5, and 6 are the most difficult to apply and their results might be less reliable. Criterion 2 (misidentification) depends on an examination of the thread (which was not always possible) and on the analyst's judgment. The application of criterion 5 (taxonomic technique) was based on clues of the accuser's beliefs about the concept of fallacy, which can be very tricky, and the value "No" was charitably assigned when there were no clues. And a similar interpretive problem affects criterion 6 (problematic theory), together with the fact that the analyst's conception of what is "problematic" might be challenged by other argumentation scholars. Criteria 1, 3, 4, and 7, on the other hand, seem more reliable, also because there was no default value in the absence of clues, so that a "No" in them will be as significant as a "Yes".

3. Results and discussion

Let us begin then by showing the results of the application of the criteria. I will give them in percentages rather than absolute values (see Fig. 1).

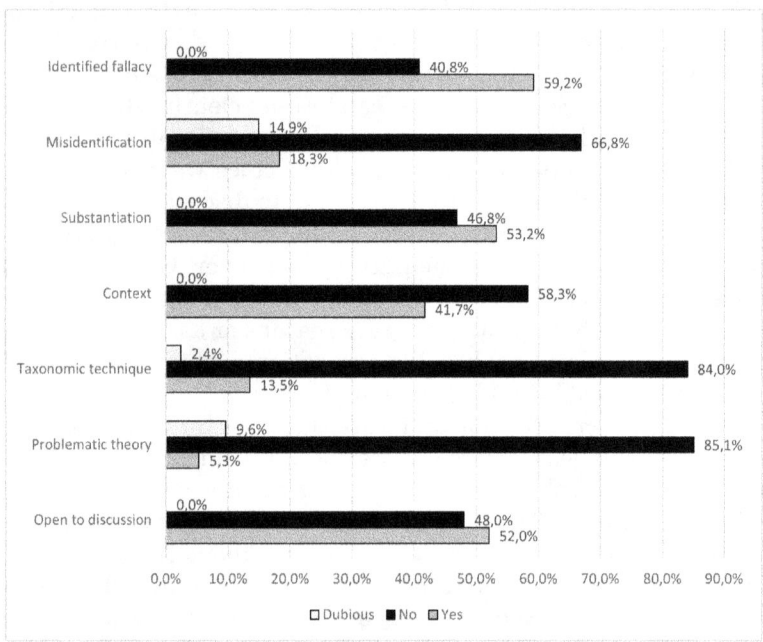

Figure 1. Analysis of accusations

Only three criteria have dubious instances: misidentification (14,9%), taxonomic technique (2,4%) and problematic theory (9,6%). These were the criteria that were based on interpretive clues and evaluation of the accused tweet and, for this reason, sometimes it was not possible to determine the value with confidence. Misidentification stands out with dubious instances because, in order to apply that criterion, it was necessary to compare the accusation with the argument that was being accused of being fallacious, and in many cases those tweets had disappeared. When that was the case, I assigned the value "Dubious" to it. In the case of problematic theory, the value "Dubious" is also high because, as I explained in the previous section, it was assigned to fallacy labels that did not fit with any list of fallacies in the scholarship but that did not contradict current scholarship either.

More than half of the accusations (59,2%) identified the fallacy, while the rest (40,8%) simply pointed out that there was a fallacy, without adding a label to it. For the purposes of deciding whether the fallacy was identified and whether the identification was wrong, and also of collecting the kinds of fallacies mentioned, some laxity was exercised with the labels of fallacies. So, for example, the following tweets were considered as correctly identifying the fallacy that I have put in square brackets:[2]

Example 1. *That's just you lying again. No opinions were offered. Gableman wasn't discredited. Name calling is a fallacy. [Ad hominem]*

Example 2. *That's another example of the bifurcation fallacy. [False dilemma]*

Among the accusations that do not specify the fallacy that was allegedly committed, we find the following:

Example 3. *That's not evidence, that's just more fallacy.*

Example 4. *I'm sure there's a logical fallacy somewhere but I'm just tired. Why?*

Example 5. *No Beth! As people are saying here the 14M people's mandate belongs to the Conservative Party, NOT Boris Johnson personally. Britain is not a Presidential political system and Boris Johnson is not President, he is a Prime Minister. It annoys me this fallacy is being promoted.*

Examples 3 and 4 might give the impression that accusations in which the fallacy is not identified tend to be very poor—without substantiation. However, as the example 5 shows, an unidentified accusation can be quite detailed and explanatory. I will come back to this in my discussion of criterion 3 (substantiation).

Given that an accusation was regarded as a misidentification only when the error was manifest, it is no surprise that the percentage of misidentifications was relatively low (18,3%). The following are some examples of accusations considered as misidentifying a fallacy:

Example 6. *Indeed. Ad absurdum is an argumentative fallacy you rely on a great deal, unfortunately, in your bad faith baseline characterizations of those you disagree with.*

Example 7. *That is an equivocation fallacy. You're taking two unrelated events and clicking them together as relatable. The feelings from Cloud are always unspecified. You're taking a book from 2021 and matching it with one from 20 years ago.*

Reductio ad absurdum is not a fallacy but a well-known logical pattern, and "equivocation" refers to the use of an expression with different meanings, not to a false analogy.

Regarding criterion 3, a little more than half of the tweets (53,2%) justified their accusation. Many of those who did not were bare accusations such as the following:

---

2 Some of the tweets quoted here have been slightly edited for orthographical and grammatical correction without altering the contents.

Example 8. *If your only argument is a logical fallacy, you have no argument.*

Example 9. *Strawman fallacy.*

Example 10. *Motte & Bailey fallacy.*

There were, however, also many accusations that were supported by a remarkably good justification. The following are some examples:

Example 11. *The argument that Boris has a mandate from the public is a fallacy. We don't operate in an American system where you elect a president. You vote for party candidates and the largest party forms a government.*

Example 12. *Really weak straw man fallacy. I've never heard anyone claim the Constitution was infallible. Further, the fact that the amendment process was included means that the guys who wrote didn't think it was infallible either!*

Furthermore, as I was analysing the tweets, I suspected that whether a fallacy was identified or not (criterion 1) influenced whether the accusation was substantiated (criterion 3). As the previous examples apparently show, when the fallacy was not identified, it seemed more likely that the accusation was justified. This would make sense, since the fallacy labels can be used as substitutes for explanations or justifications. In order to test this hypothesis, let us see the percentage of accusations that were substantiated depending on whether the fallacy was identified or not (see Fig. 2).

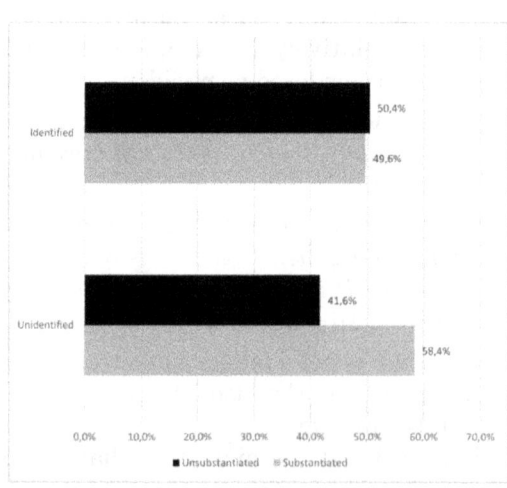

Figure 2. Identification and substantiation

There is here a noticeable difference. Whereas only 49,6% of the accusations that identified the fallacy were substantiated, the percentage of substantiated accusations rises up to 58,4% when the fallacy was not identified. If this finding is confirmed by further studies, it could mean that teaching fallacy labels is *detrimental* to argumentative practice.

We can perhaps be more optimistic about criterion 4: whether the accusations made any reference to the context in which the argument was put forward. Even though, in the global analysis presented in Fig. 1, 58,3% of the accusations made *no* reference to the context, we must bear in mind that that number includes the accusations that were *not* substantiated—and which therefore obviously did not make any reference to context either. If, however, we take into account only the substantiated accusations, the numbers are very different (see Fig. 3).

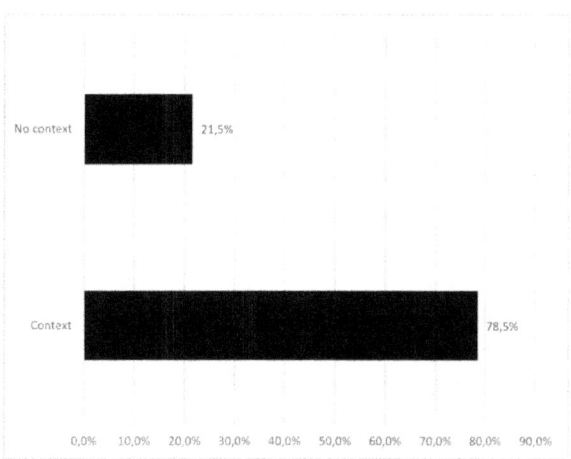

Figure 3. Reference to context in substantiated accusations

Of those accusations that were substantiated, only 21,5% made no reference to the context, whereas 78,5% did. The following are some examples of substantiations without reference to the context (notice that they give us no information about the argument that is being criticised):

Example 13. *Appeal to how many believe something is a logical fallacy. Nearly everyone thought the world was flat before. Knowledge is not a democracy.*

Example 14. *Nice try. Red Herring fallacy. A Red Herring argument is one that changes the subject, distracting the audience from the real issue to focus on something else where the speaker feels more comfortable and confident.*

Contrast those with the following remarkable examples of accusations that discuss the context or the content of the argument:

Example 15. *The "independent state legislature" 'doctrine' takes a fragment of the Constitution out of context, exaggerates the omission of an understood and implied element (the courts), and then prioritizes it over everything else the Constitution says. A massive "cherry-picking" fallacy.*

Example 16. *That's a logical fallacy. Going to the black market isn't as easy as going to Walmart. Many won't put in the extra work to get guns that way, many won't do it since it's now illegal and those that do get it you automatically know they are illegal so can take police action.*

Criterion 5 (taxonomic technique) was assigned an affirmative value mainly when the accuser explained the scheme of the fallacy without getting into the content or the context of the argument. Given that, as we have just seen, references to context are relatively high, signals of taxonomic technique were obviously low (13,5%). The following are some examples of accusations that relied on the taxonomic technique:

Example 17. *Haha! Fallacy of thinking. Slippery slope. Arguing that if an opponent were to accept some claim C1, then they would have to accept some other closely related claim C2, eventually leading to the conclusion that the opponent is committed to something absurd or obviously unacceptable.*

Example 18. *And your attack is ad hominem. Meaning you question my credibility, but not my arguments with evidence. It's a fallacy.*

There were even less signs of accusers that relied on a problematic theory (5,3%). Some of the instances were due to a reliance on the taxonomic technique—which I take as problematic—but there were also cases of questionable epistemological beliefs. The following examples show an exaggerated distrust of authorities, with the implication that no argument from authority can be reasonable:

Example 19. *No shit and I applaud these doctors but that has been obvious for a long time. Trust your own instincts, no need for an appeal to authority fallacy to be convinced.*

Example 20. *"Experts say" = What follows will be a lie, but hey, it fits the narrative. Appeal to authority is in fact a logical fallacy—especially when, like with masks, the vax, lockdowns... The Science changes every month.*

There were also cases of fallacies that were misconceived or poorly explained:

Example 21. *Pointing out logical fallacies automatically invalidates your position due to it being a fallacy fallacy.*

Example 22. *No need to bring in defeated topics which are shown to contain false premises. That is the definition of appeal to tradition logical fallacy.*

The relatively high percentage of "Dubious" results (9,6%) in this criterion is due to the appearance of several fallacies that are not included in traditional or current taxonomies, but that I was reluctant to consider problematic because they did not clearly clash with those taxonomies. Up to 41 categories of fallacies were of that kind. Some examples are "absolutism", "appeal to morality", "appeal to the future", "argument from incredulity", "bold face lie", "common sense", "God of the gaps", and "good guy with a gun". The following is a sample of such accusations:

Example 23. *"Good guys with a gun" is a fallacy. Gun restrictions would be a better way to protect your family, you, and others. Try fighting for the right to life instead of the right to guns.*

Example 24. *'Patriarchy' is a logical fallacy named golden hammer.*

Example 25. *This is just such a great example of the mind-projection fallacy. Without any irony, this dude assumes that the class discussing LGBT issues is what causes people to decide/declare/define themselves as nonbinary and bisexual.*

Example 26. *Once again: Isolation limits transmission, even if the isolation is imperfect & not up to the standards it should be. Hence why most infections were before or around quarantine. It's not my fault you abuse the nirvana fallacy to pretend otherwise.*

Finally, for a more significant result, only 52% of the accusers were open to discuss their accusations, while the rest showed an unwillingness to do so. When an accusation was reasonably substantiated, the accuser was considered as open to discussion. But there were also more manifest signs of openness, such as asking questions:

Example 27. *Then where do you get your definition? Just the "everyone knows" fallacy?*

Example 28. *"Perverting the nature of..." Why is perverting the nature of sex immoral? This is a classic is-ought fallacy. 1) Sex *is* for procreation therefore it *ought* to be used for procreation. This is exactly what Hume argued against and the primary problem with natural law.*

Also expressions of uncertainty in the accusation:

Example 29. *But I think the logical fallacy that he used was that causation equals correlation, which it does not.*

Example 30. *I'm wondering if there's an ecological fallacy here—at the group level I think you're right, but at the individual level cancellation will make a big difference. I'm not sure this is a case against debt cancellation but the point is very well taken.*

And sometimes the accuser's openness was inferred from his or her attitude thorough the thread (i.e. the conversation), which I examined for each tweet. Unwillingness to discuss the issue, on the other hand, was sometimes shown in statements that the conversation was over, such as in the following examples:

Example 31. *We're not talking about fucking Joe Biden's failures and accomplishments as a legislator. That's a Red Herring and a Strawman fallacy in one. You suck at this. Funny part is you think you're a skilled logician. All done now, bye.*

Example 32. *I hope people will realise one day that there is a huge logical fallacy with the argument that life begins after breath. If so, when did life begin on Earth? Sorry, in advance: but the truth is I most probably won't reply & it's up to you to work it out! Just think about it!*

Example 33. *1st- comparing a random citizen wearing a sports hat is such a fallacy to comparing virtue signalling by forcing a Ukrainian flag in your name without any real action that it's not worth continuing discussing. You're a phony, clear and simple, and hope you one day do more than [Ukrainian flag].*

Of course, although almost half of the accusers (48%) displayed an unwillingness to discuss their accusations, we do not know whether this has to do with their use of the concept of fallacy. They were, after all, engaged in heated discussions on contested issues—gun regulation, abortion, religion, transgender rights, vaccines, and even the existence of God! Moreover, hostility and unwillingness to listen to the other side seem to be unfortunately common on Twitter.

4. Conclusion

Some of the criteria that were chosen for the analysis turned out to be more informative than others. Some of them may even be questionable—when is an underlying theory of fallacy "problematic"? Nevertheless, even if certain results of this study should be disregarded, the relative independence of the criteria allows the rest of the findings to be of some value. In particular, it was striking to find that the percentage of accusations that were substantiated was significantly higher when the fallacy was not identified than when it was. Further research should look into this issue. If the use of fallacy labels tends to lead to making unsubstantiated accusations, then arguably we are doing a disservice by teaching them.

References

Blair, J. A. (1995). The place of teaching informal fallacies in teaching reasoning skills or critical thinking. In H. V. Hansen & R. C. Pinto (Eds.), Fallacies: Classical and contemporary readings (pp. 328–338). The Pennsylvania State University Press.

Boudry, M., Paglieri, F., & Pigliucci, M. (2015). The fake, the flimsy, and the fallacious: Demarcating arguments in real life. Argumentation, 29(4), 431–456.

Eemeren, F. H. van, Garssen, B., & Meuffels, B. (2009). Fallacies and judgments of reasonableness. Springer.

Finocchiaro, M. A. (1981). Fallacies and the evaluation of reasoning. American Philosophical Quarterly, 18(1), 13–22.

Finocchiaro, M. A. (1987). Six types of fallaciousness: Toward a realistic theory of logical criticism. Argumentation, 1, 263–282.

Hitchcock, D. (1995). Do the fallacies have a place in the teaching of reasoning skills or critical thinking? In H. V. Hansen & R. C. Pinto (Eds.), Fallacies: Classical and contemporary readings (pp. 319–327). The Pennsylvania State University Press.

Hundleby, C. (2010). The authority of the fallacies approach to argument evaluation. Informal Logic, 30(3), 279–308.

Jason, G. (1987). Are fallacies common? A look at two debates. Informal Logic, 8(2), 81–92.

Willingham, D. T. (2008). Critical thinking: Why is it so hard to teach? Arts Education Policy Review, 109(4), 21–32.

THE ARGUMENT'S THE THING ...

GEOFFREY C. GODDU
University of Richmond
ggoddu@richmond.edu

Abstract
At first blush, arguments seem akin to poems, plays, short stories, and other literary works—they are all some sort of socio-linguistic entity. But linguistic entities is the least popular ontological category amongst argumentation theorists for the nature of arguments. So how to resolve the conflict between our first blush and argumentation theory? I consider three options: deny arguments are in the same category as, say poems; accept they are in the same category, but deny that that category is socio-linguistic; accept that arguments are linguistic entities and so deny the dominant positions in argumentation theory. I shall conclude that the first option is the most likely to be the case.

1. Introduction

Consider the following limerick:

There was a young lady named Bright,
whose speed was far faster than light;
She set out one day
in a relative way
and returned on the previous night,

or Shakespeare's *Hamlet,* or the short story, "The First Time Machine", by Frederic Brown, or Anselm's ontological argument for the existence of God. For many, it might seem obvious that these are all the same sort of thing--poems, plays, short stories, arguments, etc., are all a sort of socio-linguistic entity—objects created out of words in accord with certain social norms. They all have authors, they all involve language, and they all occur with a socio/historical context that (i) in part determines what does or does not count as a poem or a play, say, and (ii) determines norms of adequacy and inadequacy. As intuitive as this similarity seems, however, treating arguments as linguistic entities is, within argumentation theory at least, probably the least popular of the three ontological options, i.e., abstract objects, acts, or linguistic entities.

The purpose of this paper is, primarily, to explore the viability of two strategies for resisting the potential argument for arguments being linguistic entities in virtue of their similarity to poems, plays, short stories, etc. One strategy is to deny that arguments are the same sort of thing as poems, plays, etc. The similarity is at best superficial and not one with ontological significance. The other strategy is to accept that they are indeed all the same sort of thing, but to deny that *any* of them are linguistic entities. For the remainder of this paper I shall, for the most part, focus on arguments and poems and leave it to the reader to determine the degree to which what I say here can be applied to short stories or plays or other works of literature.

I begin, in Section 2, with a brief discussion of why we might think that arguments and poems are, ontologically, the same sort of thing and the puzzle this causes. In Section 3, I shall present the pros and cons of three potential solutions to the puzzle raised in Section 2. The first insists that arguments and poems are not in the same ontological category. The second maintains that arguments and poems are in the same category, but are not linguistic entities. The third insists that arguments and poems are both in the same category and are in fact linguistic entities. I shall conclude in Section 4, that the most viable option is to deny that arguments and poems fall into the same ontological category.

2. Are Arguments and Poems Ontologically the Same?

Why think poems and arguments are in the same ontological category? Both have authors—our opening limerick was first published in *Punch* in 1923, but without attribution. Despite various reprintings (some with slight alterations) and numerous attributions, the evidence strongly suggests that the author was in fact A.H. Reginald Buller, a professor of Botany at the University of Manitoba.[1] Arguments also have authors—Aquinas' Five Ways, Anselm's Ontological Argument, Descartes' Cogito, Peter Unger's Argument for Skepticism, and so on. The authors all used words to construct the poem or the argument—one suspects that the authors considered several different possible word combinations before settling on the final version. [Note: for the purposes of this paper, I am putting aside the controversy within argumentation theory over the status of "visual arguments".]

The authors made these constructions within socio-historical contexts that at least gave typical boundary conditions for what is or is not a poem

---

[1] There was a Young Lady Named Bright Whose Speed was Far Faster than Light, Quote Investigator, Dec 19, 2013, https://quoteinvestigator.com/2013/12/19/lady-bright/, last accessed 9/15/2022.

(or a limerick) and an argument. Limericks, typically, have five lines with an AABBA rhyme scheme. Arguments, typically, have reasons and a conclusion.

Finally, there are things that seem to be both poems and arguments simultaneously. Many didactic poems provide reasons for their morals. Consider, for example, William Blake's, "Eternity":

> He who binds to himself a joy
> Does the winged life destroy
> He who kisses the joy as it flies
> Lives in eternity's sunrise

Any argument can be rewritten as a poem and still be an argument. Here, for example, is Unger's argument for skepticism, (see Unger, 1974, 132) written as a poem:

> 1. If someone *knows*
> Something to be so,
> Then
> It is all right
> For the person to
> Be absolutely
> Certain it is so.
> 2. It is never all right
> For anyone to
> Be absolutely
> Certain
> Anything is so.
> 3. Nobody *knows*
> That
> Anything is so.

I did not promise it would be a good poem. But if something can simultaneously be both an argument and poem, then, assuming that all arguments fall into the same ontological category as each other and all poems fall into the same ontological category as each other, all arguments and poems fall into the same ontological category.

Suppose we accept these similarities as evidence that they are all the same sort of thing. Well the unreflective view of poems and plays is that they are just the texts. This view is also a fairly standard philosophic view. Nelson Goodman and Catherine Elgin (1986), for example, argue for a quite strong version of this view. Anna Christina Soy Ribeiro writes: "In writing culture today we usually think of a poem as an unchanging sequence of words neatly printed on the page in line of various lengths, to be read in silence and solitude."(Ribeiro, 2015, 129) But if poems and plays are just texts and arguments are the same sort of thing as poems and plays, then arguments are just texts.

The problem, however, is that, given a choice amongst potential candidates for the nature of arguments such as abstract objects like propositions, or acts, or linguistic entities such as sentences, the last is the least popular amongst argumentation theorists. Here, for example, is Mark Vorobej's argument against sentences:

The following two passages

(A) 5 is a square root of 25. Therefore, 25 is not a prime number.

and

(B) 25 is the square of 5. It follows that 25 is not a prime number.

could express the same argument even though they are composed of different sentences. The author of the first passage uses certain words in order to draw an inference involving the two propositions expressed by the two sentences she employs. The author of the second passage uses two different sentences to accomplish exactly the same end. In each case, a single inference is drawn from the same premise to the same conclusion, and neither the nature of that inference nor the semantic content of the premise or conclusion are apparently affected in any way by the author's choice of words or by the passages' sentential structure. That's why arguments are composed of propositions, and not sentences. (Vorobej, 2006, 9.)

Compare also Vorobej's (A) 5 is a square root of 25. Therefore, 25 is not a prime number, with the following Google translation:

(AS): Cinco es una raíz cuadrada de veinticinco. Por lo tanto, veinticinco no es un número primo.

A and AS are different texts—they use different words in separate languages. Yet, there is only one argument here. Two texts, but one argument, so arguments are not texts.

Consider also the following text uttered at a soccer/football match:

(C) He just about scored a goal, so can get back in this game.

In American English 'just about' is usually a synonym of 'almost', so this passage is arguing that since the team almost scored a goal, the momentum may be shifting, so we can get back into this game, especially if the score is something like 0-1. In British English, 'just about' is usually a synonym of 'barely', so this passage is arguing that since the team just scored a goal, we can get back into this game, especially if the score was say 0-2 and now is 1-2. Same text, but different arguments, so arguments are not texts.

Arguments survive the replacement of coreferential terms or phrases—texts do not. Hence, arguments are not texts. Arguments can be translated from one language to another. To the degree that translation is the replacement of terms, phrases, and sentences in one language with coreferential terms in another language, texts cannot survive translation from one language to another. Hence, arguments are not texts.

On the one hand arguments seem akin to poems and plays and so on and those are most often thought of as texts and yet we have good reasons to think that arguments are not just texts. How, then, to resolve the conundrum? I turn now to three possible solutions.

3. Three Potential Solutions

Option 1: Despite initial appearances arguments and poems are not the same sort of thing. After all, unlike arguments, poems are much more sensitive to the replacement of coreferential terms and to translation. Consider, firstly, the opening six lines of Ellis Parker Butler's, The Ballad of a Bachelor:

Listen ladies, while I sing
The ballad of John Henry King.
John Henry was a bachelor,
His age was thirty-three or four.
Two maids for his affection vied,
And each desired to be his bride,

Compare that with:

Listen ladies, while I sing
The ballad of John Henry King.
John Henry was an unmarried male of marriageable age,
His age was thirty-three or four.
Two maids for his affection vied,
And each desired to be his female participant in her own marriage ceremony,

While the latter is still a poem, I am strongly disinclined to say it is the *same* poem as Butler's, even though the only changes involve replacing 'bachelor' and 'bride' with co-referring expressions.

Similarly, compare the opening limerick of this paper with the following German Google translation version:

Da war eine junge Dame namens Bright,
deren Geschwindigkeit viel schneller war als das Licht;
Sie machte sich eines Tages
relativ auf den Weg
und kehrte in der vorangegangenen Nacht zurück.

The English version is a limerick, but the German translation is not—the rhyme scheme is non-existent and the syllable counts are off. Limericks are a type of poem, so to have the same poem it must be of the same type. But the German translation is not of that type, so the poem did not survive translation into German.

Poems, like texts, do not appear to survive the replacement of coreferential terms or translation. Arguments do. Hence, arguments and poems are not the same sort of thing.

Option 2: Despite our initial take on the matter, poems and plays and stories are not texts. There are acts of arguing, there are performances of plays, and there are poetry readings. There are clearly argument texts, and play texts, and poem texts. But given that one and the same argument can be expressed by various different argument texts in either the same language or in different languages, the arguments themselves are not the texts. Argument texts express arguments; acts of arguing enact arguments. So why not say the same of poems or plays? Poetry texts express poems; poetry readings enact poems, and the poems are something else than either the text or the poetry reading. Ruby Meager, for example, calls the texts and the readings, *manifestations* of the work.(Meager, 1959, 52, emphasis in original)

But what could this thing be that is neither the text nor the enactment? The most obvious candidate is the meaning or content of the text. For arguments there is an obvious candidate to be this meaning or content—the propositions expressed by the premises and the conclusion. But plenty of poems seem to express no propositions at all, but still have a meaning, so if we wish to keep arguments and poems in the same ontological category we cannot flesh out the meaning or content of texts in terms of propositions. [Note: there is ongoing debate about whether all poems could be paraphrased in terms amenable to propositional analysis, eg. (Currie and Frascaroli, 2021)] Regardless, one strategy for resisting the argument that arguments are linguistic entities is to maintain that arguments, and poems, etc are not the texts, but rather the content or meaning of those texts. Joseph Stevenson, for example, has been described as holding the view that literary works are "roughly the meaning shared by two instances of such a work(including translations of that work into different languages and close paraphrases of it in its own language)". (Howell, 2002, 69).

But even without specifying what this content or meaning is we still have a problem—there are plenty of properties of poems that are not properties of the content or meaning of texts. Here are some examples—meter, rhyme, alliteration, consonance, assonance, euphony, cacophony, enjambment, end-stopped lines, etc.... To be a typical limerick, a type of poem, is to have an AABBA rhyme pattern. Hence, some poems have the property of having an AABBA rhyme pattern. But meanings or contents of expressions do not rhyme—words rhyme. Meanings are not euphonious or alliterative; meanings are not enjambed or end-stopped—but texts and parts of text are. Hence, if these truly are properties of poems, then poems cannot be the meanings or content of the expressions. This of course explains why it is often quite challenging, if not impossible, to translate poems in one language into another—keeping meaning constant will often lose the relevant rhyme or meter or sound properties that make it the poem it is; but keeping those properties involves an often significant change in meaning.

Okay—poems are hybrid—they are texts with certain contents. But ontologically this makes things worse, not better. Since a poem is in part the text, then the poem still has the identity conditions of texts and so if we change the text, we change the poem. Hence, we still cannot translate poems into other languages because that changes the text. But we can translate arguments into other languages—hence arguments are not hybrid entities of this sort and so arguments and poems do not fall into the same ontological category.

But suppose we channel Roy Sorensen (see especially his examples in 1991)—what then to make of this argument:

(D) 1. First degree murder is a crime.

2. The number two is the lowest prime.

3. So, some arguments have premises that rhyme.

We already have a good reason to not treat arguments and argument expressions as the same thing, even if we sometimes use the word 'argument' to refer to either one. Clearly it is the premise expressions that rhyme, so at the very least we need to read the conclusion as: so, some argument expressions have premise expressions that rhyme. Do those premises force that conclusion—no. Here is the French version of that argument expression in which the premises clearly do not support the conclusion.

(DF) 1. Le meurtre au premier degré est un crime.

2. Le nombre duex est le plus petit nombre premier.

3. Ainsi, certains arguments ont des prémisses qui riment.

Different question—should argument expression D convince you of the truth of the conclusion that some argument expressions have premises that rhyme? Yes, but that is because D is an example of an argument expression with rhyming premises. Note that DF is not, so DF will not convince us of the truth of the expressed conclusion. We can, however, express the convincing that is going on via the following:

(E) D is an example of an argument expression with rhyming premises, so some argument expressions have premises that rhyme.

E can be translated into any language and still express the same argument about D. Hence, D is not an example of an argument that involves rhyming premises, even though it is an example of an argument expression with rhyming premise expressions. But argument expressions are composed of words and so it is quite possible that argument expressions involve rhyming or alliteration or euphony or ...

This also explains Unger's Argument for Skepticism as a poem case. We are still in Option 2, so poems are somehow not the same as the expressions. Hence, what I did was turn Unger's argument expression into a poem expression. Since poem expressions and argument expressions are in the same category, namely texts, one and the same text can be both a poem expression and an argument expression.

Option 3: Despite our strong intuitions that arguments are not texts, we are mistaken—arguments, like poems are just texts. Earlier I mentioned that Goodman and Elgin (1986) take a very strong line on literary works just being texts. A consequence of their view is that any change in text generates a new literary work. Change one word and we have a new work. Hence, literary works cannot be translated into other languages, since that will generate different texts. One could take the same hard line with arguments. They too are just texts. Change one word and you have a different argument—hence, Vorobej's (A) and (B) are in fact different arguments. Arguments also cannot be translated into different languages. Arguments and literary works such as poems are in the same ontological category—it just turns out that a lot of things we thought were true of arguments (and some of the things we thought true of poems) are mistaken.

Goodman and Elgin try to ameliorate some of the consequences of their view by allowing that a revision to a poem, say, would, depending on the circumstances, make either the unrevised poem an *interpretation* of the revised version or vice versa. Similarly, translations of literary works, would be *interpretations* of the original, but not the same work as the original. One could then take a similar line about arguments—(A) and (B) might not be the same argument, but they could be interpretations of each other. Our modern day translations of Anselm's arguments are not actually Anselm's arguments, but rather interpretations of Anselm's arguments.

But what might an Option 3 advocate say about (C)—He just about scored a goal, so we can get back into this game? That was supposed to be an example of one text, but two separate arguments. Appealing to interpretations is out of place here—that appeal was meant to capture how different *texts* could be related in a way that, though not identity, would capture enough of our intuitions about 'same work' or 'same argument' to keep the hard line palatable. But there is just one text here, not two that are meant to be interpretations of each other.

Here is another case: Suppose, perchance, someone were to create a story, in Russian, that just happened to be identical with what could result if Frederic Brown's, "The First Time Machine" were translated into Russian.
Given that we are supposing that the two texts are created completely

independently of each other they are neither the same short story nor interpretations of each other. But now suppose that Brown's story is translated into Russian with the result being textually identical to the independently created Russian story. Now we have one text that both is and is not an interpretation of Brown's.

Of course, one way to try to solve both of these cases is to make a Type/Token distinction. In both we have one type of text, but two tokens and each token is a different argument or story. Goodman and Elgin cannot make this distinction since the prime motivation for their hard line was a nominalism that is equally averse to meanings and types and abstract objects, etc. In addition, they do not want arguments and literary works to exist at the level of tokens since they do think arguments and literary texts are repeatable—just write the same text again and you still have the same work.

But what of someone who was willing to resort to arguments and poems as both text types? Two things: Firstly, once one appeals to text types one still needs to explain what counts as the same type—is one text in print all in wood block letters, but the other in cursive the same type or not? I suspect that attempts to salvage that they are the same type will ultimately have to appeal to some notion of 'representation of the same letter in an alphabet' which is abstract enough the one might as well just have appealed to meanings or some such abstract object in the first place. Secondly, this move still does not solve the two cases, since the solution offered above relies on taking the tokens as the arguments/stories—after all, our intuition says different argument, but the type is the same. [I suspect our intuition in the second case is that all three texts are in fact the same story—but they share no text type.]

So where do we stand? Option 1 argues that the evidence best supports that poems are texts, but arguments are not. At the same time, the seemingly strong similarities between arguments and poems have to be dismissed as superficial and the possibility of translating poems into other languages remains problematic at best.

Option 3 accepts the strong similarity between literary works and arguments and insists that arguments are like poems, just texts. Now the problem is to provide a notion of 'same text' that both allows poems and arguments to be repeatable, i.e. written down more than once, but also accounts for cases in which it appears the same text are in fact different arguments or different literary works.

Option 2 accepts the strong similarities and accepts the reasons for holding that arguments are not texts and concludes that arguments and poems are not to be equated with the expressions, but rather some sort of content of those expressions. The challenge here, however, is to produce a notion of poem that allows both translation into other languages and accepts that rhyme and meter and enjambment are properties of poems. The question then would be whether this notion of a poem is also a notion that would carry over to arguments.

Here is but one possibility from Ribeiro, responding to Robert Howell's skepticism that there is any unified ontology category that captures all works of literature, including poems. [Note if Howell is right, then there is no category that literary works such poems, plays, etc are all in, into which we could also put arguments.] She starts out: "A poem is an intentional abstract artefact". So far so good I suspect for many of my argumentation colleagues who also think arguments are intentional abstract artefacts. But she continues, "A poem is an intentional abstract artefact; a type, consisting of an instantiation template". Unfortunately, she does not elaborate at all on what she means by an instantiation template. Argumentation theory does have something that could be an instantiation template, i.e. argumentation schemes, but no argumentation theorist I know of wants to claim that the argumentation schemes are in fact the arguments. [In Goddu (2021) I argue that argumentation schemes are best construed as texts (and the current strategy is to try to give an account of poems and arguments according to which they are both not texts.] She, however, is not done:

> A poem is an intentional abstract artefact; a type, consisting of an instantiation template whose creation is spatio-temporally located via its original token, and thus embedded in either a declamation-based or an inscription-based practice, which will dictate the kinds of ontological strictures embodied in that original token and required of all future tokens.(Ribeiro, 2015, 131-132.)

She is trying to capture both oral and written poetry traditions to rebut Howell's arguments, at least with respect to poetry. The problem is that she goes on to say the following about written poems: "inscription-type poems will be linguistically rigid (words and word order may not vary if the work is to remain the same.) (Ribero, 2015,. 132) This just reintroduces all the problems cited above since those problems arise as much for token expressions as they do for expression types delineated in terms of words and word order. Change the words, even if they have the same meaning, in either the same or different languages, and we have changed the type instantiated. Arguments survive such changes; text types do not, and so, given Ribiero's definition, neither do written poems. Hence, arguments are not text types, and not in the same category as written poems.

4. Conclusion

I have only scratched the surface of the nature of literary works—indeed if anything this initial foray has made me even less sure than I was before about what a poem is. Regardless, there still appears to be a significant problem for the claim that arguments fall into the same category as literary works—the properties of words, either how they look or sound or how they are arranged visually is not a property of arguments and yet are often a crucial part of what makes something the poem or play or story it is. Argument expressions, on the other hand, being composed of words, *can* depend on the properties of the composite words—but there are really good reasons to not confuse the argument with its expression—reasons that also count against making the expression part of some ontological hybrid thing that is the argument. Hence, were someone to argue that arguments are texts given their similarity to works of literature, it should fail to catch the conscience of the king.

References

Currie, G. & Frascaroli J. (2021). Poetry and the Possibility of Paraphrase. *The Journal of Aesthetics and Art Criticism* , 79, 428-439.

Goddu, G C.. (2021). A Simple Theory of Argument Schemes. *Informal Logic* 41(4), 539-578.

Goodman , N. & Elgin, C. (1986). Interpretation and Identity: Can the Work Survive the World? *Critical Inquiry,* 12, 567-574.

Howell, R. (2002). Ontology and the Nature of Literary Work. *The Journal of Aesthetics and Art Criticism,* 60, 67-79.

Meager, R. (1958-1959). The Uniqueness of a Work of Art. *Proceedings of the Aristotelian Society, New Series,* 59, 49-70.

Ribiero, A.C. (2015). The Spoken and the Written: An Ontology of Poems. In J. Gibson (Ed.), *The Philosophy of Poetry,* (pp 127-148(. Oxford: Oxford University Press.

Stevenson, J. (1965). *The Languages of Art and Art Criticism,* Michigan: Wayne State University Press.

Sorensen, R. (1991). 'P, Therefore P' Without Circularity. Journal of Philosophy 88, 245-266.

Unger, P. (1974). An Argument for Skepticism. *Philosophic Exchange* 5(1), 131-155.

Vorobej, M. (2006). *A Theory of Argument,* New York: Cambridge University Press.

TWO SUBTYPES OF ILLOCUTIONARY ACTS OF ARGUING

AMALIA HARO MARCHAL
University of Granada
amaliaharo@ugr.es

Abstract
In the Linguistic Normative Model of Argumentation, argumentation is characterized as a second-order speech act complex that, from an illocutionary point of view, counts as an attempt by a speaker at showing that a target-claim is correct. In this model, the successful performance of a speech act of arguing depends on the production of the illocutionary effect that consists in the securing of uptake. However, speaker's utterances that constitute speech acts of arguing also produce other kinds of illocutionary effects, namely, effects that affect not only the set of speaker's rights, obligations, and entitlements, but also that of the hearer. The main goal of this paper is to show that it is necessary to distinguish two notions of illocutionary act in order to account for this second type of effects; the first notion is related to the speech act performed by the speaker, and the second one allows us to characterize the speech act performed by the speaker that also involves the hearer.

Keywords: argumentation theory, speech act of arguing, normative effect, illocutionary act, deontic modal competence.

1. Introduction

The development of argumentation theory has been greatly influenced by speech act theory. Specifically, Pragma-dialectics (van Eemeren & Grootendorst 1984) and the Linguistic Normative Model of Argumentation (Bermejo-Luque 2011) are two of the most influential theories of argumentation that adopt speech act theory as their theoretical framework. Both theories follow Searle's (1969) account in considering that some conditions must be fulfilled for the speaker's utterance (or set of utterances) to count as a speech act of arguing. In particular, the Linguistic Normative Model of Argumentation (henceforth referred to as LNMA), provides a characterization of argumentation as a second-order speech act complex which, from an illocutionary point of view, counts as an attempt by the speaker at showing that a target-claim is correct. Bermejo-Luque adopts Searle's view in considering that there are certain conditions that, when fulfilled, make the speaker's utterance a speech act of arguing (2011, p. 70). She also shares with the Searlean account that the successful performance of a speech act (in this case, a speech act of arguing) depends, among other things, on the securing of uptake[1]. The Searlean characterization of the speech act of arguing allows LNMA to provide an adequate characterization of what does it mean for a speaker's utterance to constitute a speech act of arguing. In Bermejo-Luque's model, for this speech act to be carried out, it is only necessary for the speaker's utterance to be graspable to the potential hearer.

The characterization of the speech act of arguing offered by Bermejo-Luque responds to one of the motivations that can be pursued when analyzing argumentation, but it is not the only one. As I view it, when carrying out an analysis of argumentation, one can have two legitimate interests. On the one hand, we can be interested in accounting for what does it mean for a speaker's utterance to count as a speech act of arguing, i.e., what are the features that make one's utterance be interpreted as a single illocutionary act, which is an act of arguing. And, on the other hand, we can be interested in analyzing argumentation as a communicative exchange, an exchange that involves of course the speaker's utterance, but also the hearer's response. In other words, we can be interested in what an argumentative exchange consists in, and, relatedly, in how the world, the social context, changes by means of our argumentative acts.

In this paper I will contend that, although the Searlean approach assumed in LNMA provides an adequate characterization of what counts as argumentation in terms of what the speaker does, it leaves out both the role played by the hearer in communication, as well as the normative effects that any speech act, including that of arguing, brings about. These normative effects consist in the production of changes in what Sbisà (2006), following Austin's (1962) account, calls *Deontic Modal Competence*, i.e., the set of rights, obligations and entitlements that can be attributed to the participants of a communicative exchange and that can be modified and affected by the performance of speech acts. These normative effects concern not only the speaker's but also the hearer's obligations and entitlements. I will argue that this is so because they are linked to another speech act performed by the speaker that is not limited to what a speaker does, but that also involves the hearer in a certain way.

In addition, I will argue that, to account for these normative effects in the case of argumentation, we must distinguish between two different levels in the analysis of the speech act of arguing: one related to the

---

[1] In Bermejo-Luque's model, the securing of uptake is characterized as an effect consisting in that the potential hearer must be able to grasp the meaning and the force of the speaker's utterance. This characterization of the securing of uptake is similar to the one offered by Sluys (2018).

speaker's utterance, and another one related to the communicative exchange in which both the speaker and the hearer are involved. Furthermore, I will argue that, to account for these two levels of analysis, it is necessary to distinguish between two subtypes of illocutionary acts of arguing. The first subtype involves only what the speaker does and can be carried out regardless of the hearer's response, i.e., of what the hearer does or understands. In particular, the performance of this first subtype of illocutionary act of arguing just requires the fulfillment of the condition of securing of uptake by the speaker. By contrast, the second subtype of illocutionary act refers to the act performed by the speaker that involves and affects the hearer in a certain way. The successful performance of this subtype of illocutionary act is associated with the production of changes in the deontic modal competence. In particular, I will draw from the Austinian perspective known as the *interactional account* of speech acts (Corredor, 2021; Witek, 2015) to characterize this second subtype of illocutionary act of arguing. The distinction between the two subtypes of illocutionary acts will result in two different but related notions of *speech act of arguing*.

2. Argumentation as a speech act: LNMA

The Linguistic Normative Model of Argumentation developed by Bermejo-Luque (2011) is framed within a pragmatic-linguistic approach to argumentation that incorporates a critical re-elaboration of Toulmin's (1958) material conception of inference, and in which argumentation is characterized as a specific type of speech act. In this model, argumentation is understood as a communicative activity consisting in an attempt by the speaker at showing that a target-claim is correct, that is, at justifying a target-claim. In this model, a speech act of arguing is characterized as a second-order speech act complex formed by two speech acts, namely, the speech act of adducing (a reason) and the speech act of concluding (a target-claim). These are second-order speech acts too because they can only be carried out by means of performing first-order speech acts; in particular, in LNMA, the act of adducing and the act of concluding are carried out by means of performing two constative speech acts, which can be performed either directly or indirectly, or literally or non-literally. The constative speech acts, R and C, become speech acts of adducing a reason and concluding a target-claim because there is an implicit inference-claim which establishes a relationship between the content of both constative speech acts (Bermejo-Luque 2011, p. 60). The propositional content of this implicit inference-claim is "if r (the content of R [and its pragmatic force]), then c (the content of C [and its pragmatic force])" (2011, p. 61). Thus, according to Bermejo-Luque, speech acts of arguing are composed of three constitutive elements: a target-claim, a reason that is put forward to the target-claim, and an implicit inference-claim whose function is to turn a claim into the act of adducing and to turn another claim into an act of concluding.

Following Searle, Bermejo-Luque contends that "there are constitutive conditions that make certain performances acts of arguing" (2011, p. 70). These conditions would be the following (2011, pp. 71-72):

Preparatory conditions:
(i) S believes that a claim R, having such and such pragmatic force, may be taken to be correct by L.
(ii) It makes sense to attribute to S a conditional claim, with a certain pragmatic force, whose antecedent is "R is correct," and whose consequent is "C is correct".
(iii) S takes the correctness of a claim C to be in question within the context of the speech-act.
(iv) S takes a claim R to be a means to show a target-claim C to be correct.
Propositional content conditions:
(v) The content of the reason is that a claim R' is correct.
(vi) The content of the target-claim is that a claim C' is correct.
Sincerity conditions:
(vii) S believes the propositional content of R in a certain way and to a certain extent, namely, the way and extent that correspond to the pragmatic force of the claim R'.
(viii) S believes that R being correct is a means to show that a target-claim C is correct.
(ix) S believes the propositional content of C in a certain way and to a certain extent, namely, the way and extent that correspond to the epistemic pragmatic force of the target-claim C.
Essential conditions:
(x) Adducing R with such and such pragmatic force is a means to show that a target-claim C is correct.
(xi) S aims to show that a target-claim C is correct.

Let's illustrate this with the following example offered by Bermejo-Luque (2019, p. 664):

(1) *I promise I'll take care, so don't worry*

Bermejo-Luque characterizes this utterance as an illocutionary act of arguing in which two other acts are carried out: a speech act of adducing (a reason) and a speech act of concluding (a target-claim). In her account, the utterance of "I promise I will take care" constitutes a speech act of adducing in (1). This is so because, by uttering it, the speaker does not only promise that she will take care but, since she has also implicitly made the claim that, if (it is true that) she commits herself to take care, then (it is true that) the hearer does not have to worry, she is also adducing as a reason that she commits herself to take care (2011, p. 65). The utterance of "don't worry" constitutes an act of concluding in (1) because, in uttering it, the speaker is suggesting the hearer not to worry; given that she has implicitly claimed that if (it is true that) she commits herself to take care, then (it is true that) the hearer should not worry, she indirectly claims that the hearer should not worry, thus turning this act into an act of concluding (Bermejo-Luque 2011, p. 66).

As I view it, LNMA provides a sound account of what the speech act of arguing consists in from the speaker's perspective. In this sense, it enables us to account for the conditions that the speaker must comply with for her utterances to count as an attempt at justifying a target claim (that is, as a case of argumentation). However, what I want to stress here is that this leaves out some other important aspects of communication. Specifically, what is missing is an account of the role of the hearer and the normative effects that illocutionary acts of arguing bring about. In the following section, I will present what I consider to be the main problems of the Searlean approach that is assumed in LNMA. I will also introduce the so-called *interactional approach* to speech acts, which is the account or set of accounts that have raised the specific problems that I will point out.

3. An alternative to the Searlean account

In the previous section we saw that, in Bermejo-Luque's LNMA, the speech act of arguing is characterized as an attempt by the speaker at showing that a target-claim is correct. According to Bermejo-Luque, in order to carry out a speech act of arguing, only the speaker's utterance is necessary, provided that some conditions are fulfilled. If the speaker utters certain words that constitute an attempt at justifying which can be understood by the hearer as such, then it can be claimed that she has successfully performed a speech act of arguing. However, some important questions arise at this point. Firstly, it poses the question whether it makes sense to say that a speech act of arguing has been really performed if there is no hearer that actually understands and recognizes the speaker's utterance as a speech act of arguing. In addition, if one only takes into account what a speaker does, it seems difficult to account for the changes in the set of rights, obligations, and entitlements of the participants of the communicative process that are produced by a speech act of arguing. Finally, we can also ask if what we want to do when we argue is limited to trying to make our utterances understandable. These questions, which were firstly formulated in relation to speech acts in general[2], have been considered by many authors whose proposals are

---

[2] However, see Corredor (2021) and Labinaz (2021) for different ways of addressing this type of question in the particular case of argumentation.

framed within the so-called *interactional approach* to speech acts (Carassa & Colombetti, 2009; Clark, 1996; Sbisà, 2006, 2009; Witek, 2015).

Searle's (1969) approach to speech acts has greatly influenced the development of the speech act theory in general, as well as the study of argumentation. However, it has also given rise to important criticisms, such as the one pointed out by Clark (1996). According to Clark (1996, p. 137), in Searle's view it is irrelevant for the performance of a speech act whether there is a hearer that listens, understands or receives speaker's act. Regarding this, Clark points out that "this view is, of course, absurd. There can be no communication without listeners taking actions too - without them understanding what speakers mean" (1996, p. 138). Clark's criticisms have been subsequently taken up and developed by other interactional approaches (Carassa & Colombetti, 2009, p. 1840; Sbisà, 2009, p. 37). In particular, here I will focus on two interrelated problems of the Searlean perspective that these approaches have pointed out. The first one is related to the mere passive role that such perspective attributes to the hearer, while the second one consists in that it leaves out the normative effects brought about by illocutionary acts.

According to the speech act theory formulated by Austin, the performance of a certain speech act is associated with the production of three types of illocutionary effects (1962, pp. 115-116): (i) the securing of uptake, (ii) producing effects that change the normative facts, and (iii) the inviting of an appropriate response. Several authors have reformulated Austin's considerations about the illocutionary effects produced by an utterance focusing on the effect (i), the securing of uptake. As Sbisà indicates (2009, p. 35), Strawson (1964), motivated by his intention to make Austin's proposal compatible within the Gricean theoretical framework, played a fundamental role in advancing and promoting the subsequent consideration of the securing of uptake as the essential illocutionary effect associated with the successful performance of an illocutionary act. His considerations greatly influenced further developments of the speech act theory, including Bach and Harnish's (1979) as well as Searle's (1969) proposals.

As Sbisà indicates (2009, p. 37), Searle agrees with Strawson's view in considering the securing of uptake as the central illocutionary effect that must be intended to be brought about in order to carry out a certain illocutionary act. Here is where the two inter-related problems that I mentioned above arise. The first one, illustrated by Clark's remarks above, is that this perspective assigns a mere passive role to the hearer; if, as Searle proposes, the only relevant illocutionary effect is the securing of uptake by the speaker, then the hearer's response seems to play no role in the performance of a certain illocutionary act. Relatedly, the second problem has to do with the fact that the Searlean perspective leaves out the second type of illocutionary effect formulated by Austin, i.e., the effect consisting in the production of changes in the normative facts (that is, in the set of rights, obligations and entitlements of the participants of the communicative process). This set of obligations, rights and entitlements is what Sbisà calls the *Deontic Modal Competence* (2006, p. 158). The fact that the Searlean approach disregards the illocutionary effect (ii) in the characterization of speech acts entails an important consequence, namely, that it is not able to explain how it is possible that, when we carry out certain speech acts, the interpersonal relationship between speakers and hearers changes.

Let's illustrate this with the following example. When someone utters (2),

(2) *I will be there at 8:00 pm*

given the fulfillment of the conditions put forward by Searle (1969, pp. 57-61) (i.e., the propositional content conditions, preparatory conditions, sincerity condition and the essential condition), we can say that the speaker has carried out an illocutionary act of promising, which is successfully performed when the hearer recognizes the speaker's intention to produce an effect on her; an effect that consists in her understanding of the speaker's utterance (1969, p. 60), that is, in the securing of uptake. The characterization of the actual and successful performance of a speech act as dependent on this condition allows us to account for argumentation in terms of what the speaker does, i.e., in terms of the sentence (or sentences) uttered by the speaker which, if some conditions are fulfilled, would constitute a speech act of arguing. For this speech act to be successfully carried out, it is only necessary that the speaker's words are understandable to the potential hearer; no actual hearer (neither her response) is necessary for the performance (and successful performance) of the speaker's act. The hearer's role in communication is here reduced to just hearing and understanding the speaker's utterance. In my view, this approach neglects how actual communicative processes work, where normally the speaker and hearer participate in the conversation actively, responding in a certain way (either explicitly or implicitly), and exchanging their role along the process. As I have mentioned above, this problem was pointed out by Clark (1996) regarding the individualist view of communication put forward by Searle.

Secondly, Searle's characterization does not take into account how the performance of the illocutionary act of promising changes the deontic modal competence of both the speaker and the hearer. When the speaker utters (2), if he gets the hearer to listen and understand his utterance, he is acquiring a certain commitment, that is, that of delivering what he has promised. In this sense, his deontic modal competence has changed in a specific way. In addition, the hearer's deontic modal competence has changed too. When someone promises something, in addition to the commitment he acquires, he also changes the normative facts for his interlocutor: if the hearer responds (either explicitly or implicitly) by accepting the speaker's promise, then she acquires the legitimate expectation that the speaker will keep his promise. In the case of (2), if for instance they have agreed to go to the movies, then she will expect the speaker to fulfill his promise to not be late. Thus, we can see how both speaker's and hearer's interpersonal relationship has changed: by the speaker's utterance and the hearer's response they have introduced changes in the set of their rights, obligations, and entitlements, that is, in their deontic modal competence. And this is not an effect only associated with promises. It is an effect associated with the performance of any type of speech act, which must be taken into account in order to offer a plausible explanation of how communication actually works. Also, the consideration of this second type of illocutionary effect involves taking into account the role of the hearer as not limited to merely hearing and understanding, but also as an active actor in the communicative process.

4. Two subtypes of illocutionary acts of arguing

In this section, I will argue that, in order to retain the virtues of Bermejo-Luque's model and, at the same time, incorporate the insights of interactional approaches (thus avoiding the problems of the Searlean perspective), we must distinguish between two different levels of analysis. Furthermore, I will argue that, in order to account for these two levels of analysis, it is necessary to distinguish between two subtypes of illocutionary acts. The distinction between two levels, together with the distinction between the two subtypes of illocutionary acts, will allow us to show how the evaluation of the same act of arguing varies depending on the level which we pay attention to. Consider again example (1):

(1) *I promise I'll take care, so don't worry*

Here, as we saw in section 2, if the speaker counts as fulfilling certain conditions, which in this case would be the conditions for an act of arguing, then her utterance would count as an attempt at showing that a target-claim is correct, i.e., as an act of arguing. In this first level (the level of the speaker's utterance), the only thing that is needed for the speaker's act of arguing to be successfully performed is that she counts as trying to justify a target-claim. As I view it, LNMA allows us to account for speech acts of arguing in terms of what the speaker does. From LNMA, the utterance of (1) would constitute a speech act of arguing that, from an illocutionary point of view, counts as an attempt by the speaker at showing that a target claim is correct, i.e., an attempt at justifying a target-claim. However, this Searlean characterization of the speech act of arguing only in terms of what the speaker does entails the two inter-related problems presented in section 3. Taking into account the insights of interactional approaches, we might now ask: does the securing of uptake exhaust what a speaker does when she utters "I promise I'll take care, so don't worry"? As I view it, it does not, and something else must be said. Thus, it is necessary to distinguish a second level of analysis, where not only the speaker, but also the hearer plays a fundamental role. Here is an initial list of the possible normative effects (i.e., changes in the deontic modal competence) that can be associated with the speech act of arguing:

- The hearer's legitimate expectation that the speaker can provide more reasons to justify the target-claim.
- The hearer's entitlement to ask for more reasons.
- The speaker's commitment to the truth of the implicit inference-claim.
- The speaker's obligation to provide more reasons if required to do so by the hearer.

Now that we have established a distinction between these two levels of analysis, what I would like to argue is that a proper way to characterize the phenomenon of argumentation at each level is to distinguish between two subtypes of illocutionary acts of arguing. Regarding the previous examples, we successfully carry out illocutionary acts of arguing of the first subtype when our utterance counts as an attempt at showing that a target claim is correct, regardless of the hearer's response. However, if we analyze argumentation from the second level, from the level of the communicative exchange, we can distinguish a second subtype of illocutionary act of arguing, which would be successfully carried out if certain normative effects are produced. To see this more clearly, consider the consequences of taking into account the hearer's response:

(3) a. I promise I will take care, so don't worry
b. Every time I've been in the car with you, I've seen you using the cell phone and exceeding the speed limit. So don't make promises that you won't keep.
c. Okay. Well, goodbye then.

In this case, if we only take into account (3a), it is correct to claim that the first subtype of illocutionary act of arguing has been performed: the speaker has secured the uptake of her utterance, so it counts as an attempt at showing that a target claim is correct. However, if we pay attention to the second level and consider the hearer's response (3b) as well, can we still say that the speech act of arguing has been successfully performed? In this case, by means of (3b), the hearer is raising doubts about the commitments that the speaker is trying to acquire when she says "I promise I'll take care". The speaker has secured the uptake, which can be seen in the hearer's response (3b), but the hearer's non-acceptance of her promise and, thus, the lack of production of the relevant normative effects, would make the speaker's act of arguing unsuccessful, at least in terms of the second subtype of illocutionary act. For instance, the hearer no longer has the legitimate expectation that the speaker is able to provide more reasons to show the correctness of the target-claim, nor the speaker is obliged to do so if required to by the hearer, etc.

5. Conclusions

In this paper, I have contended that, in order to account for the normative effects that acts of arguing introduce in the set of rights, obligations and entitlements of agents, it is necessary to distinguish between two different levels of analysis. Each of these levels is linked to different subtypes of illocutionary acts. The first subtype is successfully performed if the speaker just secures the uptake, while the successful performance of the second one depends on the normative effects introduced by the hearer's response.

I would like to conclude by saying that this analysis opens the possibility of investigating other features of the evaluation of argumentation. For instance, one can ask if it is only the speaker who has a responsibility for the argument she is putting forward, or if the hearer must be taken into account when attributing responsibility. This can be seen clearly in cases of harmful argumentation, e.g., racist arguments, where not only the speaker, but also the hearer's response can play a key role in sanctioning certain things as arguments worth debating or not, with important consequences for those whom such harmful speech is directed to. In this way, it would be interesting to investigate if the distinction between the two levels of analysis and the two subtypes of illocutionary acts could allow us to account not only for the responsibility that can be attributed to the speaker regarding the correctness of her argumentation (e.g., in cases of fallacious discourse), but also for the responsibility that can be attributed to the hearer regarding her interpretation of the speaker's act of arguing.

References

Austin, J.L. (1962). *How to Do Things with Words*. Oxford: Oxford University Press.
Bach, K. & Harnish, R.M. (1979). *Linguistic Communication and Speech Acts*. Mass, MIT Press, Cambridge.
Bermejo-Luque, L. (2011). Giving Reasons: A Linguistic-Pragmatic Approach to Argumentation Theory. Dordrecht: Springer.
Bermejo-Luque, L. (2019). Giving reasons does not always amount to arguing. *Topoi, 38*, 659-668.

Carassa, A. & M. Colombetti (2009). Joint Meaning. *Journal of Pragmatics, 41*, 1837-1854.
Clark, H. H. (1996). *Using Language.* Cambridge: Cambridge University Press.
Corredor, C. (2021). Illocutionary Performance and Objective Assessment in The Speech Act of Arguing. *Informal Logic, 41(3),* 453-483.
van Eemeren, F.H., & Grootendorst, R. (1984). Speech-acts in argumentative discussions. A theoretical model for the analysis of discussions directed towards solving conflicts of opinion. Foris, Dordrech/Mounton de Gruyter: Berlin.
Labinaz, P. (2021). Argumentation as a speech act: A (provisional) balance. *Croatian Journal of Philosophy, 21(63),* 357-374.
Sbisà, M. (2006). Communicating Citizenship in Verbal Interaction: Principles of a Speech Act Oriented Discourse Analysis. In H. Hausendorf & A. Bora (Eds.), *Analysing Citizenship Talk* (pp.151-181). Amsterdam: John Benjamins Publishing Company, 151-180.
Sbisà, M. (2009). Uptake and conventionality in illocution. *Lodz. Pap. Pragmat, 5(1),* 33-52.
Searle, J.R. (1969). *Speech Acts: an Essay in the Philosophy of Language.* Cambridge, Mass: Cambridge University Press.
Sluys, M. (2018). Getting the message and grasping it: the give-and-take of discourse. *Philosophia, 47(1),* 1-18.
Strawson, P. F. (1964). Intention and Convention in Speech Acts. *The Philosophical Review, 73(4),* 439-460.
Toulmin, S. E. (1958). *The Uses of Argument.* Cambridge University Press, Cambridge.
Witek, M. (2015). An Interactional Account of Illocutionary Practice. *Language Sciences, 47,* 43-55.

FORGET THE TOULMIN SCHEME, REMEMBER THE EPICHEIREME!

MIKA HIETANEN
Lund University
mika.hietanen@kom.lu.se

Abstract

The Toulmin model is reminiscent of the classical argument scheme called the *epicheireme*. Upon closer inspection they turn out to be fundamentally different. For the contemporary argumentation analyst or rhetorical critic, both schemes are burdened with drawbacks. The Toulmin model suffers from unclarity regarding definitions of its elements and the epicheireme from divergent traditions in classical times. Neither model offer criteria for evaluation. The *epicheireme* was devised for the production of speeches and Toulmin's model is descriptive. Even though Toulmin's model for half a century has been the main method of argumentation analysis in many disciplines, including Rhetoric, I argue that for some uses, the *epicheireme* is a better choice since (1) the elements are easier to use, (2) the *epicheireme* is more flexible, and (3) it naturally connects with neo-Aristotelian rhetorical criticism whereas the Toulmin model is based on a different theoretical concept.

1. The Toulmin Model and the Epicheireme

Philosopher Stephen Toulmin's (2003 [1958]) model (henceforth 'Toulmin') for argumentation shows no signs of diminishing popularity and continues to be a staple for argumentation analysis within many fields such as Educational Sciences, Psychology, Critical Thinking, and Literacy studies (e.g. Andriessen & Baker, 2014; Inglis et al., 2007; Inch & Warnick, 2011; Weinberger & Fischer, 2006). The model has even been described as 'a standard account of the components and structure of single arguments' (Breivik, 2020, p. 11).

In contrast, the *epicheireme* is not even mentioned in most textbooks on rhetoric or argumentation. Instead, many favour Toulmin. Within the academic field of Rhetoric in Scandinavia, Toulmin is the most taught method for argumentation analysis (via Jørgensen & Onsberg, 2008 & 2017).

Although Toulmin's model or scheme is similar to the epicheireme, the latter is all but forgotten, except for some periods of re-discovery, e.g. 1964–1967 in *Western Speech* where it was discussed without reaching any consensus on its use or nature. The title of Feezel's (1967) essay is symptomatic: 'The Mystery of the Epicheireme.'

In 1968, the first study comparing the epicheireme with Toulmin appeared when Trent (1968) noted that Toulmin's concept 'is basically an enthymeme or epicheireme (a supported enthymeme) [...] described in detail by Cicero' (p. 254). The novelty in Toulmin's model, Trent noted, was the concept of probability in the qualifier and rebuttal (p. 254).

The epicheireme resurfaces when exegetes re-discover classical rhetoric. For instance, based on Cicero, Long demonstrates that the apostle Paul's argumentation is sometimes 'epicheirematic in form' (Long, 2003, p. 697) and Reid (2005) indicates Paul's use of specifically the ad Herennial epicheireme. New testament exegetes have been interested in Toulmin, also (Thurén, 1995; Hietanen, 2002).

I suggest three reasons for the limited interest in the epicheireme. First, the brief and unclear or even contradictory descriptions in the primary sources make it challenging to understand what the epicheireme is. Second, it seems difficult to use and there are not many example applications. Third, it is easier to turn to a modern model clearly presented in textbooks.

Even authors specifically introducing 'classical rhetoric,' usually resign trying to formulate a classical approach when turning to the subject of argumentation analysis and turn to modern and fairly simple models, such as Venn-diagrams or Toulmin (Lindqvist Grinde, 2008; Fleming, 2014).

My objective is to clarify the differences between Toulmin and the epicheireme, mainly from the perspective of their usefulness for rhetorical criticism and for practical argumentation analyses in general. My questions are: (1) What are the differences between the epicheireme and the Toulmin model? and (2) Should we continue to favour the Toulmin model over the epicheireme?

2. The Toulmin Model

2.1. General Introduction

Toulmin's model is a countermove against the dominating formal conceptions at the time. His interest in validity permeates his *The Uses of Argument* (2003 [1958]) where he concludes that formal validity does not reveal the soundness of argumentation. He rejects the idea of universal criteria for validity and suggests that evaluation criteria should relate to the field or subject concerned. Twenty years later, the popularity of his scheme resulted in the textbook *An Introduction to Reasoning* (1984 [1978]), offering minor revisions of the method.

Ehninger and Brockriede (1960) noted that Toulmin is better suited to practical argumentation analysis than traditional logical models. The initial popularity of Toulmin can be traced to their textbook *Decision by Debate* (1963; according to van Eemeren et al. 1996, p. 150).

The model has been influential for the Informal logic movement as well as for different strands within communication studies and argumentation theory. It is a staple on reading-lists on argumentation (van Eemeren et al., 2014, chapter 4).

In the following, I go through the elements of the model, focusing on their definitions. For the model, see Fig. 1.

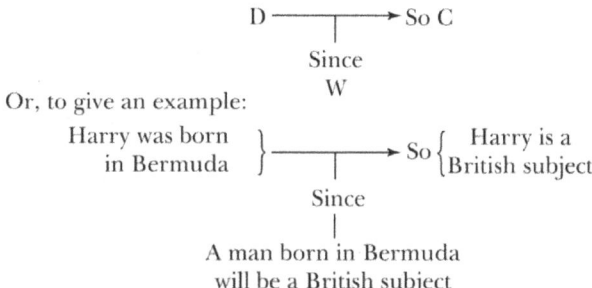

Figure 1. The Toulmin scheme and an example (Toulmin, 2003, p. 94)

2.2. The Claim

Toulmin's claim is synonymous with a standpoint: 'assertions put forward publicly for general acceptance' (Toulmin, 1984, p. 29).

2.3. The Datum (Grounds)

The datum (sg.) or data (pl.) – later labelled 'grounds' (Toulmin et al., 1984) – answer the question: 'What do you have to go on?' referring to 'the *specific* facts relied on to support a given claim;' facts 'that can be agreed on as a secure starting point acceptable to both sides, and so "not in dispute"' (p. 38).

Grounds are thus similar to *pisteis atechnoi*, non-artistic proofs, which are neither contested nor part of the orator's *inventio* (i.e. the 'art' of rhetoric; *Ars Rhetorica*, 1,1,3 in Bartlett, 2019; '*Rh.*').

Evaluation criteria for grounds centre around whether they truly fulfil the requirement of burden of proof; the 'facts' used may be 'slim and shaky' (Toulmin et al., 1984, p. 40). What grounds are acceptable depends on the field in question and on the situation. Toulmin does not develop any actual system of evaluation.

2.4. The Warrant

Toulmin et al. (1984) describe the warrant as (1) 'what entitles us to be confident that, in this particular case, the step from grounds to claim is a step of a generally reliable sort' (p. 47) and (2) helping us see the relevance of the grounds since it will otherwise not be possible to formulate a warrant that connects them with the claim. They give the following example: "'It's my turn" (*A* [refers to *G*]) – "The one whose turn it is should choose" (*W*) – "So I should choose" (*C*)' (p. 50).

A fundamental, and in 1958 radical, aspect is that the backing does not contain any of the information that appears in the conclusion, but the argumentation is still considered valid in an informal sense (van Eemeren et al. 2014, p. 226).

Two questions can be asked about a warrant: (1) 'Is that warrant reliable at all?' and (2) 'Does that warrant really apply to the present specific case?' (Toulmin et al., 1984, p. 62). Furthermore, 'the difference between grounds and warrants (facts and rules) is a *functional* difference' (p. 47). Also, warrants look different in different contexts, such as in science and engineering, in law and ethics, in medicine, and in aesthetics and psychology (pp. 50–55).

Finally, the grounds should be explicit factual information whereas the warrant should be implicit (Toulmin 2003, p. 92).

2.5. The Backing

Since data are uncontested facts, they do not need further support, but the warrant may need support. Toulmin et al. (1984) explain: 'It is one thing to state a warrant, but it is quite another thing to show that it can be relied on as sound, relevant, and weighty. This is particularly the case, if there are several possible ways to connect *G* and *C*, which support conflicting claims' (p. 62). In brief, *'warrants are not self-validating'* (p. 62). The purpose of the backing is to produce '"further, substantial support, all considerations" – and so demonstrate that [...] [the] warrant is sound and relevant' (p. 63).

Not only the warrant, but also the backing looks different in different fields. In science, the backing may point to known research-results, in law to historic rulings and statutes, and in sports to historic facts.

2.6. The Qualifier and Rebuttal

By adding a modality to the claim, and by including exceptions, a more precise claim can be presented. Although a qualifier and a rebuttal give added value to the Toulmin model, the four basic elements, claim, datum/grounds, warrant, and backing, can be used independently and irrespectively of qualifier and rebuttal.

3. The Epicheireme

3.1. General Introduction

The epicheireme has developed from the 4th century BCE onwards (Klein, 1994, pp. 1252–1256). Freezel (1967) notes that Cicero (1949) in *De Inventione* (c.85 BCE, '*de Inv.*'), Quintilian (2001) in *Institutio Oratoria* (c.95 CE, '*Inst.*'), and the author of *ad Herennium* (mid 80's BCE; '*ad Her.*;' Caplan, 1954), 'appear to be the only rhetoricians who have treated the epicheireme extensively' (p. 109).

Contrary to its current state, the epicheireme was 'the cornerstone of rhetorical argument for over fifteen centuries' (Church & Cathcart, 1965, p. 147). It was then all but forgotten and its re-discovery in the 1960's did not rehabilitate it for the post-modern world.

Deductive argumentation has three forms: *syllogismus, enthymema,* and *epicheirema* (Lausberg, 2008; 'Lausb.' §372). It is often said that the syllogism is the logically complete form of *ratiocinatio*,[1] and the *enthymeme* the incomplete form since the latter typically has one implicit premiss. More correctly, they have different uses: the syllogism for more formal contexts and the enthymeme for public argumentation (*Rh.* 1,2,13).

Some researchers describe the epicheireme as some kind of 'formal, rhetorical syllogism' (Meador, 1966, p. 154), but Braet (2004, p. 328) argues that the epicheireme originally was a stylistic device and that the syllogistic understanding only played 'a marginal role.' Rather, 'the dominant meaning was: a form of argumentation that stylistically amplified the corresponding enthymeme' (p. 328).

Freezel (1967) seems right in that there were 'two different concepts of the epicheireme standing parallel in time' (p. 111): the one of *ad Her.* and the one of Cicero. This is the conclusion of Klein (1994), also. The purpose is basically the same: to explicate an argumentation for an audience in a memorable and clear manner. The ad Herennial variant is embellished, the Ciceronian more stringent and closer to Toulmin.

---

[1] In classical sources, *ratiocinatio* can signify the syllogism, the enthymeme, or the epicheireme or rhetorical deductive reasoning in general (Lausb. §371). Aristotle and Quintilian use *epicheirema* synonymously with *enthymema* (*Rh.* 1,2,13; *Inst.* 5,10,4).

3.2. Cicero's Epicheireme

Cicero discusses the epicheireme in *de Inv.* where it has five parts, defined as follows (quotes from *de Inv.*, 1, 37; for variations in spelling and form, see Klein, 1994 and Kienpointner, 1992).

The proposition (*propositio*): 'by which a topic is briefly explained from which all the force of the ratiocination ought to proceed'; the main point of the argumentation (not to be confused with the *propositio* in the rhetorical *dispositio* where it signifies the standpoint).

The proof of the proposition (*approbatio propositionis*): 'by which that which has been briefly set forth being corroborated by reasons, is made more probable and evident'; i.e. support of the *propositio* by various reasons.

The assumption (*adsumptio*): 'by which that is assumed which, proceeding from the proposition, has its effect on proving the case'; i.e. that which must be demonstrated.

The proof of the assumption (*approbatio adsumptionis*): 'by which that which has been assumed is confirmed by reasons'; i.e. support for the *assumptio*.

The conclusion (*complexio*): 'the summing up, in which that which results from the entire argumentation is briefly explained'; a conclusion of the argumentation, possibly preceded by a repetition.

Since Cicero's discussion is concerned with practice, his description needs not be more precise than what an orator has need for. He gives a lengthy example (in English 512 words), indicating that the epicheireme is much more than a concise standpoint–arguments unit, but includes rhetorical elaborations and details.

The epicheireme can sometimes be reduced to four or even three parts. Unnecessary proofs should be withheld, e.g. 'when either the proposition is understood by its own merits, or when the assumption is self-evident and is in need of no proof' (*de Inv.* 1,39). Cicero provides examples of the omission of the proof, of the proposition, of the proof of the assumption, and of both (1,39) but discourages to omit the conclusion or the proof (1,60).

3.3. The Epicheireme's Graphical Representation

An attraction of Toulmin's model is its graphical representation. There is none available in the original sources, but I believe Kienpointner has captured the structure of the Ciceronian epicheireme (Fig. 2, my translation).[2]

---

[2] For other suggestions as to the graphical representation of the epicheireme, see Trent, 1968, p. 256; van Eemeren et al., 1996, p. 48; Santibáñez, 2010, pp. 96–97, 100–101.

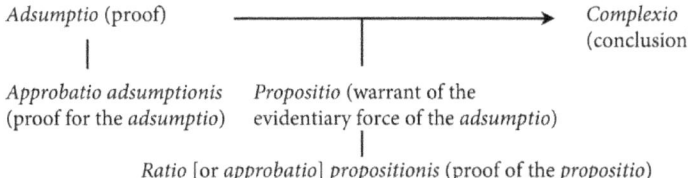

Figure 2. Kienpointner's (1992, p. 891) graphical representation of the epicheireme

In this scheme, the thesis is represented by the *complexio,* with the *propositio* placed at the heart of the argumentation. Kienpointner (1992, p. 891) explains, with a quote from *de Inv.* 1,67 (my translations):

> Within the frame of this argument scheme, Cicero characterizes the *locus* as a formula of proof *(propositio),* from which the power of the proof emanates: 'the formula of proof, through which the Locus is briefly expressed, from which the whole power of proof of the conclusion should emanate' (p. 891).

Locus here refers to Cicero's system of *loci* (Gr. *topoi*), a set of argument types for the speaker to choose from. Connecting the epicheireme with the *topoi* gives the epicheireme considerable depth.

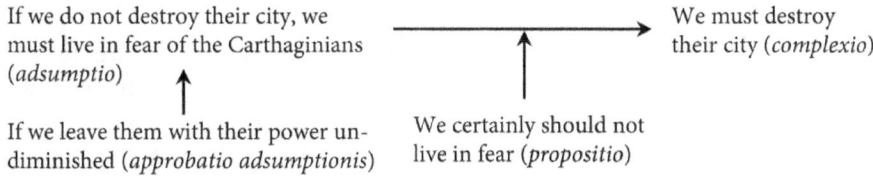

Figure 3. Example of the Ciceronian epicheireme from *de Inv.* I,39 (adjusted for clarity)

Fig. 3 is based on Cicero's example. Here, the non-artistic proof makes up the *adsumptio* that the Romans must live in fear of the Carthaginians, unless the *complexio* is acted upon. The *propositio* is based on the *topos* of the desirable/undesirable; nobody wants to live in fear.

For Cicero this is a three-part epicheireme, but it includes a support for the *adsumptio* which corresponds to an *approbatio adsumptionis* wherefore I have explicated it as a four-part epicheireme. Cicero may have seen it as a paraphrase of the *adsumptio* and not as a separate element. It is somewhat 'messy' that two parts are if-phrases, but Cicero's explication is not always clear-cut (Braet, 2004, pp. 329–334). Clear, though, is how the *propositio* is crucial; it is on the idea of living in fear that the argumentation rests.

Cicero's example for a *complete* epicheireme (*de Inv.* 1,34) contains no less than three arguments supporting the *ratio* and three supporting the *adsumptio* in what for him is 'one' argumentation, indicating the scope considered natural for the ancient epicheireme.

4. Comparing the Elements of the Two Models

4.1. Structure

I suggest that for the sake of clarity we should not talk about an epicheireme unless the *propositio* or the *adsumptio* has further support; otherwise it is rather an enthymeme. Also, a supported premiss should not be implicit since that amounts to an enthymeme. In classical literature these distinctions are not made and often enthymemes are called epicheiremes and *vice versa* (Lausb. §371).

Following my suggestion, we have six variants: with three parts: (1) *complexio* with an implicit *propositio* and a supported *adsumptio*; (2) *complexio* with an implicit *adsumptio* and a supported *propositio*; with four parts: (3) an implicit *complexio* with both *propositio* and *adsumptio* supported; (4) *complexio* with a *propositio* and a supported *adsumptio*; (5) *complexio* with a supported *propositio* and an *adsumptio*; and with five parts: (6) all parts explicit.

The epicheireme and the enthymeme are connected: if we let the *propositio* or the *adsumptio* – or both – be implicit when they are supported, we have one or two enthymemes. I find it better to break up such structures into enthymemes rather than to try to analyse them as a 'fat' epicheireme of sorts.

In Toulmin, the minimum number of parts is three (datum and claim), with an implicit warrant. When including backing, qualifier, and rebuttal, the maximum number of parts is six. The Toulmin model has eight possible variants (configurations with or without B, Q, R).

4.2. The Nature of the Parts

In rhetoric, premisses *can* be true, but in general, rhetorical argumentation rests on matters that have no truth-value, such as policy decisions. When premisses are only probable, the conclusion is that also. The fundamental difference with Toulmin is that a rhetor may use artistic or non-artistic proofs for any of the premisses or sub-premisses. The non-artistic proofs are undisputed although technically only highly probable. Since they are accepted by the audience – often part of their *doxa* – they in practice function as facts.

Rhetorical argumentation typically builds upon something acknowledged in order to move the audience towards something not originally acknowledged (*Inst.* 10,5,11–12). A rhetor uses all the possibilities of artistic proofs to achieve this: logos (*signa, argumenta, exempla*; *Inst.* 5,9,1; Lausb. §357), *ethos*, and *pathos*.

For Toulmin, data are undisputed facts and as such equivalent to non-artistic proofs. The difference is that the datum must both be explicit and in the position of the minor premise. The warrant is general, rule-like, and implicit. Although typically this holds true for the major premiss of rhetorical argumentation also, it does not have to do so.

The backing bears a resemblance to grounds: both are facts related to the case in question and both are similar to the *pisteis atechnoi*, but whereas data are the immediate facts on which the claim rests, the backing provides support for the inference taken from datum to claim.

Toulmin (1984) argues that the two extensions to his model, the qualifier and the rebuttal, 'are distinct both from data and from warrants, and need to be given separate places in our layout' (p. 93). In rhetorical argumentation both probability and modality are subtly included in the arguments themselves since rhetorical arguments are by nature probable. Trent (1968, p. 254) noted that 'the information contained in Toulmin's rebuttal would, in a traditional argument, be included in the support for the major premise (Toulmin's backing) and/or in the major premise (Toulmin's warrant).'

Consequently, Toulmin's additions do not necessarily present an advantage over traditional rhetorical models. From a rhetorical point of view: why would a rhetor indicate the amount of uncertainty in a claim by introducing a qualifier or specifically mention a rebuttal that would make the claim unsuccessful? That task falls on the antagonist.

4.3. Chaining

A speech or a text typically includes several main arguments. Regarding enthymemes and epicheiremes, conclusions can become minor premisses for the next argument. In the Toulmin model, also, the claim can become a datum for the next level of argumentation. This is not unproblematic since a datum is supposed to be undisputed, but a claim not.

With the epicheireme this is not an issue since the premisses only need to be probable. This probability can be achieved by previous argumentation: by successfully arguing a standpoint it can then be used as a premiss for another standpoint.

The problem with chaining in Toulmin is that even a successfully argued claim does not become a fact. Perhaps the change in terminology from data to grounds in part reflects an attempt to solve this problem? A successfully argued claim can certainly be used as *grounds* for another argumentation.

4.4. Original Purpose and Evaluation

The Toulmin model does not include specific criteria for evaluation. In its original form it can therefore only be used for descriptive analyses. Since evaluation is the goal of most analyses, Toulmin needs to be used in one of its further developments. Also, it is not geared towards production. The epicheireme, on the other hand, is specifically intended for production. For this reason it also does not originally include criteria for evaluation and, when used for analysis, needs to be complemented. On this point, both methods have a comparable weakness.

4.5. Goal

In rhetoric, the goal is *pistis* or persuasion. For this, argumentation needs not be logically complete. It is enough that it is compelling. When comparing the examples in Toulmin and in Toulmin et al. on the one hand, and the classical rhetorical handbooks on the other, the differences are striking. The former are neatly arranged, including both qualifier and rebuttal, which make the arguments seem complete. Toulmin's examples are textbook type ('Petersen is a Swede ...'). The epicheireme examples are different. For the most part, they are rhetorical speeches which include unnecessarily many words and therefore do not adhere to any ideal of rational clarity, but to a rhetorical mode of expression. Certainly both Cicero's and *ad Herennium's* examples are artificial textbook examples, also, but the difference with Toulmin is significant. Rhetoric and argumentation theory are different perspectives, due to different goals.

5. Concluding Discussion and Summary

Toulmin has been interpreted differently by researchers with different backgrounds: as rhetorical (van Eemeren and Grootendorst, 2004, p. 47) or as non-rhetorical (Santibáñez, 2010). A crucial difference, which makes Toulmin non-rhetorical, concerns probability. In rhetoric, arguments are probable by nature and the information in Toulmin's rebuttal would be included either in the support for the *propositio* or in the premiss itself. Toulmin considers the datum to be uncontested and places the probability in the rebuttal or qualifier.

When a claim from a lower-level argumentation becomes the datum for a higher level of argumentation, its nature changes, creating a tension in the model. It is difficult to reconcile the factual nature of the datum with the probability of the claim ('facts of the case' vs. 'assertions put forward publicly for general acceptance;' Toulmin et al., 1984, pp. 29, 38).

Furthermore, the similarity of the datum and the backing complicates analysis since it may be difficult to know what is what in the analysed artefact; this distinction is sometimes difficult to make. Van Eemeren et al. (1996, p. 159) suggest that Toulmin is not consistent in his description of the datum, warrant, and backing. The risk of confusion between datum and backing has been indicated also empirically (Voss, 2006).

As regards Toulmin's idea of field-dependency, it has not materialized into any commonly used model for evaluation based on different fields although there are suggestions (already mentioned, Ehninger and Brockriede, 1960, Trent, 1968, and Toulmin et al., 1984 [1978] as well as several in Hitchcock & Verheij, 2006).

It seems to me that the popularity of Toulmin's scheme rests on three aspects. It is one of the first modern successful attempts at describing real-life argumentation that does not rest on formal validity. It better than logical models reflects argumentative reality. Second, the graphical presentation is appealing. Third, compared to other advanced contemporary models, such as Pragma-Dialectics, the learning-curve is low. However, due to unclarities of the method, initial time savings may be lost in application.

From the point of view of rhetoricians, I conjecture that the popularity of Toulmin is in some part due to the various insufficient and confusingly varying descriptions of classical alternatives.

One of the most substantial contributions by Informal logic to the analysis of argumentation are attempts to introduce criteria for evaluation. Toulmin's suggestion is almost a repetition of Aristotle's understanding of *doxa*: to rely on experts for what is acceptable (*Topica* 1,1, Aristotle, 1960; Toulmin et al. 1984, Chapter 25). Consequently, in the one area where modern theories can assert themselves with regard to classical rhetoric, evaluation, Toulmin does not. That some contemporary textbooks have tried to complement the method with criteria for evaluation does not in theory favour Toulmin over the epicheireme since such attempts could be made with regard to classical rhetoric as well.

What, then, are the differences between the epicheireme and the Toulmin model? The differences are fundamental, see Table 1.

Table I. Differences between the epicheireme and the Toulmin model

	The Epicheireme	The Toulmin model
Theoretical framework	Classical rhetoric (further: dialectics)	Informal logic (further: Philosophy of language)
Original purpose	Production of speeches	Inquiry regarding what elements are field-dependent in an argumentation; also analysis of everyday argumentation
Minimum elements	Three: explicit thesis + two arguments *or* implicit thesis with two explicit arguments and a support for one of the arguments	Two: claim + datum (with warrant implicit)
Maximum elements	Five: thesis, two arguments each supported by an argument	Six: claim, datum, warrant, backing, qualifier, rebuttal
Truth-value	Parts that are *pisteis entechnoi* are probable, parts that are *pisteis atechnoi* are regarded as *doxa*	Warrant probable (may need backing), datum true or false (uncontested facts)
Expression of probabilities	Probability inherent in each *pistis entechnos* whereas *pisteis atechnoi* are considered uncontested. Factors that can lessen the credibility of the thesis are not expressed.[3]	Datum expresses uncontested facts. Factors that can lessen the credibility of the claim are expressed in a rebuttal and/or a qualifier.
Criteria for evaluation	No developed system. What persuades the audience is good argumentation.	No developed system. Good argumentation is accepted by experts within the relevant field.
Further developments	A few, but they are mostly not generally known or in use.	Several theoretical ones but only a few concerning the model as an instrument for analysis, and not in general use.

Second, should we continue to favour the Toulmin model over the epicheireme? The answer is contingent on two factors: the discipline in question and the purpose of the analysis. The best method is the one that lets us successfully answer our research questions in relation to the artefact we wish to study, within the limits of the study. Sometimes Toulmin is a better choice, sometimes the epicheireme.

In general, for fields within the Humanities, which retain a connection with classical rhetoric, the epicheireme is for historical reasons a good choice – sometimes in its minimum form, the enthymeme. Within the actual discipline of Rhetoric, the epicheireme and the enthymeme are

---

3 In contrast to the rebuttal, a *refutatio* raises a counterargument and refutes it while the rebuttal contradicts the claim.

obvious choices since they organically connect with a rhetorical understanding of probabilities as expressed in the system of the *pisteis*, as well as with the *topoi*. For other fields also, an enthymeme or epicheireme analysis in many cases provides the required results. If evaluation is an important consideration, none of these methods are by themselves sufficient.

Even though Toulmin's model for half a century has been the main method of argumentation analysis in many disciplines, I argue that the *epicheireme* is often a better choice since (1) the elements are easier to use because of how they are defined, (2) the *epicheireme* is more flexible, and (3) it naturally connects with neo-Aristotelian analysis whereas the Toulmin model is based on a different theoretical concept.

References

Andriessen, J. & Baker, M. (2014). Arguing to learn. In R.K. Sawyer (Ed.), *The Cambridge Handbook of the Learning Sciences* (pp. 439–460). Cambridge Univ. Press. https://doi.org/10.1017/CBO9781139519526.027

Aristotle. (1960). *Topica*. Transl. E.S. Forster. LCL 391. Harvard Univ. Press.

Bartlett, R.C. (2019). *Aristotle's Art of Rhetoric*. Univ. of Chicago Press.

Braet, A.C. (2004). Hermagoras and the Epicheireme. *Rhetorica, 22*(4), 327–347. https://doi.org/10.1525/rh.2004.22.4.327

Breivik J. (2020). Argumentative patterns in students' online discussions in an introductory philosophy course. *NJDL, 15*(1), 8–23. https://doi.org/10.18261/issn 1891 943x 2020-01-02

Caplan, H. (transl.). (1954). *Ad C. Herennium* [Anonymous]. LCL 403. Harvard Univ. Press. https://10.4159/DLCL.marcus tullius cicero-rhetorica ad herennium.1954

Cicero. (1949). *On Invention. The Best Kind of Orator. Topics*. Transl. H.M. Hubbell. LCL 386. Harvard Univ. Press.

Church, D.A. & Cathcart, R.S. (1965). Some Concepts of the Epicheireme in Greek and Roman Rhetoric. *Western Speech, 29*, 140–147.

van Eemeren, F.H., Grootendorst, R., & Henkemans, F.S. (Eds). (1996). *Fundamentals of argumentation theory*. Erlbaum.

van Eemeren, F.H. and Grootendorst, R. (2004). *A Systematic Theory of Argumentation*. Cambridge Univ. Press.

van Eemeren, F.H., Garssen, B., Krabbe, E.C.W., Henkemans, A.F.S., Verheij, B. & Wagemans, J.H.M. (2014). *Handbook of argumentation theory*. Springer. https://doi.org/10.1007/978-90-481-9473-5.

Ehninger D. & Brockriede W. (1963). *Decision by debate*. Dodd Mead.

Fleming, D. (2014). Rhetoric and Argumentation. In G. Tate, A. Rupiper Taggart, K. Schick, & H. B. Hessler (Eds.), *A Guide to Composition Pedagogies,* 2nd ed. (pp. 248–265). Oxford Univ. Press.

Freezel, J. D. (1967). The Mystery of the Epicheireme. *Western Speech, 31*, 109–115.

Hietanen, M. (2002). Profetian är primärt inte för de otrogna – en argumentationsanalys av 1 Kor 14:22b [Prophecy is not Primarily for Non-

Believers – an Argumentation Analysis of 1 Cor. 14:22b]. *Svensk Exegetisk Årsbok, 67,* 89–104.

Hitchcock, D.L. & Verheij, B. (Eds.) (2006). *Arguing on the Toulmin model.* Springer. https://doi.org/10.1007/978-1-4020-4938-5

Inch, E.S., & Warnick, B. (2011). Critical Thinking and Communication: The Use of Reason in Argument. Allyn & Bacon.

Inglis, M., Mejia-Ramos, J.P., & Simpson, A. (2007). Modelling mathematical argumentation. *Educ Stud Math, 66*(1), 3–21. https://doi.org/10.1007/s10649-006-9059-8

Jørgensen, C. & Onsberg, M. (2008). *Praktisk argumentation. [Practical Argumentation].* 3rd ed. Nyt Teknisk Forlag.

Jørgensen, C. & Onsberg, M. (2017). *Praktisk argumentation. [Practical Argumentation.].* Transl. O. Vigsø. 2nd ed. Retorikförlaget.

Kienpointner, M. (1992). Argument. In G. Ueding with W. Jens et al. (Eds.), *HWRh, 1* (pp. 889–904). Max Niemeyer.

Klein, J. (1994). Epicheireme. In G. Ueding with W. Jens et al. (Eds.), *HWRh, 2* (pp. 1251–1258). Max Niemeyer.

Lausberg, H. (2008). *Handbuch der literarischen Rhetorik.* 4th ed. Franz Steiner.

Lindqvist Grinde, J. *Klassisk retorik för vår tid.* [Classical Rhetoric for Our Times]. Lund: Studentlitteratur, 2008.

Long, F.J. (2003). 'We Destroy Arguments ...' (2 Corinthians 10:5): The Apostle Paul's Use of Epicheirematic Argumentation. In F.H. van Eemeren, J.A. Blair, C.A. Willard, & F.S. Henkemans (Eds.), *Proceedings of the Fifth Conference of ISSA* (pp. 697–703). Sic Sat.

Meador, P. A. Jr. (1966). The Classical Epicheireme: A re-examination. *Western Speech, 30,* 151–155.

Quintilian. (2001). *The Orator's Education.* Transl. D.A. Russell. LCL 126. Harvard Univ. Press.

Reid, R.S. (2005). Paul's Conscious Use of the Ad Herennium's 'Complete Argument.' *RSQ, 35*(2), 65–92.

Santibáñez Y.C. (2013). Retórica, dialéctica o pragmática: a 50 años de 'Los usos de la argumentación' de Stephen Toulmin [Rhetoric, dialectic, or pragmatic. 50 years after 'The Uses of Argument' by Stephen Toulmin]. *clac, 42,* 91–125.

Thurén L. (1995). Argument and Theology in 1 Peter. JSOT Press.

Toulmin, S.E. (2003 [1958]). *The uses of argument.* Cambridge Univ. Press, 2003. Orig. publ. 1958. https://doi.org/10.1017/CBO9780511840005

Toulmin, S., Rieke, R., & Janik, A. (1984 [1979]) *An introduction to reasoning.* Macmillan.

Trent, J.D. (1968). Toulmin's model of an argument: An examination and extension.' *QJS, 54*(3), 252–259. https://doi.org/10.1080/00335636809382898

Voss, J. F. (2005). Toulmin's Model and the Solving of Ill-Structured Problems. *Argumentation, 19,* 321–329. https://doi.org/10.1007/s10503-005-4419-6

Weinberger, A. & Fischer, F. (2006). A framework to analyze argumentative knowledge construction in computer-supported collaborative learning. *Computers & Education, 46*(1), 71–95. https://doi.org/10.1016/j.compedu.2005.04.003

CULTIVATING NORMATIVE TERRAINS IN INFORMATION ECOLOGIES

BETH INNOCENTI
University of Kansas
bimanole@ku.edu

Abstract
How do arguers cultivate information ecologies in order to get addressees to acknowledge experiential truths, and why do they expect the strategies to work? Addressing the question is important because socially dominant groups may derail discussions of experiences by historically marginalized groups. Getting social actors to recognize truths outside of their own experiential purview may make information ecologies healthier, e.g., expand considerations in deliberation. How? By cultivating normative terrains that contour ongoing public deliberation.

1. Introduction

How do arguers cultivate information ecologies in order to get addressees to acknowledge experiential truths, and why do they expect the strategies to work? Addressing the question is important because experiences can ground argumentation (Hundleby, 2021) and because social groups whose experiences define the "mythical norm" (Lorde, 1984, p. 116) may derail discussions of experiences by historically marginalized social groups (Innocenti, 2022; Peach, 2020). I submit that arguers use strategies to cultivate normative terrains that contour ongoing public deliberation.

To make the case, I first broadly and schematically explain how social actors may cultivate normative terrains in information ecologies. I then use the explanatory framework (i.e., normative pragmatics) to analyze a case where a member of a historically marginalized group—Black women in the United States—cultivated a normative terrain designed to get a socially dominant group—including middle-class White women—to recognize and acknowledge Black women's experiences in broader public deliberations about women's political movement. I conclude that normative pragmatic accounts can deepen cognitive accounts of argumentation.

2. Cultivating normative terrains

I use the phrase "normative terrain" (e.g., Goodwin, 2007) to refer to normative materials—e.g., principles, obligations, norms, responsibilities—that are "out there" and brought to bear in specific times and places by using communication strategies. For example, installing a Stop sign at an intersection cultivates the local normative terrain. How? A principle to not endanger others' lives is already "out there"; for example, car drivers should not drive on sidewalks or while intoxicated. Installing a Stop sign cultivates a local normative terrain where all can hold each other accountable for living up to the principle in a specific way at a specific time and place, i.e., by stopping at the intersection.

The keystone is the accountability for all—addressees and message sources—to live up to normative materials brought to bear just by using strategies. First, by installing the Stop sign, city officials (i.e., the message source) put themselves in a position to be held accountable for not endangering lives. Installing a Stop sign regardless of traffic patterns could produce dangerous conditions, such as faster driving between Stop signs, ignoring the Stop sign, and a false sense of security for pedestrians. So installing the Stop sign creates a reason for drivers to stop: drivers can reason that city officials would not risk criticism for endangering lives unless they had made responsible efforts to determine that installing the sign at the location was prudent. Second, installing a Stop sign puts drivers in a position to be held accountable for not endangering lives. In the absence of a Stop sign, or if the sign is obscured by tree branches or is poorly designed (e.g., difficult to read due to the small size of the letters or low color contrast), drivers could not easily be held accountable for failing to live up to the principle of not endangering others' lives by stopping. So installing a well-designed Stop sign creates a second reason for drivers to stop: to avoid criticism for endangering others' lives.[1] Of course the

---

[1] Driving contexts supplement this ethical reason with formal sanctioning mechanisms, such as paying a fine, and formal accountability mechanisms, such as needing to pass a driver's license examination. Context matters. For example, consider a public health context and specifically a global pandemic. A Mask Required sign on a train may not reliably secure the intended response. First, train passengers may calculate that the local government is vulnerable to public criticism for not having made responsible efforts to determine that wearing a mask on public transportation is prudent. Second, train passengers may calculate that they can disregard the sign with impunity because they will not be held accountable for living up to a principle to not endanger others' lives specifically by wearing a mask on public transportation. Indeed, in the absence of formal sanctioning mechanisms, and in the presence of widespread public skepticism about mask mandates and distrust in government, they may calculate correctly.

message designs could be expected to work only if a principle to not endanger others' lives is "out there" and salient.

This account of the persuasive force of message designs has been described as "normative pragmatic" because it accounts for the practical efficacy of messages in normative terms (e.g., Goodwin, 2000; Jacobs, 2000; Kauffeld, 1998).[2] It covers both macro-level designs, such as communication systems and structures, and micro-level designs of specific messages (Aakhus, 2013; Goodwin, 2003; Innocenti & Miller, 2016; Jackson, 2015; Jacobs, 2006; Kauffeld, 1994; Kauffeld, 2009a; Vasilyeva, 2015, 2021). Like any account, it is partial. However, as philosophers of social science have noted, "a host of studies show that the main variable affecting behavior is not what one personally likes or thinks he should do, but rather one's belief about what 'society' (i.e., most other people, people who matter to us, and the like) approves of" (Bicchieri, 2017, p. 10; see also Brennan et al., 2013, p. 37).

In what follows I use the normative pragmatic framework to explain potentially efficacious communication designs for cultivating normative terrains in information ecologies.

3. Analysis

3.1. Text and context

To address the question of what strategies Black women in the U.S. use to constrain White middle-class women from ignoring or discounting Black women's experiences and perspectives in deliberations about women's political movement, and why they can reasonably be expected to work, I analyze an exemplary message designed to do just that: Toni Morrison's 1971 essay in the *New York Times Sunday Magazine* entitled "What the Black Woman Thinks about Women's Lib." Morrison's essay has been described as an "unequivocal" and "damning" critique of White middle-class feminism and the women's liberation movement (Roynon, 2013, p. 98; Lumsden, 2009, p. 120). Morrison's (1971) essay covers both "the question of color" and "the question of the color of experience" that explain Black women's worldview. Although I have chosen an historical case, scholars have documented that the rhetorical problem of getting White women to recognize and acknowledge experiential truths of Black women is ongoing (e.g., Cooper, 2018; Coles & Pasek, 2020; Lindsey, 2020; Johnson-Bailey, 2003; Strongman, 2018; Taylor, 1998; Taylor, 2020).

---

2 See Kauffeld (2009a) for an account of the Gricean philosophy of language that underpins normative pragmatic accounts, and Gray & Lennertz (2020) for discussion of the need for Gricean analysis to attend to sociopolitical contexts.

Morrison's communication designs merit critical attention. Her essay displays qualities that would later earn her Pulitzer and Nobel Prizes for Literature. The *New York Times,* and the Sunday edition in particular, was then and continues to be a widely-circulated, leading national newspaper, marketed as a serious, legitimate alternative to news sources that include sensationalized stories (Campbell, 2012).[3] It is certainly addressed to readers who want to lay some claim to holding well-informed beliefs and opinions and staying current on various topics. However, as was true for other major newspapers, the perspective of the *New York Times* as a whole was White, male, and middle-class.[4]

Morrison's essay contributed to widespread deliberations about women's political movement. Morrison concludes the essay by referring to the inaugural meeting of the National Women's Political Caucus (NWPC), an organization designed to deliberate about and get more women into positions of political leadership. At the meeting, Black women successfully pushed for the group to disavow racism and not endorse racist candidates.[5] However, news sources reported the U.S. President and top administration officials at the time joked about a picture of NWPC organizers looking like a "burlesque" and more (e.g., Kilpatrick, 1971; "Never Underestimate," 1971). Certainly there were needs for women's political movement to disavow racism and be taken seriously.

About macro-level design I only note that publishing the essay as a cover story in the *New York Times Sunday Magazine* puts a wide swath of middle-class White women in a position to be held accountable for recognizing and acknowledging Black women's view of women's liberation. The national significance and reach of Morrison's essay can be gleaned from the fact that press releases about the essay were published in Black newspapers across the U.S.[6] Of course Morrison's essay was not the only

---

3 In late October 1971 the New York Times reported Sunday circulation of 1,412,017, and a daily circulation of 814,290 (New York Times, 1971). By comparison, in March 1970 the daily circulation of the Washington Post "topped half a million for the first time" (Washington Post Staff, n.d.).

4 In 1973 the New York Times, published "in a city with 1,850,000 or 23.4% non-white citizens, employs 20 minority group members on its editorial staff of 557. In a city with almost half a million more women than men, it employs but 62 women" (Chisholm, 1973, p. 21). As one news bureau head put it, people in the news business "'look at everything through white, middle-class eyes'" (Chisholm, 1973, p. 21).

5 The New York Times provided front-page reporting on the efforts and successes of the Black women's caucus as part of the NWPC's first meeting (Shanahan, 1971; see also Rovner, 1971).

6 E.g., Atlanta Daily World (August 24), Amsterdam News (August 28), Oakland Post (September 2), and Los Angeles Sentinel (September 9).

one promulgating Black women's view of women's liberation. Morrison cultivates, rather than creates ex nihilo, a normative terrain.[7]

3.2. Micro-level design

I focus on just one strategy Morrison uses to constrain middle-class White women from ignoring or discounting Black women's view of women's liberation: irony. Morrison displays irony in how Black women see themselves in contrast to how White people see them; and in how White women see themselves in contrast to how Black women see them. In order to "get" the irony, addressees must recognize and acknowledge Black women's experiential truths. Failing to "get" the irony puts addressees at risk of criticism for not living up to norms such as holding well-informed opinions and beliefs, knowing and understanding U.S. history and current events, and acting in anti-racist, anti-sexist ways. In what follows I illustrate four ways Morrison displays irony and, by doing so, cultivates a local normative terrain.

First, Morrison (1971) describes the irony of historical efforts to mark Black women as inferior to White women. When racial segregation was legal in the U.S., signage designed to reify social hierarchy read "White Ladies" and "Colored Women." The qualities that then made "ladies" "worthy of respect" were "softness, helplessness and modesty," but feminists now "see them as characteristics which served only to secure their bondage to men." Morrison amplifies: "the word 'lady' is anathema to feminists," "there is no such thing as Ladies' Liberation," and the qualities of women then considered by White people as "unworthy of respect"—e.g., "tough, capable, independent and immodest"—are now recognized by self-proclaimed feminists as sources of respect. To "get" the irony that the segregation signage inadvertently marked the superiority of Black women to White women—as Morrison (1971) puts it, "Not racially superior, just superior in terms of their ability to function healthily in the world"—addressees need to recognize and acknowledge the veracity of Black

---

[7] For example, several months earlier Charlayne Hunter (1970) wrote a story for the New York Times covering the topic. She quoted several Black women leaders, including Aileen Hernandez, leader of the National Organization of Women (NOW) at the time; Eleanor Holmes Norton, chair of New York City's Commission on Human Rights; and Frances Beal, member of the Third World Women's Alliance. Further, later that year "The Great American Dream Machine," a nationally-distributed television show broadcasted on Public Broadcasting Service, aired an episode on the "Black experience" that included interviews with four Black women about women's liberation. Press releases about the episode quoted one of the four Black women who spoke about women's liberation quoting the part of Morrison's (1971) essay where Morrison quoted a line published in Essence from Ida Lewis about Black women seeing women's liberation as a family quarrel (e.g., Michigan Chronicle, December 4; Milwaukee Star, December 4; Chicago Daily Defender, December 6; Philadelphia Tribune, December 11).

women's view of historical and current events. Not "getting" it is a fallible sign that addressees do not share Black women's worldview, perhaps valuing qualities such as softness over toughness. But Morrison uses strategies that make ascribing to old value hierarchies risky.

For example, and second, Morrison amplifies the perspective of Black women as responsible human beings by describing how Black women see White women. She asserts that Black women regarded White women as children rather than "competent, complete people" and "real adults capable of handling the real problems of the world"; as "ignorant of the facts of life—perhaps by choice, perhaps with the assistance of men"; as "totally dependent on marriage or male support"; and as people who "can escape the responsibilities of womanhood and remain children all the way to the grave." It would be difficult for middle-class White women to disclaim the truth of this perspective given its consonance with the information ecology cultivated by the publication of Betty Friedan's book *The Feminine Mystique* (1963). Friedan was one of the leading organizers of the NWPC. Her book was a best-seller documenting that White middle-class women as a social group were treated as children by magazine editors, physicians, educators, and more.[8] Further, Morrison's descriptions augment risks of criticism for continuing to ascribe to the old value hierarchy of, say, softness over toughness because her descriptions vividly display how old value hierarchies perpetuate both sexism and racism by bolstering institutions and practices that infantilize White women and denigrate Black women.

Third, Morrison displays irony in mass-mediated stereotypes and caricatures of Black women. I mention just one Morrison discusses: Geraldine, a character on one of the highest rated television shows in the U.S. at the time and the first variety show in the U.S. starring a Black man. Morrison (1971) notes Geraldine is offensive to Black people because "the virtues of black women are construed in her portrait as vices. The strengths are portrayed as weaknesses—hilarious weaknesses." But Morrison (1971) notes that "one senses even in the laughter some awe and respect." Morrison describes Geraldine's "accuracy: for defensive read survivalist; for cunning read clever; for sexy read a natural unembarrassed acceptance of her sexuality; for egocentric read keen awareness of

---

[8] The critique stated by bell hooks in 1984 was relevant in 1971 and earlier: "Friedan was a principal shaper of contemporary feminist thought. Significantly, the one-dimensional perspective on women's reality presented in her book became a marked feature of the contemporary feminist movement. Like Friedan before them, White women who dominate feminist discourse today rarely question whether or not their perspective on women's reality is true to the lived experiences of women as a collective group. Nor are they aware of the extent to which their perspectives reflect race and class biases, although there has been greater awareness of biases in recent years. Racism abounds in the writings of white feminists, reinforcing white supremacy and negating the possibility that women will bond politically across ethnic and racial boundaries" (hooks, 1984, p. 3).

individuality; for transvestite (man in woman's dress) read a masculine strength beneath the accoutrements of glamour." "Getting" the irony again requires recognizing and acknowledging Black women's experiences and virtues, and recognizing that Black women see that White feminists recognize them as virtues.

Fourth, Morrison concludes the essay by writing, "Now we have come full circle: the morality of the welfare mother has become the avant-garde morality of the land. There is a good deal of irony in all of this." She lists examples: "common-law marriage (shacking); children out of wedlock, which is fashionable even now if you are a member of the Jet Set (if you are poor and black it is still a crime); families without men; right to work; sexual freedom, and an assumption that a woman is equal to a man."

In sum, what do addressees get out of "getting" the irony? They can lay some claim to being well-educated (because understanding the examples of segregation signage requires knowledge or experience of U.S. history); well-informed and current (because understanding references to popular culture and the women's liberation movement requires interaction with news and entertainment media); and acting in anti-racist, anti-sexist ways (because understanding the irony requires recognizing that Black people recognize White people's—and White feminists'—recognition of Black women's superiority). At the same time, irony constrains addressees from ignoring or discounting Morrison's take on Black women's view of women's liberation, because doing so would be a fallible sign that addressees are not well-educated, well-informed, current, and acting in anti-racist, anti-sexist ways.

4. Conclusion

Morrison effectively cultivated a normative terrain by using strategies that created conditions for all to hold each other accountable for recognizing and acknowledging Black women's experiential truths. On a macro-level, Morrison places her essay in a highly-visible location in the information ecology. On a micro-level, Morrison vividly, concretely displays irony in how White people see Black women, and how Black women see White women. The effective cultivation of the normative terrain is also based on Morrison's success in living up to reciprocal normative materials that her strategies bring to bear. Addressees can reason that Morrison would not risk criticism for ignorance or misunderstanding historical and current events, or for distorting Black women's worldview, unless she had made responsible efforts to see and think carefully about the experiences. The normative materials are already "out there," but Morrison's strategies bring them to bear in vivid, concrete ways that create ethical, practical reasons to acknowledge the veracity of Black women's view of women's liberation. However, these designs are effective only to the extent that the normative materials they bring to bear are salient for addressees.

It would be possible to provide a more cognitively-oriented account of Morrison's message designs. For example, repetition of irony could induce belief due to increased familiarity or processing fluency (Dechêne et al., 2010; Koch & Zerback, 2013). However, generally, cognitive accounts would not reflect the sorts of reasoning about the efficacy of strategies that would pass muster in social interaction. Imagine a political leader telling supporters to believe what he says *because* he keeps repeating it, and *because* when humans hear something repeated multiple times, they may begin to think it is true because it feels familiar (e.g., Dreyfuss, 2017; Paschal, 2018). This sort of reason smacks of manipulation so is easy to dismiss on just those grounds. In contrast, imagine a political leader telling supporters to believe what he says *because* he keeps repeating it and *because* he expects to be held accountable (e.g., Schwartz, 2020). This sort of reason is an open undertaking of an obligation and an explicit statement about how the strategy of repetition is designed to work in ethical terms. Normative pragmatic accounts reflect the sorts of reasoning that can pass muster in human communication as social actors talk about what they are doing in terms of ethical principles such as fairness, objectivity, and open-mindedness (e.g., Innocenti, 2022; Kauffeld, 2009b; Spring & Gastil, 2013; Townsend, 2009).

In short, Morrison (1971) cultivates a normative terrain from materials already out there and, by doing so, creates and amplifies ethical reasons for addressees to acknowledge Black women's experiential truths. Her essay and other messages designed by Black women over decades have effectively constrained White middle-class women who want to lay some serious, legitimate claim to acting in anti-sexist, anti-racist ways from blithely focusing exclusively or primarily on their own interests with respect to race, class, gender, ability, sexuality, and more. Nonetheless, in the U.S. White women continue to march out of step with women in the global majority (Maharaj & Campbell-Stephens, 2021). For example, from the year 2000 forward the majority of White women have voted for the Republican candidate for U.S. President while large majorities of Black, Latinx, and Asian women have voted for the Democratic candidate (Center, n.d.). Efforts to get at least White women to recognize the experiential truths of Black women and act in solidarity are ongoing and necessary.[9]

---

9 I thank ECA Programme Committee members and ECA participants including Scott Jacobs, Henrike Jansen, Dima Mohammed, and Fabio Paglieri for comments and questions that helped to shore up some arguments.

References

Aakhus, M. (2013). Deliberation digitized: Designing disagreement space through communication-information services. *Argumentation in Context, 2(1)*, 101-126.

Bicchieri, C. (2017). *Norms in the wild: How to diagnose, measure, and change social norms.* New York: Oxford University Press.

Brennan, G., Eriksson, L., Goodin, R. E., & Southwood, N. (2013). *Explaining norms.* Oxford: Oxford University Press.

Campbell, W. J. (2012, February 10). Story of the most famous seven words in U.S. journalism. *BBC News.* https://www.bbc.com/news/world-us-canada-16918787

Center for American Women and Politics. (n.d.). *Gender gap: Voting choices in presidential elections.* https://cawp.rutgers.edu/gender-gap-voting-choices-presidential-elections

Chisholm, S. (1973). The White press: Racist and sexist. *The Black Scholar, 5(1)*, 20-22.

Coles, S. M., & Pasek, J. (2020). Intersectional invisibility revisited: How group prototypes lead to the erasure and exclusion of Black women. *Translational Issues in Psychological Science, 6(4)*, 314-324.

Cooper, B. (2018). *Eloquent rage: A Black feminist discovers her superpower.* New York: St. Martin's Press.

Dechêne, A., Stahl, C., Hansen, J., & Wänke, M. (2010). The truth about the truth: A meta-analytic review of the truth effect. *Personality and Social Psychology Review, 14(2)*, 238-257.

Dreyfuss, E. (2017, February 11). Want to make a lie seem true? Say it again. And again. And again. *Wired.* https://www.wired.com/2017/02/dont-believe-lies-just-people-repeat/

Goodwin, J. (2000). Comments on "Rhetoric and dialectic from the standpoint of normative pragmatics." *Argumentation, 14(3)*, 287-292.

Goodwin, J. (2003). Manifestly adequate premises. *OSSA conference archive.* https://scholar.uwindsor.ca/ossaarchive/OSSA5/papersandcommentaries/30/

Goodwin, J. (2007). Argument has no function. *Informal Logic, 27(1)*, 69-90.

Gray, D. M., & Lennertz, B. (2020). Linguistic disobedience. *Philosophers' Imprint, 20(21)*, 1-16.

hooks, b. (1984; 2015). *Feminist theory: From margin to center.* New York: Routledge.

Hundleby, C. E. (2021). Feminist perspectives on argumentation. In E. N. Zalta (Ed.), *The Stanford Encyclopedia of Philosophy.* https://plato.stanford.edu/archives/spr2021/entries/feminism-argumentation/

Hunter, C. (1970, November 17). Many Blacks wary of "women's liberation" movement in U.S. *New York Times,* 47, 60.

Innocenti, B. (2022). Demanding a halt to metadiscussions. *Argumentation, 36(3)*, 345-364.

Innocenti, B., & Miller, E. (2016). The persuasive force of political humor. *Journal of Communication, 66(3)*, 366-385.

Jackson, S. (2015). Design thinking in argumentation theory and practice. *Argumentation, 29(3)*, 243-263.

Jacobs, S. (2000). Rhetoric and dialectic from the standpoint of normative pragmatics. *Argumentation, 14(3)*, 261-286.

Jacobs, S. (2006). Nonfallacious rhetorical strategies: Lyndon Johnson's Daisy ad." *Argumentation, 20(4)*, 421-442.

Johnson-Bailey, J. (2003). Everyday perspectives on feminism: African American women speak out. *Race, Gender & Class, 10(3),* 82-99.

Kauffeld, F. J. (1994). Veracity, accusation and conspiracy in Lincoln's campaign for the senate. *Rhetoric Society Quarterly, 24(1/2),* 5-21.

Kauffeld, F. J. (1998). Presumptions and the distribution of argumentative burdens in acts of proposing and accusing. *Argumentation, 12(2),* 245-266.

Kauffled, F. J. (2009a). Grice's analysis of utterance-meaning and Cicero's Catilinarian apostrophe. *Argumentation, 23(2),* 239-257.

Kauffeld, F. J. (2009b). What are we learning about the pragmatics of the arguers' obligations. In S. Jacobs (Ed.), *Concerning argument* (pp. 1-31). Washington, DC: National Communication Association.

Kilpatrick, C. (1971, July 14). Nixon and Kissinger begin series of talks: Between us boys. *Washington Post,* A2.

Koch, T., & Zerback, T. (2013). Helpful or harmful? How frequent repetition affects perceived statement credibility. *Journal of Communication, 63(6),* 993-1010.

Lindsey, T. (2020, July 22). Black women have consistently been trailblazers for social change. Why are they so often relegated to the margins? *Time.* https://time.com/5869662/black-women-social-change/

Lorde, A. (1984). Sister outsider: Essays and speeches of Audre Lorde. Berkeley: Crossing Press.

Lumsden, L. (2009). "Women's lib has no soul"? Analysis of women's movement coverage in Black periodicals, 1968-1973. *Journalism History, 35(3),* 118-130.

Maharaj, S., & Campbell-Stephens, R. (2021, February 9). We are not visible minorities; We are the global majority. *Toronto Star.* https://www.thestar.com/opinion/contributors/2021/02/09/we-are-not-visible-minorities-we-are-the-global-majority.html

Morrison, T. (1971, August 22). What the Black woman thinks about women's lib. *New York Times.* https://www.nytimes.com/1971/08/22/archives/what-the-black-woman-thinks-about-womens-lib-the-black-woman-and.html

Never underestimate... (1971, July 26). *Newsweek,* 29-30.

New York Times Company. (1971, October 28). Corporations report figures covering their sales and earnings. *New York Times,* 70.

Paschal, O. (2018, August 3). Trump's tweets and the creation of "illusory truth." *The Atlantic.* https://www.theatlantic.com/politics/archive/2018/08/how-trumps-witch-hunt-tweets-create-an-illusory-truth/566693/

Peach, H. (2020). Dissent, disadvantage, testimony and the ideological "truth" of presumptions. In C. D. Novaes, H. Jansen, J. A. van Laar, & B. Verheij (Eds.), *Reason to dissent: Proceedings of the 3rd European Conference on Argumentation, Volume 1* (pp. 607-619). London: College Publications.

Rovner, R. (1971, July 17). Black women capture one-third of elected seats at convention. *Philadelphia Tribune,* 6.

Roynon, T. (2013). *The Cambridge introduction to Toni Morrison.* Cambridge: Cambridge University Press.

Schwartz, M. S. (2020, September 19). "Use my words against me": Lindsey Graham's shifting position on court vacancies. *NPR.* https://www.npr.org/sections/death-of-ruth-bader-ginsburg/2020/09/19/914774433/use-my-words-against-me-lindsey-graham-s-shifting-position-on-court-vacancies

Shanahan, E. (1971, July 11). Women organize for political power. *New York Times,* 1, 22.

Sprain, L., & Gastil, J. (2013). What does it mean to deliberate? An interpretive account of jurors' expressed deliberative rules and premises. *Communication Quarterly 61(2)*, 151-171.

Strongman, S. (2018). The sisterhood: Black women, Black feminism, and the women's liberation movement. Dissertation, University of Pennsylvania.

Taylor, U. Y. (1998). Making waves: The theory and practice of Black feminism. *The Black Scholar, 28(2)*, 18-28.

Taylor, K.-Y. (2020, July 20). Until Black women are free, none of us will be free. *The New Yorker*. https://www.newyorker.com/news/our-columnists/until-black-women-are-free-none-of-us-will-be-free

Townsend, R. M. (2009). Town meeting as a communication event: Democracy's act sequence. *Research on Language and Social Interaction, 42(1)*, 68-89.

Vasilyeva, A. L. (2015). Identity as a resource to shape mediation in dialogic interaction. *Language and Dialogue, 5(3)*, 355-380.

Vasilyeva, A. L. (2021). Constructing disagreement space: Talk show host's actions. *Language and Dialogue, 11(3)*, 379-404.

Washington Post Staff. (n.d.), Washington Post company history. https://www.washingtonpost.com/company-history/

Qualitative and Quantitative Evidence for Linguistic and Discursive Features of Rephrase

Konrad Kiljan
University of Warsaw (Poland)
& Laboratory of The New Ethos
Warsaw University of Technology (Poland)
konrad.kiljan@uw.edu.pl

Marcin Koszowy
Laboratory of The New Ethos
Warsaw University of Technology (Poland)
marcin.koszowy@pw.edu.pl

Abstract

Rephrasing a message to alter original content for rhetorical gain serves multiple, often overlapping purposes, such as clarifying the point, summarising the content of the message, escaping towards safer deniability options, showing off skills or increasing the emotional load of the message. This rich repertoire of goals for performing rephrases implies that researching the persuasiveness of specific rephrase types promises to yield valuable insights. We propose a systematic inquiry into the linguistic features of rephrase types along with their typical role in a discourse. As there is no robust methodological framework to systematically deal with such a variety of texts, in this paper we propose an initial toolset to capture, qualitatively and quantitatively, the diversity of rephrasing acts. To this end, we employ corpus methods which allow us to introduce and test an annotation scheme for building corpora for rephrased arguments. We collected initial linguistic evidence for rephrase patterns displayed in attested uses of rephrase. To that aim, we establish a pilot corpus of argument maps annotated with rephrase types. As our resources, we took the corpora annotated with the OVA+, Online Visualisation of Arguments software, for (i) the US 2016 Presidential Debate transcripts, (ii) a Reddit thread on US 2016 Presidential Debates and (iii) a Reddit thread dedicated to conspiracy theories. This pilot study allowed us to propose a roadmap to design a fully-fledged corpus analysis of rephrase types.

1. Introduction

Rephrase is a dynamic discursive phenomenon through which a speaker reformulates what has been previously said, simultaneously altering the content for rhetorical gain (Koszowy et al., 2022, pp. 50-51). The two text spans rephrase is built from (which we will call rephrasandum and rephrasans) have the same argumentative function but different linguistic surface. Thus they should not be considered as two arguments for or two arguments against an idea (Konat et al., 2016, p. 33). Despite having the same argumentative role in a discourse, the said specific linguistic surface of rephrase potentially has a rich repertoire of discursive tasks. Among them there are functions such as clarifying the point, summarising the content of the message, helping audiences memorise it, demonstrating argumentative density, escaping towards safer deniability options, showing skills and thus establishing or strengthening speaker's ethos, as well as increasing the emotional load of the message to establish an emotional relation between the speaker and the audience.

This paper summarises the recent work done in the area of rephrase types in order to explore in greater detail the variety of ways in which rephrase is used. It does so by (i) using a different classification criterion that allows for distinguishing new (and more diverse) rephrase types, (ii) designing an annotation scheme and guidelines to capture those new rephrase types, and (iii) employing agile corpus creation methods (Voorman & Gut, 2008) for the sake of an iterative process for generating annotation guidelines.

The last task included two different approaches to annotating rephrase kinds. Firstly, we took the work of Konat et al (2023, forthcoming) who treated the compendium of argumentation schemes (Walton et al., 2008, Ch. 9) as a source of inspiration for annotating typical emotional appeals. We applied a similar approach with respect to annotating rephrase types and looked for the argument schemes from the compendium, as they are likely to constitute a ground for proposing matching rephrase patterns. For instance, argumentation schemes as argument from expert opinion, argument from authority, and argument from popular opinion were combined as one rephrase category, namely rephrase ad alia, capturing all those instances of rephrasing that add a reference to "others" in the rephrasans.

Although this approach allowed us to explore some interesting connections between selected patterns of reasoning (such as argument from analogy) and the phenomenon of rephrasing in communication, it has shown some difficult linguistic and classificationary issues preventing us from establishing sharp boundaries between the proposed rephrase categories. That led us to dropping the idea of taking argument scheme theory as a conceptual framework to deal with the variety of rephrase phenomena. Making use of insights gained while working with the argumentation scheme approach, we proposed a new conceptualisation of rephrase kinds that reflected speakers' main intentions behind rephrasing, e.g. illustrating, clarifying or intensifying the message.

In recent studies (Koszowy et al., 2022), the idea of collecting linguistic and cognitive evidence to study rephrase types led scholars to distinguish rephrase-specification and rephrase-generalisation as two major types of rephrase that have proven to be quite significant in the pilot corpora as they covered over 80% of all instances of rephrase. Preliminary study of rephrase frequencies discussed in (Koszowy et al., 2022) proved that rephrase is not marginal when compared to other propositional relations such as inference or conflict. Out of almost 7k elementary discourse units and over 4k propositional relations between those units across different discourse genres, it proved to be 1.5 times more frequent than conflict, namely 23% vs. 15% (see Koszowy et al., 2022, pp. 62-64). Although rephrasing is on average 2.7 times less dense than inferences (23% vs. 62%, respectively), its importance for conversation dynamics seems unquestionable, justifying research on its various types.

It is worth noting that extensive research on rephrase can be motivated not only by its abundance in natural language. Multiplicity of forms rephrase takes is an equally interesting aspect, implying questions on its essence and possible categorisations. The difficulties arising from the variety of the discursive and linguistic features of rephrase become apparent when we try to compare examples such as the following pair (whereas the single underlining emphasises the key part of a rephrasandum, bolding and underlining marks the main part of a rephrasans):

Example 1:
Hillary Clinton: We've got to do several things at the same time. **We have to do two things.**

Example 2:
Trump: My tax cut is the biggest since Ronald Reagan. **I'm going to cut taxes big league.**

The common features of these two pieces of discourse is that they repeat certain words ("things", "tax") and modify the original content ("two things" in Ex 1, and "big league" in Ex. 2). However, they are also somewhat different as each of them seems to aim at achieving different rhetorical goals. As those similarities and differences have not yet been systematically explored, we treat them as a challenge to face in the present study of rephrase types. For although both elaborations in Ex 1 and Ex 2 are recognisable as rephrase, since the second sentences repeat and strengthen the persuasive effects of the first ones without adding new argumentative value content, their diversity brings about important questions, such as: (1) which theoretical perspectives would agree that they serve the same function? (2) which functions are then secondary or unimportant? (3) how should the differences between them be analysed to enable useful classification? (4) can such a classification explain the differences between their persuasive strength? Our study attempts to answer these questions in an iterative process providing empirical evidence for the validity of most common theoretical approaches.

In order to emphasise and explain the steps of our iterative procedure of designing annotation scheme for a study of rephrase types, in Sect. 2, we took selected argumentation schemes (Walton, Reed, Macagno, 2008) that appeared relevant for the analogous patterns for rephrase, and designed guidelines for annotating them. Having experienced issues with this take, we next took some intuitions from our study to design a new annotation format based not on argumentative patterns, but rather on the intentions speakers have when performing speech acts of rephrasing, such as making the previous statement clearer or increasing the strength of a claim. The results of studying the diversity of rephrase are discussed in Sect. 3. The conclusions reached in Section 4 allow us to formulate directions of future inquiry aimed at obtaining a model of rephrase types that would, through reliable annotation, help us capture the richness of rephrase uses across different discourse genres.

2. Annotation of Rephrase Types Inspired by Argumentation Schemes

The difference between rephrase patterns, although potentially interesting in terms of varied persuasiveness of particular rephrase types, has not been extensively verified empirically. Initial research in this area indicated some persuasive differences between two kinds of rephrase, namely specification and generalisation (Koszowy et al., 2022). Most importantly, the pilot annotation of the corpora containing 161 instances of rephrased arguments has proven that these two kinds of rephrase patterns are most frequent across the studied corpora with their total frequency of 86%, with the dominance of specification over generalisation, 63% vs. 23%, respectively (Koszowy et al., 2022, pp. 68-69). The fact that rephrase-generalisation and rephrase-specification dominate the corpora shows on the one hand that the intentions of making a previously stated content more specific or more general is quite frequent in discourse (regardless of the different discourse areas which were US 2016 Presidential Elections, BBC Moral Maze radio programme, and the BBC Question Time TV programme), and, on the other hand, does not do full justice to the diversity and richness of the examples of rephrase present in the data. Thus our motivation to further explore rephrase types was to employ different distribution criteria for various rephrase uses, such as what particular content has been added in a rephrasans (the annotation scheme presented in this section). Testing it and perfecting via multiple iterations between which we compared the findings of two annotators was meant to help produce a more diverse annotation scheme. The lessons from this part of the process were meant to improve subsequent annotation, focusing on the intentions of rephrasing speakers (see the annotation scheme proposed in Section 3). Limiting the application of an unspecified 'other' type (making up 14% of cases in previous study) was introduced as an indicator of annotation success.

In this section, we present our initial idea of an annotation scheme for rephrase types inspired by selected argumentation schemes. The rationale behind this approach was based on the observation of the overlap between rephrase and arguments. As (Konat et al., 2016, pp. 34-35) have shown, in some cases of rephrasing, there are good reasons for interpreting the same structure in parallel as an argument. The indication of such overlaps drove us to employ an argumentation scheme approach (Walton et al, 2008) in search for argumentation schemes likely to model some types of rephrase. This idea resulted in us designing a test annotation of rephrase instances present in existing corpora of annotated discussions inspired by rephrase-related argumentation schemes. As demonstrated in the "Challenges" section (2.2), this approach met with complicated linguistic issues, making it difficult to disentangle between the proposed rephrase types. It was nonetheless a necessary part of the process, as it provided us with ideas for a more robust annotation scheme discussed in Section 3.1.

2.1. Process and results

For the purpose of designing an annotation scheme for rephrase types inspired by the list of common patterns of reasoning, we applied and modified "A User's Compendium of Schemes" (Walton et al., 2008, Ch. 9). We identified a subset of argumentation schemes that seemed likely to overlap with rephrase types (such as argument by example or argument by analogy) and annotated the corpora with specific rephrase types inspired by argument-scheme theory. As already mentioned in Section 1, the use of the Compendium was based on recent work of (Konat et al., 2023, forth.), which utilised pathotic argumentation schemes to design annotation guidelines for emotional appeals. Building upon the similarities between rephrase and arguments (Konat et al., 2016; Koszowy et al., 2022), we adapted a similar take as in (Konat et al., 2023). Argumentation schemes were used to propose the taxonomy of rephrase to be further incorporated into the rephrase types annotation. This approach promised to allow for a more nuanced annotation than the one differentiating between generalisation and specification types. In so doing, we therefore aimed at consolidating annotation tasks in corpus studies of rephrase.

Looking for a possible typology, we went through the schemes compendium on arguments developed by Walton et al. in 2008 (Ch. 9). This classic publication consists of 104 argument schemes categorised on the basis of their content (such as "argument from expert opinion", "argument from sunk costs" and so on), which are supposed to "capture stereotypical patterns of human reasoning"[1]. For the basis of analysis checking their correctness, they can be formalised and tested by a set critical questions the authors provide for each scheme.

With these limitations in mind, we intuitively selected the arguments that may have a rephrase counterpart, constructing a set with elements such as rephrase by analogy (based on argument by analogy) or Ad hominem rephrase (based on argument ad hominem). From the very beginning, the scheme was intended to be experimental, the annotation results and the annotators 'notes being the material to be tested in an iterative process. After initial guidelines with provisional definitions of types were constructed, we aimed to run an initial pilot annotation and iteratively modify it in 3 iterations.

The annotation scheme with analogical rephrase types analogical to Waltonian arguments. The first iteration used the following rephrase types:

Rephrase by example / analogy / illustration / specification
Rephrase by positive consequences
Rephrase by negative consequences
Rephrase ad alia
Ad hominem rephrase
Pro homine rephrase
Other instances of rephrasing

Almost all the names used in this categorising were taken from the argumentation scheme compendium (e.g. rephrase by negative or positive consequences or rephrase by analogy – based on arguments from consequences or arguments by analogy, respectively), with a sole exception of rephrase ad alia. This category denoted all instances of rephrasing referring to 'others 'in a rephrasans. This category encompasses such usages as rephrasing by appealing to expert opinion, authority, popular opinion and other kinds of external sources and is inspired by the recent work by Budzynska and Reed (2023) who discuss argumentum ad alia as a similarly broad category and we intended to test its usefulness.

The rephrase types list was modified in subsequent iterations. With 2 annotators working independently, a reflective phase focused on

---

1 Walton, D. N., Reed, C., & Macagno, F. (2008) Argumentation Schemes. Cambridge: Cambridge University Press, p. 1.

discussing and comparing their results as well as working notes was carried out after each iteration to reflect on the process and options for its improvement. Usage of working notes was particularly helpful in building as it offered insight into such issues with functionality of proposed typologies as such as borderline cases (e.g. 'my decision was to annotate type X, but I also considered Y as this text might also mean Z"), similarities (e.g. "X seems to overlap with Y") or alternative ideas for guidelines and definitions. After each iteration, the list was adjusted with some types merged and other offered names that would better direct the annotators. This evolutionary process and results of each phase are represented on Figure 1.

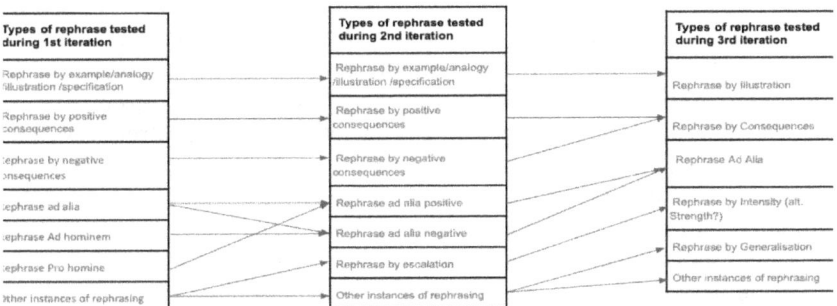

Figure 1. The process of testing rephrase typologies in Iterations

As the inter-annotator agreement after the first iteration reached only 30,77%, it was obvious that the typology has to be improved. The biggest main points of disagreement, Ad Hominem, Pro Hominem and Ad Alia rephrases were merged. Also, a separate category of Escalation was constructed as multiple cases attributed to separate categories included some kind of rise appearing between rephrasandum and rephrasans (e.g. Trump: It 's about putting money—more money into the pockets of American workers).

The second iteration ended with 57,69% inter-annotator agreement. The increase seemed promising and made us think about further points for improvement. Some categories were merged, as Rephrases by Positive and Negative Consequences as well as by Positive and Negative Ad Alia became simpler Rephrase by Consequences and Rephrase Ad Alia. We also decided to simplify the naming, as cases including analogies, examples, illustrations and specifications were merged into a broader Rephrase by Illustration category. As some cases of rephrasans changing the scale of phenomena previously appearing in rephrasandum also included scaling these qualities down, 'escalation 'was renamed as 'intensity/strength'

The simplified categorisation was tested in the third iteration with unsatisfactory results. 26,92% inter-annotator agreement made it the lowest score of any. This called for a critical reflection on the differences behind disagreement and questioned a possibility of building a content-based categorisation. After the 3 iterations followed by rounds of error analyses we came to the conclusion that the proposed take, despite bringing in some instructive ideas (that will be used in Sect. 3) does not allow us to obtain reliable annotation results. The insights gathered during the 3 iterations were later used to build a new conceptualisation.

2.2. Challenges

Low and inconsistent inter-annotator agreement proved very evidently the limitations of the approach based on content. Lack of certainty remained an important issue. Rounds of changes in names and blurred borders between the types suggested that the applied content-based typology was to a large extent arbitrary and could not serve as a mutually exclusive nor collectively exhaustive set of categories. Even when agreement was reached, the explanations given by annotators were often motivated differently.

Linguistic cues also proved to be rarely evident. The initial guidelines included a set of potential expressions that were supposed to help the annotators identify specific rephrase types

Exemplary cues for rephrase Ad Alia would include such phrases as "...which is what everybody would agree with" (as a reference to a popular opinion of others) or ...which is what Europol data indicate (as a reference to authority of an institution in a position to know). As a result, including them in guidelines would even be misleading as annotators have to attribute specific types relying on the meanings of whole expressions and not try to find distinctive linguistic markers in them.

The challenges further justified the shift into intentional features in the next phase. Focusing on the discursive function and analysing where the speaker wants their input to lead, though more context dependent and less clear in terms of linguistic evidence, seemed like a more promising direction

3. Annotating rephrase types with speaker's intentions

Inference-wise schemes for rephrase did not work in terms of obtaining an operationalizable annotation scheme to grasp rephrase types. In this section we discuss a different conceptualisation of types of rephrase that allowed us to obtain better results. The replacement of rephrase-related inferential schemes with speakers' intentions to rephrase an original content led us to obtain four major categories: P-I-C-D, which when supplied with the 'others 'category forming the PICDO scheme. The pilot annotation and error analysis of the micro-corpus of data annotated allowed us to construct methodology for a more systematic inquiry.

The difficulties encountered in 1-3 iterations were typically caused by the multiplicity of meaning generated by the same linguistic cues which served various argumentative functions. In natural language utterances can rarely be segregated by content (as Walton suggested), as speakers care for economical and persuasive quality of their speech results in utterances loaded with meanings. For instance, an appeal to unwanted results of cooperation with an anti-hero figure described with emotional adjectives would confuse criteria for ad hominem and negative consequences. As proven by low agreement in 1-3 iterations, the same text spans could be categorised with (argumentatively defendable) criteria for various content types, suggesting it an unfit way of determining specific patterns of reasoning behind them. To counter this effect, looking not on a static product but on a process occurring with a single act of rephrase was adapted.

3.1. New attempt – process and results

The above-described motivation, combined with experience gained in 1-3 iterations, was used to construct a new annotation scheme. It was built on lessons from previous attempts that reflected similarities between various argumentative types (such as analogy, example) and interpretative differences between annotators. Though categorisation driven by argumentative proved unfit to build a working annotation scheme, it, and the working notes gathered during the tests, served as guidelines for distinguishing processual patterns for rephrase.

To categorise speakers' intentions, we decided to impose strict and simple definitions of processual functions with the aim of avoiding multiple interpretations of the same utterances. Classifying the utterances from a training corpus resulted in a following typology of rephrase types:

Pure - Rephrasans only repeats words used in Rephrasandum
Illustrating - Rephrasans adds new qualities to phenomena described in Rephrasandum
Correcting - Rephrasans amends the Rephrasandum
De-Intensifying - Rephrasans changes the degree of qualities already mentioned in Rephrasandum
Other - unfit for above categories (to be analysed if similar instances might construct a new category)

The results of this attempt were quite promising – inter-annotator agreement went up in 4th and 5th iteration, reaching 53,85% and 66,67% respectively. Importantly, the similarity between annotators 'comments and justifications also increased. Intuitive quality of the scheme is manifested clearly in cases such as following:

TRUMP : there's some bad things going on **Some really bad things**
Annotator 1: "really" as a cue for intensifying
Annotator 2: "really" stronger than 'some' so it's intensifying.

or

TRUMP : [they are leaving] instead of bringing it back and putting the money to work. **Instead of that , they 're leaving our country to get their money**.
Annotator 1: "to get their money" as an instantiation, so I
Annotator 2: I, because there is more specific action described

This experience suggests that the scheme with more training can be a reliable tool for analysis of various rephrase types. Interesting observations can be also drawn from analysing the distribution of the PICDO rephrase types. After the 5 iterations, it was used for analysis of 2 larger corpora built from 780 instances of rephrases found from Reddit threads on US 2016 Presidential Debates and Conspiracy Theories. The distribution of the types within them was very similar (Pure; 15,47% and 2,00%, Illustrating 48,62% and 63,11%, Correcting 6,63% and 2,17% , De(Intensifying) 14,36% and 15,86%, Other 4,42% and 9,52%), suggesting that the scheme is reliable and allows for grasping some dynamics occurring in the natural language.

The unequal distribution of rephrase types is an interesting phenomenon to be analysed in the future research. It is clear that the most minor types (such as Pure rephrasing) were best-defined and allow for little misunderstanding. It is also possible that the most popular Illustrating category is too broad and became a new 'other 'category[2]. Future work could also analyse if it should be split into various types with different persuasive strength.

3.2. Challenges

Although the rise in inter-annotator agreement was significant, it still remains unclear whether PICDO will be reliable across all future corpora. Such examples as the following seem to elude clear classification with it;

CLINTON: we'll guess what [Trump] is hiding CLINTON: we'll keep guessing at what it might be that he is hiding

---

[2] We thank Daniel Cohen for this insight on PICDO categorisation during panel discussion at 4th European Conference on Argumentation in Rome 2022

Even when a narrow definition of Pure rephrasing excludes its usage, the annotators would give conflicting answers. Depending on the tone of the speaker, adding 'keep guessing 'can be read as a sign of intensification or correcting the original message. At the same time 'what he might be hiding 'is a milder version of the original expression, suggesting that De-Intensification is another process occurring in the same rephrase. Multiplicity of process occurring at the same time will limit reliability of the scheme, though much less frequently than in the case of content-based approach.

Some utterances annotated as 'rephrase '(or 'MA') in OVA did not fit into our definition of rephrase. Usually it happens because they do not add anything to the content of the rephrased bit but change the subject or even disagree with it. We hypothesise that MA is annotated too hastily as a link between sentences to demonstrate some continuity between them, when upon closer examination they might be independent ideas expressed to keep talking (for instance to get to desired point by jumping from one idea to another without real link between them). It might also mean that our definition of rephrase is too rigid, but it does not seem like we narrowed understanding used in contemporary literature.

4. Conclusions and future work

Some critics of Waltonian classification of arguments (Lumer 2016) pointed that its richness does not absolve this approach from a particular weakness, namely that any content-based list can never be exhaustive, as content types can be potentially endless. Our results give reasons for another problem with content-based categorisation: as the resulting types are non-homogenous and can overlap, their functionality is very limited. Shifting our attention into the processes rather than products of inferences allowed us to construct a less ambiguous scheme, as indicated by the shrinking size of the Other category. The results also demonstrated the validity of agile iterative methodology with a note-intensive reflection phase for studies in corpus linguistics.

This pilot study let us formulate the following hypothesis: distinguishing rephrase types may contribute to a broader study of speakers' and social media users 'dialogical profiles and "dialogical fingerprinting". The data-based knowledge about linguistic features of particular rephrase kinds (that signal e.g. clarification or intensification) along with their typical roles in discourse (such as employing a certain rephrase type when summarising a sequence of argumentative moves or using a different kind of rephrase for a response to an *ad hominem* argument) should be used for analysis of speakers 'individual rephrasing styles. The insights gathered during this research will add an interesting component to broader dialogical or argumentative profiles. Two possible research directions in this area emerge:

The linguistic and discursive features of rephrase types can be used for exploring speakers 'rhetorical profiles. In line with recent corpora-based studies on argumentation analysing such features as speakers 'propensity for personal attacks (Visser et al. 2021), schemes for rephrasing would provide a fuller picture of specific profiles. The recognition of ways in which a particular speaker uses rephrases may demonstrate how they strengthen their ethos with e.g. illustration, clarification or intensification, and which rephrase types are more likely to either support or attack other speakers 'ethos. Specifically, correcting others may have a flavour of attacking ethos or may not be aimed at weakening someone's ethos. Elaborating on such dynamics will be an interesting line of future inquiry.

More generally, distinguishing rephrase types can be useful for analysis of conversation dynamics. With speech act theory as a theoretical point of reference, it would provide insight into dialogical moves made by the speakers in specific types of conversations. The results of analysis testing which rephrases are more frequently used as ethotic attacks/supports, but also which start, prolong or end the conversation will be applicable for construction of dialogical protocols for both human- and computer-run dialogues. In this way, the descriptive studies using methodology described in this paper will provide empirically-validated foundations for construction of normative models.

Acknowledgements

The work reported in this paper has been supported by the Polish National Science Centre under Grant 2020/39/I/HS1/02861.

References

Koszowy, M., Budzynska, K., Pereira-Fariña, M., Duthie, R. (2022). From Theory of Rhetoric to the Practice of Language Use: The Case of Appeals to Ethos Elements. *Argumentation, 36*, 123–149.

Koszowy, M., Oswald, S., Budzynska, K., Konat, B., Gygax, P. (2022). A Pragmatic Account of Rephrase in Argumentation: Linguistic and Cognitive Evidence. *Informal Logic*, Vol. 42, No. 1, pp. 47–8.

Konat, B., Budzynska, K., & Saint-Dizier, P. (2016). Rephrase in Argument Structure. In: P. Saint-Dizier & M. Stede (Eds.), *Foundations of the Language of Argumentation. COMMA 2016 Workshop* (pp. 32-39). University of Potsdam.

Konat, B., Gajewska, E., & Rossa, W. (2023, forthcoming). Computational pathos: argument schemes and emotional reactions in natural language.

Lumer, C. (2016). Walton's Argumentation Schemes. (2016). OSSA Conference Archive. 110. https://scholar.uwindsor.ca/cgi/viewcontent.cgi?article=2286&context=ossaarchive

Visser, J., Lawrence, J., Reed, C., Wagemans, J., & Walton, D. (2021). Annotating Argument Schemes. Argumentation, 35, 101–139. https://doi.org/10.1007/s10503-020-09519-x

Voormann, H., & Gut, U. (2008). Agile corpus creation. *Corpus Linguistics and Linguistic Theory* 4(2), 235–251.

Walton, D., Reed, C., & Macagno, F. (2008). Argumentation Schemes. Cambridge: Cambridge University Press. https://doi.org/10.1017/CBO9780511802034

CONNECTIONS BETWEEN AGE AND INTERPERSONAL ARGUING IN UKRAINE, WITH SPECULATIONS ABOUT WAR'S EFFECTS

IRYNA KHOMENKO
Taras Shevchenko National University of Kyiv
khomenkoi.ukr1@gmail.com

CRISTIÁN SANTIBÁÑEZ[1]
Universidad Católica de la Santísima Concepción

DALE HAMPLE
Western Illinois University

Abstract

Interpersonal arguing is a routine element of social life, from childhood to old age. Quite a lot of educationally-related argumentation research has been done on children, but the vast majority of other work has examined college students. The present study reports results on a study conducted in Ukraine just prior to the present war, and shows how orientations to interpersonal arguing change (or not) through the adult life span. Our design and data also permit some speculations as to the effects of imminent war on orientations to interpersonal arguing.

1. Introduction

Here we report results of a survey conducted in Ukraine, designed about a year before the advent of the current Russian invasion. A main objective of the study was to give us information bearing on the possibility that people's attitudes and understandings of interpersonal arguing change as they progress through the adult life span, a hypothesis that first acquired clear support due to a study conducted in Chile (Santibáñez, Hample, &

---

[1] The Chilean data reported in this paper were produced in the Fondecyt 1170492 research project, ANID, Chile.

Hample, 2021). We gathered our data in July, 2021, and Russia invaded Ukraine in February, 2022, having already taken Crimea by force in 2014. Russian troops had been massing near Ukrainian borders in December, 2021, and tensions had been rising before then. Unintentionally, our design permits some indirect testing of the effects the threat of war might have on people's views of interpersonal argument, and the results of an earlier study (Khomenko & Hample, 2019) allow some comparisons. As a consequence, this report about arguing orientations has two main foci: age and tensions accompanying the prospect of invasion.

2. Age

As mentioned, the Chilean study was the first to give evidence on how people's experiences and reflections about interpersonal arguing change as they age (Santibáñez, Hample, & Hample, 2021). That study contrasted Chilean seniors (mean age = 72 years) with Chilean undergraduates. Those two groups differed argumentatively in several respects, with the seniors generally being less eager to argue, but also more insistent on respect for social hierarchy and being less civil and cooperative when they did engage. Being cross-sectional (as is the present study), attributing these differences to age was uncertain, but various literature connecting aging to less aggression and less interpersonal arguing made the interpretation plausible (e.g., Costa, & McCrae, 1988; Martin & Anderson, 1996; Reiter, Janske, Eppinger, & Li, 2017; Schullery & Schullery, 2003). However, the Chilean study did not examine middle adult ages, and the present study does.

Data collected in Argentina (Mamberti & Hample, 2022) and from immigrants into Chile (Santibáñez & Hample, 2022), did allow examination of middle-aged respondents as well as younger and older adults. Consistently with the original Chilean study, those investigations also indicated that eagerness to argue declined with age; that older people were less interested in arguing for various reasons, such as utility, play, and identity display; and that seniors were more pessimistic that face-to-face arguing would impair personal relationships. However, a few differences emerged: the original Chilean study showed seniors to be interested in arguing to display dominance over the other person; the original findings about seniors being less civil and cooperative were not supported in the two later studies; and several emotional registrations of conflict (e.g., direct personalization, persecution feelings) gave varied results. Naturally, our first impulse is to explain these differences by reference to the differing cultures from which data were drawn, but certainly having more data from even more nations has obvious value. One's national culture affects the social and economic experiences that accompany aging, but aging is also, and perhaps most fundamentally, a pan-species biological process as well.

All of the investigations in this loosely-organized global project make use of essentially the same self-report scales (see Hample, 2018, for a review of the instruments and a summary of the earliest studies in the project). The argumentation scales, which all originated in the USA, are translated into the local languages. A first instrument is power distance. This is a cultural measure, not specifically an argumentation variable. It measures one's acceptance of power differentials in society, and bears on who is allowed to disagree with whom. Motives for arguing are assessed with two traditional argumentation measures, argumentativeness and verbal aggressiveness. Each has two subscales, which should be reported separately: argument-approach and argument-avoid for argumentativeness, and VA-prosocial and VA-antisocial for verbal aggressiveness. Two recently developed scales also fit here conceptually. These are willingness to argue with a supervisor at work, and willingness to argue with a peer coworker. Argument frames are a group of instruments, which asses why people might want to argue (for utility, to exhibit dominance, to display personal identity, or to play), the degree to which the other person is important in one's arguing behaviors (blurting, cooperating, being civil), and one measure of reflective insight into the nature of arguing (professional contrast, scored so that high numbers indicate sophistication). Taking conflict personally reflects people's emotional experiences with conflict throughout their lives; specific measures here are direct personalization, stress reactions, feelings of persecution, expectation of positive or negative relational effects, and general valence for interpersonal conflict. All these, collected together, offer a general description of how a person or group of people views the prospect of interpersonal arguing.

These particular measures give detail to our first research objective:

Research Question 1: In Ukraine, what are the associations between age and our measures of arguing orientations?

3. War

As we explained earlier, we had not foreseen the Russian invasion when we began this project. However, the timing of this data collection made the project unique to some degree. The fact that we had done a similar project in Ukraine earlier (those data were collected in Spring, 2018; Khomenko & Hample, 2019) allows some comparison of results. It will be obvious to readers that securely interpreting differences between those data collections as having been caused by the prospect of war in the second instance would be hopelessly fallacy-ridden.

A key element of both Ukrainian data collections was that we allowed respondents to choose whether to take the survey in the Russian or Ukrainian language. Bilingualism is very common in Ukraine. We had included this feature in both studies' designs as a courtesy to respondents, but we wondered – even in the first study – if language choice might be something of a proxy for national attachment. This, of course, became a much more pressing question for the second study. To the degree that we can address this matter at all, we will be relying on differences in results from those who self-selected the Ukrainian survey versus those who preferred the Russian-language version. In 2022, two surveys were conducted on this topic in Ukraine (Сімнадцяте, 2022; Мова, 2022). Of the respondents, 76% said Ukrainian was their native language and only 19% preferred Russian. Further, 80% believed that Ukrainian should be the national language, and only 15% advocated bilingualism.

We have paid attention to nations' political systems and histories in many of our studies (e.g., Debowska-Kozlowska & Hample, 2022; Hample & Njweipi-Konger, 2020; Rapanta & Hample, 2015). We noted how long functioning democracies had been in place, and considered whether generations had come to adulthood when their nation had been under the control of the former Soviet Union, military rule, or dictatorship. We felt that this information had a bearing on whether people had grown up feeling free to argue publicly about politics in particular, and whether people (or their parents and grandparents) had experienced suppression of free speech. These considerations were merely background information for those studies, however, simply part of our introductory cultural descriptions. Political matters were obviously more urgent in Ukraine.

Although our own studies, even with their varied national contexts, do not clearly point to any conclusions about the effects of democracy, war, and political danger on interpersonal arguing, other research does suggest some relevance. Bauer et al. (2016) reported a meta-analysis of 16 studies examining the aftereffects of recent wars in several nations, and found that cooperative behaviors actually increased. This effect was particularly evident for interactions within in-groups. These results, while encouraging, are for post-war life, which is of course not the situation here. Using data collected in Ukraine in 2017, Timmer et al. (2023) showed that exposure – even indirect or virtual exposure – to warfare increased civilians' willingness to commit interpersonal violence and aggression. That paper's extensive literature review indicates that this sort of coarsening is common throughout the world, and not restricted to Ukraine at all.

Apart from these war-related circumstances that would make interpersonal life (including arguing) more aggressive, another relevant factor is whether things can be freely discussed at all. Governments at war commonly take control of national communication media and content, as far as they can (Jowett & O'Donnell, 2015). They may also enforce restrictions on what can be said in public. Radziszewski (2013) studied interpersonal conversation in Poland regarding the nation's possibilities of participating in the 2004 Iraq War. She found that those discussions (surely argumentative ones) strongly influenced participants' political views, and the consequent public opinion supporting participation in the war contributed to Poland's eventual policy. In Ukraine during the build-up to invasion, neighbors and social acquaintances might well have different views about national independence versus a return to Russian control. Adults would surely have been aware that easy expression of opinions might be costly – socially, economically, or physically. This external suppression could be internalized or might stimulate psychological reactance among people responding to our survey. These pressures might have been felt differently by Ukrainians of contrasting allegiance. Remembering that we have only an indirect proxy for national commitment (the choice of language for the survey), we explore these possibilities:

Research Question 2: Did the choice of language (Russian or Ukrainian) for the survey indicate different orientations to interpersonal arguing?

4. Method

4.1. Procedure and Sample

Data were all self-reports requested via online surveys in Ukraine. Instruments were translated from English into both Ukrainian and Russian, and then back-translated to assure valid content. 429 respondents elected to complete the Ukrainian-language version, and 127 chose the Russian materials. Altogether, 194 respondents were male (35%) and 362 were female (65%). The male/female splits for language choice were Ukrainian (146/283; 34%/66%) and Russian (48/79; 38%; 62%). The average age of respondents was 32.7 years (SD = 18.8). Average ages for the language subsamples were 30.5 years (SD = 17.8) for Ukrainian language choosers, and 40.2 years (SD = 20.1) for those doing materials in Russian. A considerable portion of each subsample was composed of college-related people, but distributions of the two groups were similar, as Figure 1 indicates.

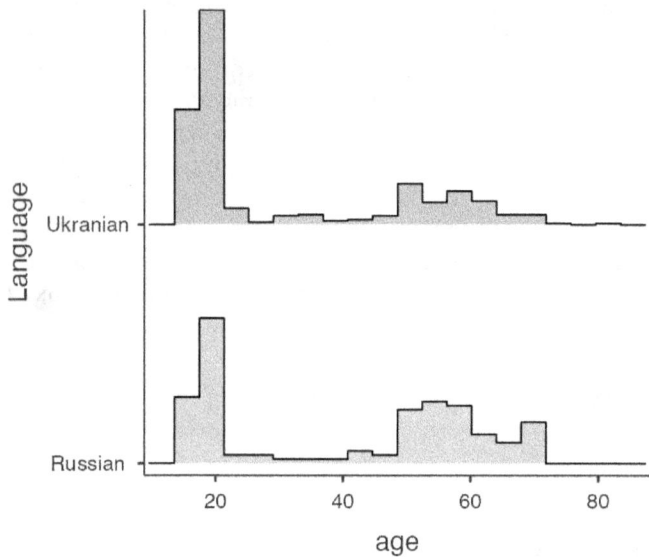

Figure 1. Age distributions by Language of Choice

4.2. Instrumentation

Instruments were those used throughout the global arguing project. Unless otherwise noted, English versions are in the Appendix of Hample (2018).

Power distance is the tolerance for social hierarchy and the deference given to those of higher status than self. We used the scales in Yoo, Donthu, and Lenartowicz (2011). Cronbach's alpha reliabilities for Ukrainian/Russian language groups were .77 and .79, respectively.

Argument motivations were measured by means of scales for argumentativeness (Infante & Rancer, 1982), verbal aggressiveness (Infante & Wigley, 1986), and two new scales assessing willingness to argue with one's supervisor and one's coworker (Mamberti & Hample, 2022). Reliabilities for Ukrainian/Russian subsamples are as follows: argument-avoid (.84/.86), argument-approach (.88/.87), VA-antisocial (.80/.85), VA-prosocial (.80/.82), argue with boss (.88/.90), and argue with coworker (.88/90).

Argument frames are in three categories. The first assess self's motives for arguing. Reliabilities were utility (.76/.78), dominance assertion (.78/.83), identity display (.70/.80), and play (.72/.79). The second group measures whether and how one integrates the other arguer into the episode. These reliabilities were blurting (.75/.82), cooperation (.73/.74), and civility (.51/.51). The civility result was unacceptable, and we do not regard this scale as providing very useful information about this data set. The last group is argument sophistication, and is assessed by one scale, professional contrast. This measures the degree of contrast between argument professionals and ordinary actors, and is scored so that high scores indicate high sophistication. Its reliabilities here were .87 and .86.

Taking conflict personally is a battery of instruments asking respondents to summarize their emotional and cognitive reactions to being in interpersonal conflicts. These scales and their reliabilities were: direct personalization (.82/.77), persecution feelings (.76/.76), stress reactions (.65/.60), expectation of positive relational effects (.85/.83), expectation of negative relational effects (.80/.87), and overall positive valence for conflicts (.85/.88). The stress reaction scale's reliability was low, but might still give some useful information. As with civility, we will still report stress results, but only for the sake of comparison to other studies.

Generally, the scores showed good internal consistency, implying that the items "hung together" in a way comprehensible to our respondents. This affords some confidence that the constructs have validity in Ukraine.

5. Results

5.1. Age

We correlated self-reported age with each of the variables in our survey. Table I shows the results for the current study in the first column. It also shows results from other studies that had enough age range to be somewhat comparable. These results combine both language groups.

Table I. Correlations Between Age and Other Variables

Study:	Ukraine	Argentina	Immigrants to Chile	Chilean Seniors UGs	Notes on Consistencies
Power Distance	.18	.03	-.13	Seniors higher	
Arg Appr	-.33	-.12	-.16	UGs higher	Older people less eager to argue
Arg Avoid	.32	.09	.14	Seniors higher	Older people more avoidant
VA Antisocial	.08	-.22	-.26	ns	
VA Prosocial	.07	.07			
Argue Boss	-.04	.07			

Argue Cowork	.05	0.09			
Utility	-.13	-.04	-.16	ns	Older people less interested
Dominance	-.18	-.18	-.21	Seniors higher	
Identity Display	-.21	-.25	-.23	ns	Older people less interested
Play	-.30	-.23	-.27	UGs higher	Older people less interested
Blurting	.03	-.16	-.16	ns	
Cooperation	.11	.09	.05	UGs higher	
Civility	-.05	.03	.11	UGs higher	
Professional Contrast	-.08	.05	-.03	UGs higher	
Direct Personalization	.13	-.20	-.17	Seniors higher	
Stress Reaction	.09	.10	.05		
Persecution Feelings	.03	-.13	-.16	Seniors higher	
Pos Relational Effects	-.30	-.18	-.10	ns	Older people less optimistic
Neg Relational Effects	.29	.11	-.08	ns	Older people more pessimistic
Positive Valence	-.37	-.15	.06	UGs higher	

Table I shows the correlations between age and other measures in the current study. With sample sizes over 500, nearly all the correlations were statistically significant. As age increased (mainly from college age into middle age and beyond – see Figure 1), we observed several substantial correlations. Older Ukrainians were more accepting of power distance than younger ones; they were more avoidant of arguments and less eager to approach them; they were less interested in arguing for any of the specific motives assessed here (utility, dominance assertion, identity display, or play); they were more worried that interpersonal conflicts would damage personal relationships and less optimistic that conflict could strengthen a relationship; and overall, they valued disagreements more negatively than younger people.

Table I also notes where these results were consistent across all the studies for which we have relevant data. In spite of national and cultural differences, some of the age results seem to cross borders and languages. These results can be summarized as showing that older people are less eager to engage others in disagreement, and they are also more concerned that interpersonal arguments may have relationship costs. These results show a reluctance to argue, but they do not indicate that older people won't or can't argue. It may be that they are simply more selective about it when they do.

We also caution readers that this is cross-sectional data. It is possible that the observed differences are due to cadre effects. That is, perhaps people born in the 1950s, 1970s, and 1990s were different in the first place and have never changed in the measured respects. However, literature about aging briefly mentioned earlier in the paper give us some confidence that we are seeing effects of moving forward in the lifespan, both biologically and socially.

5.2. War

Our second research question inquired about the possible effects of an imminent war on our respondents' views about arguing with one another face to face. Again, we caution readers that we do not have very compelling measures of people's positions on the possible war. Some wanted to maintain and strengthen Ukraine's independence, some felt that the old Soviet life was better, and some simply felt more allegiance to Russia than to Ukraine. We have no direct measures of these positionings; nor do we have any measure of enthusiasm or horror at the prospect of invasion. The only data we have concerns language preference. With all these apologies for our measure, however, we should point out that it does have one strength, its unobtrusive nature.

To the degree that we can approach the question with these data, we would like to know whether the impending war changed people's psychology about arguing. Our first result of relevance is that in this study, of our 556 respondents, 23% chose the Russian language survey. In our previous study (Khomenko & Hample, 2019), using essentially the same sampling frame and design, but collecting data in 2018, 25% chose Russian. So we saw essentially the same language preference here as we saw before the war rhetoric built up once more. A second finding is that language choice varied significantly by geographical region (χ^2 (5, N = 556) = 79.4, p < .001, Cramer's V = .38). Preference for Russian was lowest in western regions (5%) and highest in the east (49%) and the south (71%). In the most populous regions represented in our survey, Kiev had a 22% preference for Russian, and the central region had 24%.

Perhaps the most informative results we can provide on the general question of war effects are those that compare mean scores for the two language groups. Table II displays those effects. All measures were taken on a 1 - 10 metric, using a standard Likert format (except for professional contrast, which uses a semantic differential form). The t-tests were corrected for unequal group variances where needed, and Cohen's d is used as a measure of effect size.

Table II. Mean Comparisons Between Language Subsamples

	Ukrainian			Russian			t	d
	N	M	SD	N	M	SD		
PowerDist	308	3.38	1.79	90	4.33	2.11	-3.89***	.51
Arg-Avoid	429	5.61	1.87	127	5.16	1.96	2.38*	.24
Arg-Appr	429	5.86	1.93	127	5.87	1.87	-0.06	
VA-Anti	391	4.28	1.67	116	4.64	1.86	-1.98*	.21
VA-Pro	391	6.87	1.66	116	6.57	1.65	1.67	
ArgBoss	308	6.65	1.95	90	6.75	2.18	-0.37	
ArgCow	308	7.35	186	90	7.23	2.00	0.53	
Utility	366	5.16	1.63	108	5.10	1.70	0.32	
Domin	366	4.50	1.96	108	4.58	2.15	-0.41	
Identity	366	6.22	1.45	108	6.25	1.75	-0.18	
Play	366	4.84	2.16	108	5.06	2.40	-0.91	
Blurting	366	5.31	1.47	108	4.78	1.76	3.20***	.35
Prof Contrast	359	5.56	2.31	107	5.82	2.10	-1.11	
Direct	308	5.66	1.95	90	5.66	1.81	0.01	
Stress	308	5.35	1.85	90	5.19	1.80	0.70	
Persecution	308	4.72	1.84	90	4.77	1.89	-0.23	
Pos Rel Eff	308	5.49	1.97	90	5.36	1.96	0.55	
Neg Rel Eff	308	6.84	1.86	90	7.00	2.11	-0.68	
Pos Valence	308	3.99	2.08	90	4.27	2.25	-1.10	

* p < .05   ** p < .01   *** p < .001

Only four comparisons generated statistically significant mean differences, but all of those had reasonably substantial effect sizes. And those differences form a fairly clear pattern, as we will explain momentarily. Those who preferred Russian materials were far more tolerant of power disparities in society, were more argument-avoidant, and were more inclined to use ad hominem attacks during interpersonal conflicts. They were less likely to blurt while arguing.

The earlier study (Khomenko & Hample, 2019) did not include power distance in the set of measures, and did not report results bearing on language differences. However, reanalysis of that data indicated significant language-use differences for three measures. For dominance,

Russian language users were higher (Cohen's d = -.26); for blurting, Ukrainian language users were higher (d = .27); and for stress reactions, Russian choosers were higher (d = -.27). Except for blurting, these are different results from the present study. Since the sampling frames were the same and the samples were both of substantial size, we suspect that the imminence of an allegiance-heightening war might be involved in the differences.

Returning to the current study's outcomes, our reading of the observed pattern in Table II highlights our growing understanding of the centrality of power distance in these profiles. Combining our two language subsamples, power distance correlated significantly with many of our other variables: with VA-antisocial (r = .28), with VA-prosocial (-.17), with arguing with one's boss (-.11), with arguing for utility (.19), with arguing to assert dominance (.23), with being cooperative (-.22), with professional contrast (i.e., sophistication about arguing; -.12), with persecution feelings (.16), and with positive valence for conflicts (.15). People high in power distance are aggressive when they are permitted to disagree or give commands, but it is the situation – in particular, their power relationship with the other person – that determines whether they can argue at all.

Acceptance of power hierarchy is also an acceptance of who can give orders, who is entitled to disagree with the other person, whether those conversational moves need to be polite or can be aggressive, and whether the person needs to swallow more remarks in certain situations. Those who preferred Russian materials were more comfortable accepting discrepant power differentials, were aware of fewer situations in which they should avoid arguing, saw fewer problems with insulting remarks (remember, only the superior is arguing), and were less likely to blurt out their thoughts. This may be a profile of how one voluntarily lives in a society dominated by power rather than the better reasons.

6. Discussion

These results offer some insight into the effects of age on people's understandings of interpersonal arguing. The design and timing of the study also yield some hints about how people might reform their social perceptions when they and their nation are under urgent threat.

6.1. Age

Our data are informative with regard to how adults might be changing their orientations to face-to-face arguing as they age. The vast bulk of empirical research on argument-relevant personality and cognitive constructs has drawn its data from university students, and quite a bit of the work has been done in the USA. The global argumentation project contrasts on both these points.

Here, we found, as we have before, that older adults are less aggressive about arguing and more cautious about engaging in it. The original study of Chilean seniors did not include people of ages between college students and the senior sample. Here and in other recent studies our samples have included university students and somewhat older community members, most often middle-aged. The substantial concentration of seniors from the Chilean study has not reappeared in the present or other recent work. This may be part of the reason for some patterns in Table I where only the first Chilean study was clearly inconsistent: the dominance, cooperation, civility, and persecution results. Passing from middle age into retirement is a discrete stage-of-life change, and so we should be open to the possibility that somewhat abrupt rebalancing of one's perceptions about social life might be occurring at that point.

Still, we have some confidence that undergraduate-aged respondents are more aggressive about arguing than older adults. The undergraduates seem to act as though they can argue freely on many topics, and in many situations, where their elders would stay quiet. Older people are more sensitive to the chance that interpersonal arguments can do interpersonal harm. We are inclined to view those latter views as reflecting more life lessons than the student-aged respondents have been exposed to. We suggest that when research about university samples is being studied, readers should take these considerations into account.

6.2. War

We hope we have made plain that we regard this study's suggestions about the effects of war worries as being inadvertent, serendipitous, and uncertain. Even knowing this, we cannot help being interested in certain results.

Few things can throw a society into unanchored unpredictability more than war. The economy becomes erratic, one's tiny children have to learn what to do when missiles are in the air, cranky old people start to sound traitorous, and you worry that healthy friends and loved ones will become casualties without warning. You walk around piles of rubble to get vegetables and understand that police may have better things to do than investigate simple crimes. In what was once the simple realm of ordinary social interaction, you have to guard your words, test out other's people's views before you express your own, and notice carefully what other people say, or omit to say, to you.

Great swaths of our social lives are conducted by means of arguing. We give reasons for this and that, and consider other people's reasons in planning out our own actions. The quantity of uncertainty and worry that war brings can introduce randomness into anything we think or depend upon. In the case of interpersonal arguing, the changes may not be entirely random. Making an argument is usually to (try to) impose your will on another person. Sensitivity to force, power, and aggression permit those considerations to slide into ordinary interactions. Two people will absorb these matters into their daily lives differently, depending on whether they are afraid, whether they think their "side" will prevail, and what place they think they now occupy in society. Here, we found that people who preferred to take a survey in Russian were more committed to the idea of social hierarchy, whereas the Ukrainian language choosers were more egalitarian. The Russian-users were more willing to use personal attacks where reasons might have appeared. The Ukrainian-users blurted more, suggesting that they were less worried about saying the wrong thing to the wrong person; this may be an indirect expression of egalitarianism.

7. Conclusions

Over 500 Ukrainians completed a survey of their orientations to interpersonal arguing at a time when their nation was on the verge of invasion. Given the opportunity, about a quarter of our respondents chose to complete the survey in Russian, a figure nearly identical to the results of a somewhat earlier study. While many Ukrainians are competent in both Ukrainian and Russian, it is possible that some respondents could only have done the survey in Russian. Nonetheless, we are inclined to think that language choice may have been a proxy for national attachment. We found interesting differences between the language groups, perhaps most centrally a greater tolerance for power hierarchy among the Russian language users. We also examined age differences in responses within our sample, and discovered that older adults (mainly middle-aged) tended to be more careful in engaging in arguments, compared to college-aged respondents.

8. References

Bauer, M., Blattman, C., Chytilová, J., Henrich, J., Miguel, E., & Mitts, T. (2016). Can war foster cooperation?. Journal of Economic Perspectives, 30(3), 249-274.

Сімнадцяте загальнонаціональне опитування: Ідентичність. Патріотизм. Цінності. 17 - 22 серпня 2022. (Seventeenth National Survey: Identity. Patriotism. Values. August 17 - 18, 2022.)https://ratinggroup.ua/files/ratinggroup/reg_files/rg_ua_1000_independence_082022_xvii_press.pdf

Costa, P. T., Jr., & McCrae, R. R. (1988). Personality in adulthood: A six-year longitudinal study of self-reports and spouse ratings on the NEO Personality Inventory. Journal of Personality and Social Psychology, 54, 853-863.

Debowska-Kozlowska, K., & Hample, D. (2022). Agreement builds and disagreement destroys: How Polish undergraduates and graduates understand interpersonal arguing. Argumentation. Online first. doi: 10.1007/s10503-022-09570-w

Hample, D. (2018). Interpersonal arguing. New York: Peter Lang.

Hample, D., & Njweipi-Kongor, D. (2020). How do people feel about arguing in Cameroon? OSSA Conference Archive. 12. https://scholar.uwindsor.ca/ossaarchive/OSSA12/Friday/12

Infante, D. A., & Rancer, A. S. (1982). A conceptualization and measure of argumentativeness. Journal of Personality Assessment, 46, 72-80.

Infante, D. A., & Wigley, C. J. (1986). Verbal aggressiveness: An interpersonal model and measure. Communication Monographs, 53, 61-69.

Jowett, G. S., & O'Donnell, V. (2015). Propaganda & Persuasion, 6th Edition. Los Angeles: Sage.

Khomenko, I., & Hample, D. (2019). Comparative analysis of arguing in Ukraine and the USA. In B. Garssen, D. Godden, G. R. Mitchell, & J. H. M. Wagemans (Eds.), Proceedings of the ninth conference of the International Society for the Study of Argumentation (pp. 628-639). Amsterdam: Sic Sat.

Mamberti, J. & Hample, D. (2022): Interpersonal arguing in Argentina. Argumentation and Advocacy. Online First doi: 10.1080/10511431.2022.2137984

Martin, M. M., & Anderson, C. M. (1996). Communication traits: A cross-generalization investigation. Communication Research Reports, 13, 58-67.

Мова та ідентичність в Україні на кінець 2022 року (Language and identity in Ukraine at the end of 2022) https://www.kiis.com.ua/?lang=ukr&cat=reports&id=1173&page=2

Radziszewski, E. (2013). Interpersonal discussions and attitude formation on foreign policy: The case of Polish involvement in the Iraq war. Foreign policy analysis, 9(1), 103-123.

Rapanta, C., & Hample, D. (2015). Orientations to interpersonal arguing in the United Arab Emirates, with comparisons to the United States, China, and India. Journal of Intercultural Communication Research, 44, 263-287.

Reiter, A. M., Kanske, P., Eppinger, B., & Li, S. C. (2017). The aging of the social mind-differential effects on components of social understanding. Scientific reports, 7 (11046), 1-8.

Santibáñez, C., & Hample, D. (2022, Sept.). Sharing a language, sharing the argumentative attitude? The case of South American immigration in Chile today. Paper presented to the European Conference on Argumentation, Rome.

Santibáñez, C., Hample, D., & Hample, J. (2021). How do Chilean seniors think about arguing? Journal of Argumentation in Context, 10(2), 203-226.

Schullery, N. M., & Schullery, S. E. (2003). Relationship of argumentativeness to age and higher education. Western Journal of Communication, 67, 207-223.

Rahwan, I., Ramchurn, S. D., Jennings, N. R., McBurney, P., Parsons, S., & Sonenberg, L. (2003). Argumentation-based negotiation. The Knowledge Engineering Review, 18(4), 343–375.

Timmer, A., Antonaccio, O., Botchkovar, E. V., Johnson, R. J., & Hughes, L. A. (2023). Violent conflict in contemporary Europe: specifying the relationship between war exposure and interpersonal violence in a war-weary country. The British Journal of Criminology, 63(1), 18-39.

Yoo, B., Donthu, N., & Lenartowicz, T. (2011). Measuring Hofstede's five dimensions of cultural values at the individual level: Development and validation of CVSCALE. Journal of International Consumer Marketing, 23, 193-210.

DISAGREEMENT ON REDDIT: AN EMPIRICAL STUDY

ZLATA KIKTEVA
University of Passau
zlata.kikteva@uni-passau.de

ANNETTE HAUTLI-JANISZ
University of Passau

Abstract

In this paper, we report on the results of the annotation of Reddit conversations with a disagreement typology by Rees-Miller (2000). The study indicates that annotators find it to be a challenging task due to the often implicit and subjective nature of disagreement. For similar reasons, linguistic features do not always indicate an individual disagreement type. In order to explore the underlying nature of the disagreement, we, therefore, conduct empirical research on a number of strategies associated with it.

1. Introduction

Disagreement is an inherent part of interpersonal relations in real life as well as online[1]. While not all disagreements are intrinsically negative (Sifianou, 2012), hateful speech, offensive language and harassment are frequently observed on social media and have attracted increasing attention from researchers in the field of computational approaches to language (for instance, Workshop on Online Abuse and Harm). Since disagreements tend to stay unresolved (Kakava, 2001) and can easily escalate at any point, it is important to refrain from viewing them as binary phenomena. Instead, a more nuanced approach is required, for instance through a graded notion of disagreement.

In this paper, we showcase that the disagreement classification by Rees-Miller (2000) which we initially found to be the most suitable for viewing disagreements on a scale from softened to aggravated, is not robust enough for annotation purposes by both experts and non-experts. In fact, it turns out that judging disagreement is a highly subjective task,

---

[1] The work reported on in this paper was funded by the VolkswagenStiftung under grant Az. 98544 "Deliberation Laboratory".

evidenced by inter-annotator agreement varying from slight to moderate as well as empirical investigation of the data. Although our inter-annotator agreement is comparable with current state-of-the-art for disagreement annotation (de Kock et al., 2022) and other work in argument mining, our analysis leads us to conclude that the annotation might fail to capture the underlying phenomenon in a meaningful way. Similarly, even though theoretical literature suggests that individual types of disagreement correlate with specific linguistic structures, we do not find strong evidence for this in our data. While we observe that some features such as expressions of uncertainty or downtoners are likely to be associated with a particular disagreement category, for instance, softened disagreement, other features like questions, are not so straightforward. Even the use of polite words and phrasings is not always indicative of the intention to soften disagreement as it can be used in a mocking or sarcastic manner. As a result of that, we adopt a bottom-up approach by exploring the data and identifying common strategies for disagreement, namely, the use of questions, repetitions and "yes, but" structure.

2. Related work

There are several typologies in theoretical linguistics that characterise types of disagreement. Muntigl and Turnbull (1998) develop a ranking typology of disagreement from the most to least face aggravating, while Scott (2002) distinguishes between foreground, background, and mixed as well as collegial, personal challenge and personal attack disagreements. Rees-Miller's classification of disagreements (2000) categorizes them into softened, neither softened nor strengthened and strengthened categories. She broadly defines the types by the presence or absence of linguistic features. For instance, softened disagreement is distinguished by the use of personal pronouns, while aggravated disagreement is characterized by judgmental vocabulary. Unlike the other two classifications, it allows viewing types of disagreement on a scale rather than as a closed class typology.

In the field of computational argumentation, Walker et al. (2012) in their Internet Argument Corpus identified disagreements as well as attacks and cases of negative attitude. Habernal et al. (2014) distinguish between rebuttal and refutation, while more recently de Kock et al. (2022) investigate rebuttal and coordination strategies, e.g., name-calling and counter argumentation. While only the latter includes a notion of escalation, what all these approaches have in common is a level of inter-annotator agreement that is comparable to ours. However, as we will show later, this does not provide us with enough evidence that the categories employed represent the underlying patterns in the data. In the following, we briefly show the initial categories of Rees-Miller as one approach to annotating disagreement.

3. Types of disagreement on Reddit

3.1. Disagreement typology

For annotation of the data, we used Rees-Miller's typology which distinguishes between softened, neither softened nor strengthened and strengthened disagreement. The annotators were asked to judge the intensity of disagreement between two commentators using this typology. Softened disagreement is marked by the intention of the speaker to mitigate the conflict to some extent, often resorting to various politeness techniques. Aggravated disagreement is used when speakers intend to escalate the conflict by using derogatory or judgmental language.

Overall, based on the annotation conducted by expert annotators, about 37 per cent of all disagreements are softened, almost 31 per cent are neither softened nor strengthened and around 32 per cent are aggravated. The distribution observed by Rees-Miller (2000) in her original paper is different from the distribution in our data with almost twice as many instances of softened and significantly fewer cases of aggravated disagreement. This may be explained by the different nature of the data in the two cases: social media data as opposed to spoken data from an academic environment.

3.2. Data

One of the largest corpora on argumentation is the US2016 corpus (Visser et al., 2019). The corpus is annotated with Inference Anchoring Theory (IAT) (Budzynska et al., 2014, 2016), a widely-used framework for analysing dialogical argumentation across languages. IAT captures how argumentation is created, fuelled and referred to by dialogical structure, a key requirement for our investigation.

We use a subset of the US2016 corpus with Reddit reactions to one of the 2016 Democratic Party presidential debates. We identify the cases of disagreement in the data and organize them in pairs: the first element in the pair is an initial statement and the second is a statement disagreeing with the previous one. In total, we identify over 130 candidates, from which we then select the disagreement pairs for our corpus by conducting a round of annotation. Two expert annotators and one of the authors of the paper independently selected the disagreements from the candidate list. The final dataset consists of 89 pairs that were unanimously agreed upon by all three annotators.

3.3. Annotation procedure

We experiment with two slightly different annotation setups for the expert and non-export annotators in order to identify whether both groups are similarly aware of destructive and constructive disagreement. University students in the field of humanities are considered to be experts, while non-experts were recruited using Prolific. All non-experts are US citizens and native speakers of English. Both expert and non-expert annotators were presented with disagreement pairs and asked to identify the type of disagreement based on Rees-Miller's (2000) classification.

There were two rounds of annotation conducted by experts with three different annotators in each round. In the second annotation round, the annotators were also provided with context for each disagreement pair. The context consisted of up to two comments before the first disagreement turn, up to two comments after the second disagreement turn and all comments between the disagreement turns.

We also conducted two rounds of annotation with non-experts, in both cases, the context was included. The data was separated into batches of about 30 disagreement pairs and 10 different annotators were hired for each batch. In the second round of annotation, non-experts had to rank the levels of disagreement (from mild to intense) instead of using the typology.

3.4. Annotation results

In order to assess the reliability of Rees-Miller's (2000) disagreement typology, we measure the inter-annotator agreement using Fleiss' kappa. Both rounds of annotation performed by experts yielded Fleiss' kappa scores of about 0.5 (0.51 and 0.48) indicating moderate agreement between the annotators. The category of neither softened nor strengthened disagreement yielded the highest kappa scores of 0.57 and 0.52. The addition of the context in the second round of annotation did not lead to improvement in the inter-annotator agreement.

Non-expert annotators achieved a significantly lower inter-annotator agreement when compared to the scores achieved by expert annotators with a Fleiss' kappa score of 0.19 indicating slight agreement. Since in the last round of annotation we worked with ordinal data, we used Krippendorff's alpha as a measure of inter-annotator agreement and observed the score of 0.29 indicating fair agreement. This is a slight but not significant improvement from the first round of non-expert annotation, which, however, suggests that using a scale for measuring disagreement may yield a higher inter-annotator agreement.

These results indicate that disagreement is a highly subjective notion. Annotators' judgments vary greatly as individuals hold different opinions on what kind of disagreement they consider to be destructive. While a certain level of familiarity with the research field improves the agreement between the annotators, achieving a significantly higher agreement is a challenging task. Error analysis reveals that there are some disagreement patterns which make it harder for the annotators to interpret the intention of the speaker. Questions and in some cases, ethotic attacks are frequently found among these patterns.

In Example 1, SgtDowns prefaces their disagreement with an ethotic attack in a form of a question implying that Wormhog is uninformed. However, the rest of the SgtDowns' response is polite, and even includes a concession. As a result, some of the annotators interpreted the disagreement as aggravated because of the ethrotic attack and others considered it to be either neither strengthened nor softened or even softened due to the polite speech following the ethotic attack.

Example 1. *Wormhog: [...] Does it matter to women in this country if they are represented in a government that continually tries to dictate what we can and can't do with our uteruses and tries to defund clinics providing birth control and reproductive health screenings to low-income women? Yes. Does she have my vote? Not in the primary.*

SgtDowns: Did you watch the debate? There were numerous times where she alluded to her gender regardless of the question. I agree it's important they have representation but many issues were across genders but she referenced it regardless.

4. Linguistic features for disagreement type identification

There are various linguistic features associated with disagreement (Rees-Miller, 2000; Scott, 2002; Langlotz & Locher; 2012; Baym, 1996). Some of those features, like intensifiers, questions, humour or sarcasm appear across the frameworks. Taking into account the features already identified in previous work as well as our own empirical exploration of the data, we are able to classify the disagreement features into four categories: graphic signals of intensification of disagreement, lexical features of disagreement, rhetorical devices and other features. The last category includes the features which we were not able to put into any of the other categories. The full list of features as per category is presented in Table I.

Table I. Features associated with disagreement	
Lexical features of disagreement	Negations, discourse particles/markers, second/first-person pronouns, questions, judgmental vocabulary, emotionally loaded vocabulary, name-calling, naming with familiarity, downtoners, absolutes, imperative
Rhetorical devices	Repetitions, sarcasm
Graphic signals of intensification of disagreement	Verbalization of emotional reaction, visual intensification
Other features	Reasoning, elaboration, acknowledgement of the perspective of the other, apology, expression of personal position, expression of personal experience

The two most common types of features that we observed in the data are negations and discourse markers, which are not associated with any particular type of disagreement but are indicative of it in general. Emotionally loaded vocabulary, judgmental vocabulary, 2nd person pronouns, absolutes and imperatives are also frequently found in the disagreement structures. These features are more likely to be associated with aggravated disagreement. Features which can indicate softened disagreement are similarly frequent, among them are verbs and expressions of uncertainty as well as downtoners. We also identified a number of questions which can be used in both aggravated and softened disagreement categories. Fig. 1 depicts a more detailed graph with the distribution of the most common features observed in the data.

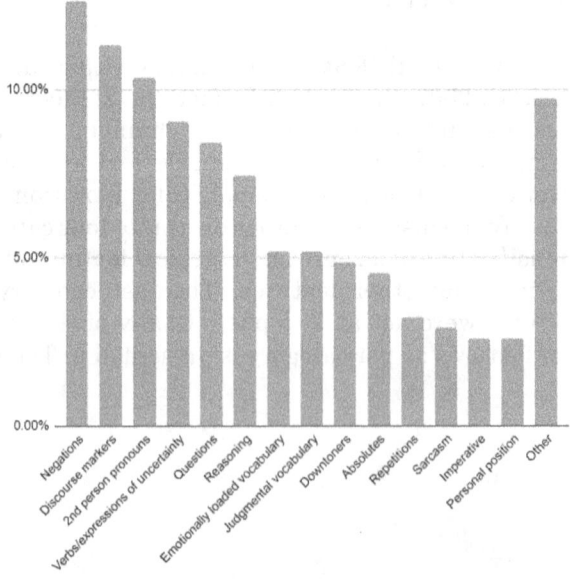

Figure 1. Distribution of the most common features

In practice, however, the use of certain features is not always indicative of the disagreement types. Even a seemingly straightforward case of naming with familiarity can be left up for interpretation. In some contexts, familiarity is endearing and suggests softened disagreement. Example 2, however, is not so unambiguous. Mitch47 uses "bruh" and "fam" to express familiarity with the previous speaker which, on the one hand, due to a certain level of anonymity on Reddit can be interpreted as rude. On the other hand, it might reflect mitch47's particular speech patterns and not an intention to aggravate the previous commentator.

Example 2. *Askew123: Better when he isn't cross-examined...*

mitch47: Bruh!!!! have a little faith fam! What exactly you mean?

Considering the results of the disagreement annotation and the observed lack of a one-to-one mapping between features and disagreement types, we conduct a deeper exploration of some of the commonly observed patterns of disagreement expression such as questions, repetitions and "yes, but" strategy in the following section.

5. Disagreement strategies

5.1. Types of questions

Questions are among the most common strategies for disagreeing. Hautli-Janisz et al. (2022) and Kikteva et al. (2022) investigate the role of questions in argumentative dialogue and observe that different types of questions play an important role in driving the conversation forward, particularly when it comes to conflicts. Assertive questions are often found to either trigger conflict or are used to disagree with the preceding content same as rhetorical questions. In our disagreement corpus, questions were found in about 23 per cent of all comments and were used to express different types of disagreement.

Due to the nature of the questions, they are often used to express indirect disagreement and can both mitigate or escalate conflict. In Example 3, the disagreement is implied when tschandler71 repeats the claim of the previous commentator in a form of a rhetorical question. By doing so they question the validity of the idea to invest 10 per cent of the GDP into solar energy. They then further escalate the conflict by asking whether the previous commentator is "nuts" if they are suggesting investing such a large percentage of GDP into renewable energy.

Example 3. *rowdyroddypiperjr: Cost of solar is less expensive than coal. If we invested 10 percent of our GDP into that we will be entirely solar by 2025. 2050 is going to happen anyway. Sooner the better*

tschandler71: Invested 10 percent? Are you nuts?

Unlike Example 3, the commentator in Example 4 seems to attempt to express their disagreement in a more polite or neutral way. While they still make an assertion about the fact that Bernie Sanders starting all of his answers in the same way is not problematic, they soften it by turning it into a question.

Example 4. *TheFaceo: I feel like Bernie starts every answer the same way*
adamdevigili: Let us understand what we're talking about"...does not sound like a bad opening does it?

Both of these cases reflect the intensity of the disagreement and the commentators' intention by means of an insult in the earlier example and careful phrasing in the latter. However, the use of questions can also make it almost impossible to correctly interpret the type of disagreement. In Example 5, for instance, it is unclear whether the commentator is simply asking for evidence regarding the lack of substance in Bernie Sanders' campaign or is challenging the statement regarding the lack of substance in the campaign.

Example 5. *Bigtuna546: [...] I feel like these debates are going to expose the lack of substance in his [Bernie Sanders'] campaign.*
SlowMotionSprint: How is there a lack of substance in his campaign?

5.2. Repetitions

Very often speakers use the exact same words as previous speakers in order to express an opposing opinion. Although counterintuitive, this strategy can be quite successful by emphasizing the target of the disagreement. Kotthoff (1993) referred to this technique as "opposition format" while other researchers described similar phenomena as "format tying" (Bolander, 2012; Muntigl & Turnbull, 1998) or "verbal shadowing" (Rees-Miller, 2000).

While this strategy is not as common as the one incorporating questions, we can still find a number of examples of repetitions being used to disagree. In Example 6, Kagawaful claims that people do not like Bernie Sanders because of his popularity on Reddit. This statement is mirrored by Velshtein who rephrases the claim with "we" instead of "you" and gives an alternative premise for it. Both premises are not in direct opposition with one another, i.e., Bernie Sanders being a bad candidate is in no way incompatible with him being popular on Reddit. The disagreement expressed via mirroring and rephrasing ("we do not like him because he's a shitty candidate") is further strengthened with the ethotic attacks on both Bernie Sanders and the first speaker ("shitty candidate" and "you morons").

Example 6. *Kagawaful: So many people are so desperate for Bernie to fail. Anyone who thinks Hilary is a better candidate is an idiot. We get it, you do not like him because he is popular on Reddit...*

Velshtein: We do not like him because he's a shitty candidate who can't substantiate any of the bullshit he spouts. The fact that all you morons lap it up only compounds on it.

5.3. The "yes, but" strategy

Even more common than questions and repetitions is the "yes, but" strategy. In our disagreement corpus, about 27 per cent of all disagreements incorporate this element in one way or another. Very often the intention behind this is to lead up to disagreement by expressing partial agreement first. The disagreement, in that case, is most likely to be formulated in a form of an alternative claim to the preceding statement. It is also important to note that "yes" does not always have to indicate concession but can also be used for the sake of politeness or even dismissal.

"Yes, but" can be expressed in a few ways with different elements either stated explicitly or implied. The most straightforward example of the strategy is when both "yes" and "but" are stated explicitly with an alternative claim following after "but". The "yes" can take different forms, for example, expressions of agreeing such as "I agree", "True" etc. are commonly used in that role. "But" can also be indicated via the use of different discourse markers such as "however" although they were rarely observed in our data. In addition to explicit "yes", agreeing can also be expressed via a rephrase of the claim made by the previous speaker or by providing a premise to the claim which can be interpreted as implicit agreement.

The three elements of the "yes, but" structure (explicit "yes", rephrase/premise" and explicit "but") are optional with the alternative claim carrying most of the implied meaning behind the strategy. Given the optionality of most of the elements of the structure, they can be combined in different ways allowing for the seven different variations of it. The only constraint is that the alternative claim cannot be preceded by an explicit "yes" with "but" being omitted.

Finite state automata are one of computer science's core concepts and are often used in computational linguistics, in particular computational morphology. Given that the "yes, but" structure is templatic, an automaton is a convenient way for formalizing the seven variations of the strategy. Finite state automata consist of states and transitions between them. In Fig. 2, circles represent the states, with 0 being the initial state and 4 being the final one. The arrows between the states indicate the transition between them which correspond to different elements of the "yes, but" structure described earlier. By following the arrows, we build a path from the initial to the final state.

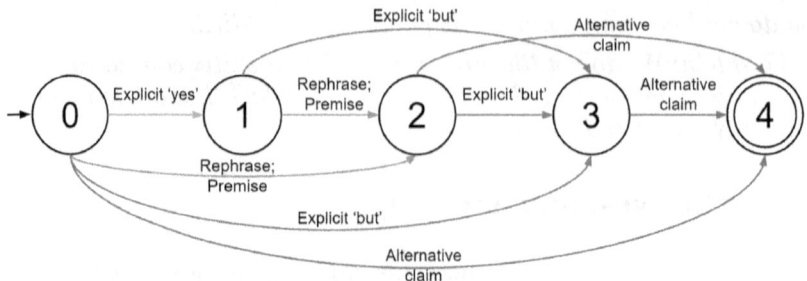

Figure 2. Finite state automaton for the "yes, but" strategy

In Fig. 2, the first path includes an explicit "yes" with either rephrase or premise which is followed by an explicit "but" and concluded with the alternative claim. Another path allows skipping the rephrase or premise while the rest of the structure is the same. Alternatively, both explicit "yes" and rephrase or premise can be skipped.

One such structure is presented in Example 7 with the finite state automaton represented in Fig. 3. In the example, fortfive implicitly agrees with the previous speaker by giving a reason for why it looks like Hillary Clinton is winning the debate, which is followed by explicit "but" and an alternative claim as to why Bernie Sanders might have more support. However, while fortfive does agree with the fact that it seems like Hillary Clinton is winning, they do not agree with what LordMacDonald implies by saying this, i.e. that Hillary Clinton is actually winning.

Example 7. *LordMacDonald: sure seems like it [Hillary Clinton is winning the debate]*

fortfive: The camera focuses on smiling faces for Clinton, but the applause sounds louder to me for Sanders.

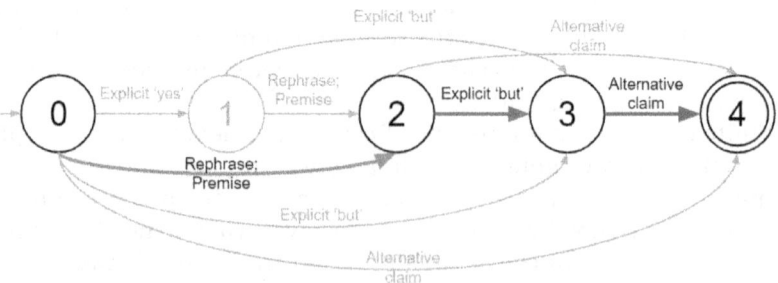

Figure 3. Finite state automaton for Example 7

One of the most interesting cases of the "yes, but" strategy is when all elements except for the alternative claim are not explicitly stated. In

Example 8, while SinisterPaige does not disagree with the fact that the participants of the debate agreed to its particular format, they reject the relevance of the claim for the argument and offer an alternative one in return. The finite state automaton for Example 8 is demonstrated in Fig. 4.

Example 8. *Soylent_Orange: True but he agreed to the format of the debate as they all did.*

SinisterPaige: The major media outlets should not be the ones dictating who wins the primaries.

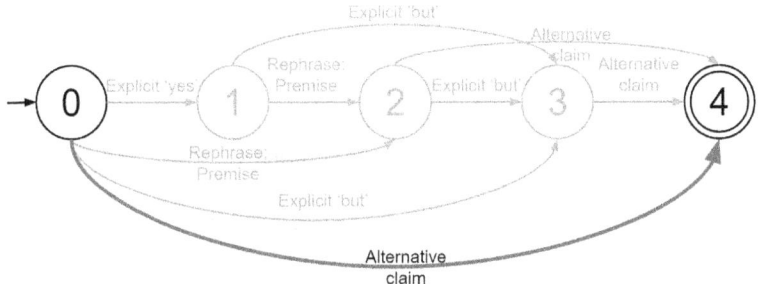

Figure 4. Finite state automaton for Example 8

Further exploration of the "yes, but" strategy and its potential use can lead to a better understanding of the nature of disagreements. However, it has the potential to pose a problem from a computational perspective due to the optionality of most of its elements.

6. Summary

In this paper, we investigate the applicability of a disagreement typology to social media data. The study results suggest that disagreement annotation is a subjective task which is significantly influenced by individual annotators' personal biases and experiences. While expert annotators achieve moderate agreement, we believe that by adopting such an approach we might fail to capture the nuanced nature of the disagreement.

We also investigate whether certain features have a clear association with different categories of disagreement. There are features which are more representative of some categories than others, however, the way they are used can influence the interpretation of the disagreement type, for instance, when politeness is used in a sarcastic manner or expressions of agreement are used to dismiss the preceding claim rather than to concede with it.

We observe and discuss several strategies used by speakers to signal disagreement, namely questions, repetitions and the "yes, but" strategy. The two, questions and "yes, but", are frequently employed, each appearing in the data in roughly 25% of all disagreement cases. Questions serve different functions in the disagreeing process, from softening opposition to triggering stronger conflict which warrants further exploration. What is interesting about the "yes, but" strategy is its templatic nature which allows for a combination of different elements in a number of unique ways and the role it plays in disagreeing process including but not limited to mitigation.

References

Baym, N. K. (1996). Agreements and disagreements in a computer-mediated discussion. Research on Language & Social Interaction, 29(4), 315–345.

Bolander, B. (2012). Disagreements and agreements in personal/diary blogs: A closer look at responsiveness. Journal of Pragmatics, 44(12), 1607–1622.

Budzynska, K., Janier, M., Kang, J., Reed, C., Saint-Dizier, P., Stede, M., & Yaskorska, O. (2014). Towards Argument Mining from Dialogue. In Proceedings of the Fifth International Conference on Computational Models of Argument (COMMA 2014) (pp. 185–196). IOS Press.

Budzynska, K., Janier, M., Reed, C., & Saint-Dizier, P. (2016). Theoretical Foundations for illocutionary structure parsing. Argument & Computation, 7(1), 91–108.

De Kock, C., & Vlachos, A. (2022). How to disagree well: Investigating the dispute tactics used on Wikipedia. In Proceedings of the 2022 Conference on Empirical Methods in Natural Language Processing (pp. 3824–3837). Association for Computational Linguistics.

Habernal, I., Eckle-Kohler, J., & Gurevych, I. (2014). Argumentation mining on the web from information seeking perspective. CEUR Workshop Proceedings, 1341.

Hautli-Janisz, A., Budzynska, K., McKillop, C., Plüss, B., Gold, V., & Reed, C. (2022). Questions in argumentative dialogue. Journal of Pragmatics, 188, 56–79.

Kikteva, Z., Gorska, K., Siskou, W., Hautli-Janisz, A., & Reed, C. (2022). The Keystone Role Played by Questions in Debate. In Proceedings of the 3rd Workshop on Computational Approaches to Discourse (pp. 54–63). International Conference on Computational Linguistics.

Kotthoff, H. (1993). Disagreement and concession in disputes: On the context sensitivity of preference structures. Language in Society, 22(2), 193–216.

Langlotz, A., & Locher, M. A. (2012). Ways of communicating emotional stance in online disagreements. Journal of Pragmatics, 44(12), 1591–1606.

Muntigl, P., & Turnbull, W. (1998). Conversational structure and facework in arguing. Journal of Pragmatics, 29(3), 225–256.

Narang, K., Davani, A. M., Mathias, L., Vidgen, B., and Talat, Z. 2022. Proceedings of the Sixth Workshop on Online Abuse and Harms (WOAH). Association for Computational Linguistics, Seattle, Washington (Hybrid), edition.

Rees-Miller, J. (2000). Power, severity, and context in disagreement. Journal of Pragmatics, 32(8), 1087–1111.

Scott, S. (2002). Linguistic feature variation within disagreements: An empirical investigation. Interdisciplinary Journal for the Study of Discourse, 22(2).

Sifianou, M. (2012). Disagreements, face and politeness. Journal of Pragmatics, 44(12), 1554–1564.

Visser, J., Konat, B., Duthie, R., Koszowy, M., Budzynska, K., & Reed, C. (2019). Argumentation in the 2016 US presidential elections: Annotated corpora of television debates and social media reaction. Language Resources and Evaluation, 54(1), 123–154.

Walker, M., Tree, J., Anand, P., Abbott, R., & King, J. (2012). A Corpus for Research on Deliberation and Debate. In Proceedings of the Eighth International Conference on Language Resources and Evaluation (LREC'12) (pp. 812–817). European Language Resources Association (ELRA).

THE CLASSIFICATION AND RECONSTRUCTION OF AUDITORY ARGUMENTS

GABRIJELA KIŠIČEK
University of Zagreb
gkisicek@ffzg.hr

Abstract
The focus of this paper is developing a classification of auditory arguments and examining its connection to argument models. A classification of auditory arguments could depend on sound producers and make a distinction between human and non-human sounds. These two main categories can be further divided into subcategories. This paper argues that differences in sound type could result in a different role in an argumentative discourse, and a different function revealed in an argument reconstruction. To test this hypothesis, auditory arguments are reconstructed using several argument models.

KEYWORDS: auditory arguments, argument reconstruction, multimodal argumentation, sound

1. Introduction

In the last decade there has been a lot of work expanding the scope of argumentation and developing research of multimodal argumentative discourse. Due to the development of technology and multimedia, argumentation scholars rightfully realized that argumentation in contemporary public discourse is no longer limited solely to verbal activity. Multimodal argumentative discourse can consist of visual images, but it could also include different nonverbal features, such as sound, smell or taste. However, it has been a challenging task to fully understand the role of the nonverbal parts of an argumentative discourse. Traditionally, arguments are understood as reasons for a specific standpoint that is expressed verbally. For centuries, argumentation scholars have been developing models and schemes for analyzing arguments, criteria for their classification, and tools for their assessment. However, basically everything in the field of argumentation is adapted to the understanding, construction, analysis and evaluation of verbal arguments. Only recently have there been initiatives to develop methods for analyzing non-verbal parts of an argumentative discourse, but mostly visual images (Groarke, 2019; Tseronis 2013; Dove 2012; Roque 2012; Feteris et al. 2011; Kjeldsen 2007; Slade 2002; Forceville, 1996).

The work in visual argumentation opened the doors for other researchers interested in other modes of argumentation, who can follow in their predecessor's footsteps in explaining how different modes operate as a part of an argumentative discourse and, thus, initiate research in multimodal argumentation.

As Godden (2015, p. 235) writes:

The study of images within argumentation theory has brought together scholars from disciplines as disparate and complementary as aesthetics and fine arts, cultural and media theory, semiotics, communication theory, rhetoric, philosophy, formal and informal logic, computer science, and mathematics. This confluence of scholarship has produced noticeable progress with, if not resolution to, some of the questions and controversies that initially surrounded the issue.

Before proceeding with a specific argument mode analysis, it is important to understand what argument "mode" stands for. Groarke (2015, p. 140) provided a figurative explanation of multimodality in argumentation.

I will define modes in terms of the ingredients (the 'material', the 'stuff') an arguer uses and arranges when they engage in an act of arguing. One might roughly compare the distinction between different modes of arguing to a distinction between different modes of building which defines the latter in terms of the ingredients (concrete, stone, bricks, wood, etc.) builders use.

Analyzing public discourse: in public policy debates, the advertisement industry, and also in "everyday argumentation", we can notice that people use more modes than just words in the attempt to prove their point, or to justify their standpoints. Kjeldsen (2015, p. 120) summarized a wide body of research conducted by numerous scholars interested in the variety of genres and forms of expression like: image events (Delicath and DeLuca 2003), flag waving (Pineda and Sowards, 2007), slide show presentations (Kjeldsen 2013), brain imaging (Gibbons 2007), folk arts (Roberts 2007), fashion (Torrens 1999), prison tattooing (2007), yarn bombing (Hahner 2014), needlework (Goggin 2003), place, architecture and memorials (Fleming 1998; Blair et al. 2011), museums (Balter-Reitz 2003), artworks (Groarke 1996; Chryslee et al. 1996), news magazine covers (Tseronis 2015), and photography (Finnegan 2001, 2003; Gronbeck 2007; Hahner 2013).

However, auditory arguments are not yet as understood nor generally accepted. This paper attempts to contribute to this uncharted territory, and strive to include auditory arguments as an important (sometimes even essential) part of an argumentative discourse. In order to be able to fully understand the role and importance of an auditory argument, it is necessary to develop methods for its interpretation, to be able to differentiate between sounds with and without argumentative relevance i.e. to detect an auditory argument, to (re)construct it and, ultimately, to assess it.

2. Auditory arguments – types and categories

If we understand modes as an "ingredient" of an argumentative discourse, then we notice that sound can be treated as one of the ingredients alongside with words, images, etc. As well as images, sound can be of different types with different roles. The two main categories of sound are human and non-human, which can be divided into subcategories. According to their role, the most important distinction is between sounds with and without argumentative relevance. Those with argumentative relevance can also be divided into two categories: those that strengthen the verbally expressed argument - usually prosodic features of the spoken language (many of the examples discussed in Kišiček, 2014) – and sounds which can serve as a reason to support a standpoint. The latter one is referred to as auditory arguments and can be defined as "an attempt to provide rational evidence for a conclusion using non-verbal sounds instead of or (more frequently) in addition to words." Groarke (2018, p. 1)

Table I. Type of sound

Type of sound	Role of sound	Argumentative relevance
human sounds connected to language – prosodic features (voice quality, intonation, rhythm, tempo etc.)	purely aesthetic and/or possible rhetorical function without argumentative relevance (many voiceovers in commercials, public speaking appearances, etc. which might contribute to the speaker's ethos) (see, e.g. Kišiček, 2018)	-
human sounds independent of language - laughing, crying, screaming, whistling...	possible rhetorical function contributing to the speaker's pathos	-
non-human sounds – artificial sounds	auditory symbols for certain actions (bell for ending a talk, etc.)	-
human sounds – prosodic features (voice quality, intonation, rhythm, tempo, etc.)	may contribute to the strength of a verbal argument (see. Kišiček, 2014, Groarke and Kišiček, 2016)	+/-

human sounds - accent	may have an essential role in an argument (see, e.g. Kišiček, 2020)	+
human sounds – laughing, crying, screaming, whistling…	may have an essential role in an argument (see, e.g. Groarke and Kišiček, 2018)	+
non-human sounds – animal sounds	may have an important role in an argument but need further examination (Groarke, 2018)	+
non-human sounds – artificial sounds (sirens, engines, alarms…)	may have an important role in an argument but need further examination (see e.g. Kišiček, 2019)	+

Evidently, some of the sound types and their role still need to be examined. One step closer to achieving this goal is determining the way in which sounds can be identified, analyzed, and classified as argumentatively relevant. Luckily, argumentation theory provided us with different models for argument reconstruction. Although none are specifically designed for auditory arguments, this paper will borrow some of the argument reconstruction models. An argument reconstruction of auditory arguments could provide an answer to several questions: can sound be a conclusion or does it function only as a premise? Can sounds have a role of a warrant or a locus? Which possible model of verbal (or visual) argument reconstruction can be applied to auditory arguments? Can an auditory argument reconstruction provide a direction to an argument evaluation?

The focus of this paper will be on presenting auditory argument reconstruction using three different methods; two recent, contemporary models: ART method developed by Groarke (2019), AMT or Argumentum Model of Topics developed by Riggoti and Greco (2019), and the well-known Toulmin model (Toulmin, 1969) which presents a traditional approach to argument reconstruction.

3. Auditory arguments reconstruction

3.1. Groarke's ART approach

Groarke (2019, p. 333) developed a method which consists of three parts: A is designed to ACKNOWLEDGE visual and other non-verbal (multimodal) argument components. R is a method that can be used to REPRESENT the components and the structure of visual arguments. T is a set of tools that can be used to TEST visual arguments in a way that determines whether they are weak or strong. Groarke explains the ART method:

Approach to visual arguments is expressly designed in a way that reduces the role that the verbal interpretation of visuals needs to play in argument analysis. It does so by emphasizing a visual account of the visual rather than verbal interpretations of their visual content. (p. 337)

In his paper, Groarke (p. 346) conducts an ART analysis on a "real life" example of an argumentative discussion. Husband and wife discuss the possibility of visiting a famous castle in Germany. One of the interlocutors uses a visual image of a castle and claims: we should visit the Castle. Before going into the analysis of the specific example, Groarke (p. 342) explains how to conduct an ART analysis. It consists of two parts: a "Key Component" (KC) table which identifies the argument's premises and conclusions; and an argument diagram that depicts its structure.

Following his example, we will add an auditory mode and change the vacation destination from Germany to The Maldives. It can easily be imagined that the interlocutors, or even tourist agencies, include sounds to attract tourists. Nowadays, we are all used to seeing photographs of sandy beaches, the clear blue sea, and breathtaking scenes of sunsets. These images successfully support claims that one should visit The Maldives. Is it difficult to imagine that sounds could also be included? Although different from visuals, sounds (of waterfalls, waves, nature, animals, etc.) might also be used to attract tourists using lines like: Imagine waking up with this? {sound of waves and birds} (https://www.youtube.com/watch?v=vPhg6sc1Mk4) The difference between images and sounds is in their verifiability. You know what the castle looks like, you know how the Eiffel Tower looks like. Sounds are less distinctive and less specific and, ultimately, could be less persuasive, but they can nevertheless have their role as one of the ingredients in a multimodal argumentation.

So, let us examine how this might work in Groarke's approach and KC tables.

Key components	Role	Explanation
Photo of beach in The Maldives	Premise (p)	Visual representation of the vacation location
Sounds of waves and birds singing https://www.youtube.com/watch?v=vPhg6sc1Mk4	Premise (a)	Auditory representation of mornings in The Maldives
We should go to The Maldives for a vacation	Conclusion (c)	Verbal claim

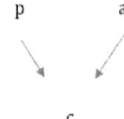

Figure 1. KC Table and Diagram for vacation at The Maldives

This example vividly presents an argument imaginable in "real-life" situations where translating images or sounds into words would seriously decrease the power of an argument. When a person decides on a vacation destination, the decision frequently depends on the visual representation of that destination (hotel, beach and other attractions). The decision can be additionally supported with sound. Or refuted! Let us examine another completely opposite situation, when a person rejects an idea of moving to another city based on a sound.

A couple is trying to decide on whether they should move to Chicago or not. The wife, who is strongly against moving, uses sounds supporting the claim that Chicago is a dangerous city. And they do not want to live in a dangerous city. She then plays an audio recording of Chicago by night which consists of gunshots, yelling, police sirens (https://www.youtube.com/watch?v=rOuyEZMKIic). The audio recording can be additionally strengthened by emphasizing the source. If the person asserting the claim was also the person who made the recording. Or if the recording is done by some neighborhood watch members. This might overcome the weaknesses of the verifiability of a sound. So, a recording might consist of a verbal part, not translating the sound into words, but naming the source (usually we see this on videos used in news reports).

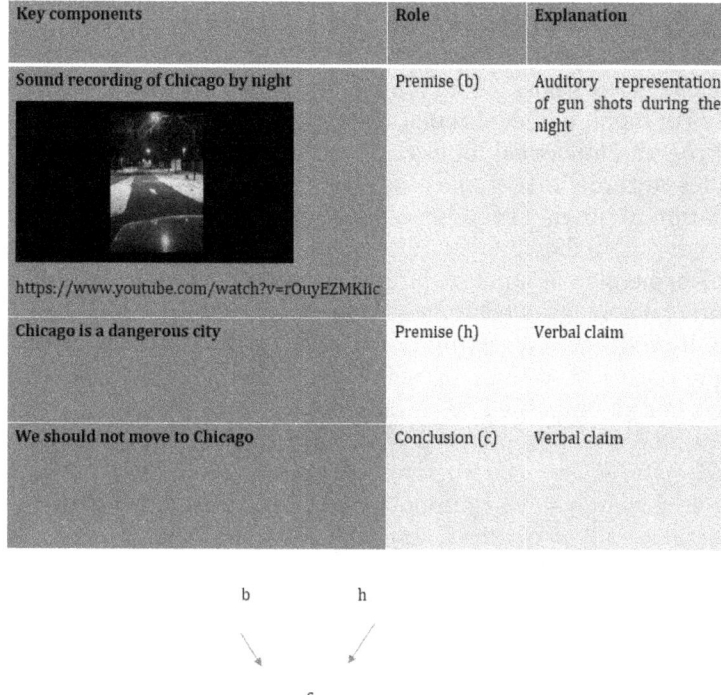

Figure 2. KC Table and Diagram against moving to Chicago

In this example, the sound of gun shoots which are connected with gunfights, criminals, gangs, etc. support the danger people might be exposed to.

In real-life argumentation, the combination of visuals, sounds, and words is often used because technology nowadays enables us to do so. If you want to persuade someone to go to the forest more often and you support your claim with calming sounds of nature and animals, it is more persuasive than using words to describe how birds sing or what waterfalls sound like.

Or, for instance, if I want to persuade someone to join me for a concert of the violin virtuoso Itzhak Perelman, I could describe the beauty of this music, or I could just play a recording of it. How can someone explain the sound of music or nature or animals?

It is certainly possible, language gives us numerous opportunities, but the argument loses its persuasiveness regardless of the speaker's eloquence.

3.2. Riggoti and Greco AMT approach

One of the novel approaches to argument analysis and evaluation is The Argumentum Model of Topics, or AMT (Riggoti and Greco in 2019). The model, as its authors claim, presents a combination of the traditional concept of topoi, and contemporary approaches to argumentation (mainly the Pragma-Dialectical approach). The authors emphasize that AMT provides argumentation analysts with a perspective for the analysis and evaluation of single arguments, (p. 208) and further on state that "AMT neatly distinguishes between premises, as maxims, which represent the logical principles of support of arguments, and premises that reflect the arguers' knowledge, worldviews and cultural expectations." (ibid, p. 209)

So, it might be interesting to see if this model would be appropriate for argumentative discourse in which the verbal part of an argument is replaced or accompanied by sound.

The model of argument consists of three levels (see (Rigotti & Rocci 2001, Rigotti & Greco 2010, Rigotti & Greco, 2019): first, the locus itself as the source from which arguments are taken; second, a series of inferential connections called maxims; and the third is that each of the maxims activates a logical form, such as the modus ponens or the modus tollens. An important part of the AMT is the *endoxon*, which is an opinion that is accepted by the relevant public or by the opinion leaders of the relevant public.

Having this in mind, the authors describe AMT as a model which "intends to explain argumentation as it happens in communicative interactions… (…) which take place within social relationships" (*ibid*. xiii). We can examine where (and if) the sound fits into the model. So, let us examine an example in which we conclude that an accident of some sort has happened based on the sound of a siren. Sirens are most commonly a sign of emergencies and are used by the ambulance, fire departments, and police. If we are sitting at home and we hear siren on the street, we conclude that someone is in need of emergency help and that, possibly, an accident has happened. The AMT reconstruction could be as follows:

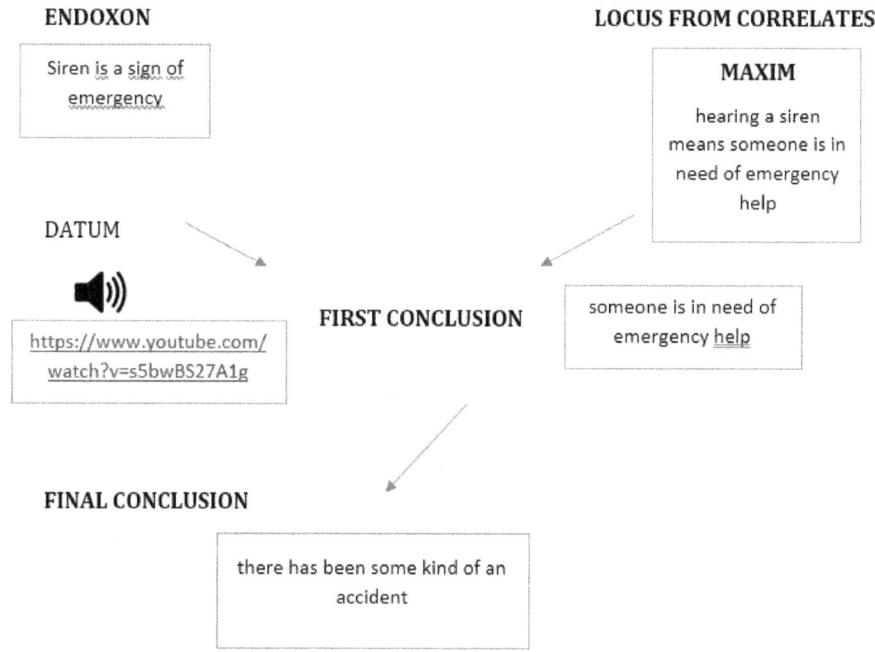

Figure 3. ATM reconstruction of a multimodal argument

This example shows that the auditory mode can be a datum (or a minor premise) in an argument reconstruction and that, in this example, the locus from correlates can explain how the reasoning works in this specific situation. Rigotti and Greco (2019, p. 253) provide a taxonomy of loci and explain the locus from correlates: The locus from correlates identifies a very specific class of arguments based on entities that logically (or semantically) correlate.

The type of maxims that is associated to this locus always draws from the habitudo of correlation, giving rise to this type of arguments: "if he is married, he must have a wife"; "if there has been a theft, there must be a thief"; if he is a father, his child must be somewhere".

Following that explanation, it can be said: "if there is a siren for emergency, there must be an emergency somewhere"

Of course, based on the specific tone of a siren, one might argue that it is not an accident but a crime scene because it is not an ambulance siren but a police siren. But, it nevertheless falls into the category of emergency situations. Furthermore, one might recognize a specific tone of a siren which indicates the police or a security convoy in situations like visits of important foreign politicians and conclude that no accident nor emergency situation occurred. However, similarly to visual or verbal arguments, auditory arguments can also be misinterpreted and, therefore, could and should be evaluated. However, for this purpose, it is important to emphasize that they can be analyzed as a part of an argumentative discourse using the AMT model.

3.3. Toulmin model approach

Perhaps the most well-known approach to argument reconstruction and the analysis of an internal structure of an argument is the co-called "Toulmin model", or the Toulmin scheme (Toulmin, 1958).

As Hitchcock and Verhrij (2006, p. 2) write: this scheme differed radically from the traditional logical analysis of a micro-argument into premises and conclusion.

Furthermore, they explain the model according to the steps corresponding to the argument (re)construction. The model consists of a Claim (C), Data (D) or facts we use as a ground for our claim, a Warrant (W) or the general principle which connects the Data with the Claim. A Backing (B) of a claim, a Rebuttal (R) which challenges this general principle, and a Qualifier (Q). Warrants, as Toulmin notes, confer different degrees of force on the conclusions they justify, which may be signaled by qualifying our conclusion with a qualifier, such as 'necessarily', 'probably' or 'presumably'.

After roughly explaining the principles of this well-known model, the question is whether sound can have a role in some of the phases of argument construction.

As Toulmin (1959, p. 90) writes: Let it be supposed that we make an assertion, and commit ourselves thereby to the claim which any assertion necessarily involves. If this claim is challenged, we must be able to establish it—that is, make it good, and show that it was justifiable.

Let us examine the following example. A mother in The Ukraine claims that her family needs to seek shelter in order to save their lives. One of the children challenges this claim saying that there is no danger. So, the mother needs to provide a foundation for her claim. Toulmin (p. 90) explains the data: "If our challenger's question is, 'What have you got to go on?', producing the data or information on which the claim is based may serve to answer him". So, can sound be the information produced by the one asserting the claim?

The mother could answer "Listen!" and the challenger would hear the sound of a military aircraft.

As Toulmin continues to explain (p.2): "Supposing we encounter this fresh challenge, we must bring forward not further data, for about these the same query may immediately be raised again, but propositions of a rather different kind: rules, principles, inference licenses or what you will, instead of additional items of information.

This principle is called a warrant. So, the mother calls on a principle: if you hear military aircraft, there is a possibility of an airstrike.

This general principle can be rebutted by saying: aircrafts are just conducting exercises or demonstrations. And not every single military aircraft will drop bombs.

To test the warrant as well as its rebuttal, it is important to know the context. If a military aircraft flies over Croatia on some random day in 2023, it might be a case of military exercise. When it flew over Croatia in 1992, people sought shelter. So, if this discussion takes place in The Ukraine in 2023, there is high probability of an airstrike and, therefore, it is necessary to seek shelter.

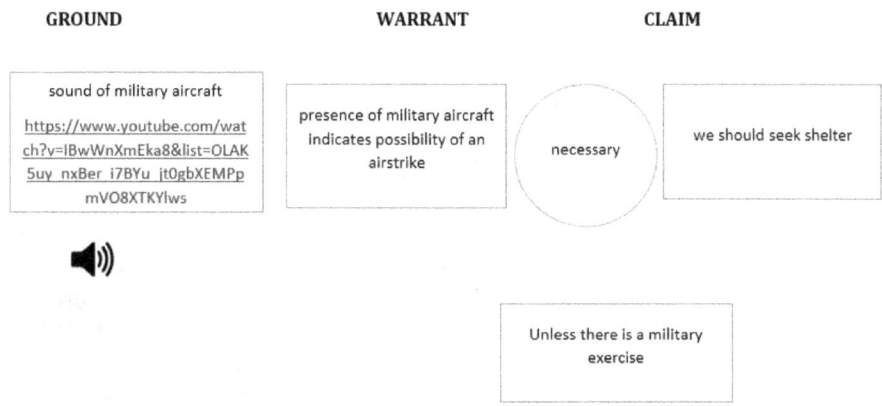

Figure 4. Toulmin model of auditory argument reconstruction

In this sense, auditory arguments are no different than verbal or visual arguments. As Blair writes (2015, p. 218):

Visual arguments can, like their verbal counterparts, have implied unexpressed or implicit components and rely on presumptions; and, like verbal arguments, visual arguments normally rely on an understanding of their contexts (what historical event occasions them, who their intended audiences are, shared background knowledge, etc.) in order to be meaningful.

The challenger can also question the interpretation of a sound and say: "This is not a sound of a military aircraft carrying bombs but only a radar recon airplane".

And again, similarly to words, sounds can be misinterpreted. And again, that calls for an elaborated model of auditory argument evaluation.

Importantly though, sounds can certainly serve as grounds for asserting a claim. Sounds we hear can be connected to different claims based on general principles backed by science, empirical research, and statistics. For instance, sounds of thunder are good indicators of storms and rains, specific sounds of engines are connected with different malfunctions of machines, different human sounds such as crying or laughing are connected with specific emotional states.

283

Therefore, I believe that auditory arguments can be an essential part of an argument construction or patterns of an argument in the Toulmin approach to argument analysis.

4. Conclusion

The field of multimodal argumentation is still relatively new and unexplored. But in a world of evolving technology, it cannot be ignored. Changes in technology introduce changes in communication and argumentation. People frequently use different modes to assert their claims, to defend their standpoints and to support their conclusions. Although words are, and always will be, the primary means of human communication, visuals, sounds and other nonverbal features are becoming more frequent in public discourse. Social media, the advertisement industry, but also political campaigns and policy debates use all available means to achieve their goals. This paper focused on the importance of sound in multimodal argumentation. However, to determine its importance, an analysis needs to distinguish between different sound types. There are human sounds, which are connected with speech i.e. prosodic features (voice quality, intonation, tempo, accent etc.), and those which are independent of speech such as laughing, whistling, crying, and non-human sounds such as sirens, alarms, animal sounds, sounds of engines, planes, etc.

Sounds also differ according to their importance and argumentative relevance. Some sounds just accompany verbal messages and contribute to their persuasive effect, but others, as elaborated in the paper, can stand alone and function as a premise to support a conclusion.

In this paper we have examined the reconstruction of an argument consisting of a sound using three different models for argument analysis: ART model, AMT model, and Toulmin model. Although different, all of the models serve the same goal: to reveal the argument structure, to differentiate between premise and conclusion. One step further in this task would be to examine whether there can be a specific model for the reconstruction of auditory arguments which might include all the particularities and specific qualities of sound, which could then contribute to the more detailed layout of an auditory argument.

This would then contribute to the next endeavor – auditory argument evaluation.

References

Balter-Reitz, S.J. (2003). She blinded me with science: Argumentation in the Indianapolis Children's Museum. In F.H. van Eemeren, J.A. Blair, C.A.

Willard, and A.F. Snoeck-Henkemans (Eds). *Proceedings of the Fifth Conference of the International Society for the Study of Argumentation* (pp. 63-68). Amsterdam: Sic Sat.

Blair, C., V.W. Blathrop, N. Michel (2011). Enthymemes of relation and national legitimation:
argument and tombs of the unknown. In F.H. van Eemeren, B. Garssen, D. Godden & G. Mitchell (Eds). *Proceedings of the Seventh International Conference of the International Society for the Study of Argumentation* (pp. 365-367). Amsterdam: Sic Sat.

Chryslee, G.J., S.K. Foss, and A.L. Ranney (1996). The construction of claims in visual argumentation. *Visual Communication Quarterly 3(2)*, 9–13.

Delicath, J.W., and K.M. Deluca. (2003). Image events, the public sphere, and argumentative practice: The case of radical environmental groups. *Argumentation 17(3)*, 315-333.

Dove, I. (2012). Image, evidence, argument. In F. H. van Eemeren, & B. Garssen (Eds). Topical themes in argumentation theory (pp. 223–238). Amsterdam: Springer.

Feteris, E, Groarke, L., Plug, J. (2011). A pragma-dialectical analysis of the use of topi that are based on common cultural heritage. In E. Feteris, B. Garssen, & F. Snoeck Henkemans (Eds). Keeping in touch with pragma-dialectics (pp. 59–75). Amsterdam: John Benjamins.

Finnegan, C.A. (2003). Image vernaculars: Photography, anxiety, and public argument. In F.H. van Eemeren, J.A. Blair, C.A. Willard, & A.F. Snoeck Henkemans (Eds). *Proceedings of the Fifth Conference of the International Society for the Study of Argumentation* (pp. 522-534). Amsterdam: Sic Sat.

Fleming, D. (1998). The space of argumentation: Urban design, civic discourse, and the dream of the good city. *Argumentation (12)*,147–166.

Forceville, C. (1996). *Pictorial metaphor in advertising*. New York: Routledge.

Gibbons, M.G. (2007). Seeing the mind in the matter: Functional brain imaging as framed visual argument. *Argumentation & Advocacy 43(3 & 4)*, 175–188.

Godden, D. (2015). Images as Arguments: Progress and Problems, a Brief Commentary. *Argumentation (29)*, 235–238.

Goggin, M.D. (2003). Arguing in "Pen of Steele and Silken Inke": Theorizing a broader base for
argumentation. In H. van Eemeren, J.A. Blair, C.A. Willard & A.F. Snoeck Henkemans (Eds). *Proceedings of the fifth conference of the international society for the study of argumentation* (pp. 383-392), Amsterdam: Sic Sat.

Groarke, L. (1996). Logic, art and argument. *Informal logic (18)*, 105–129.

Groarke, L. (2015). Going multimodal: What is a mode of arguing and why does it matter? *Argumentation (29)*, 133–155.

Groarke, L., Kišiček, G. (2016). Compassion, Authority and Baby Talk: Prosody and Objectivity. In L. Beanacquista & P. Bondy (Eds). *Proceedings of the Ontario Society for the Study of Argumentation Conference, Vol. 11* (pp 324-337). Windsor: University of Windsor.

Groarke, L., Kišiček, G. (2018). Sound Arguments: An Introduction to Auditory Argument. In S. Oswald & D. Maillat (Eds). *Argumentation and Inference: Proceedings of the 2nd European Conference on Argumentation* (pp. 177-199). London: College Publications.

Groarke, L. (2018). Auditory Arguments: The Logic of 'Sound' Arguments. *Informal Logic, 38 (3)*, 312–340.

Groarke, L. (2019). Depicting Visual Arguments: An „ART "Approach. In F. Puppo (Ed). *Informal Logic: A „Canadian "Approach to Argument* (pp. 332-375). Windsor: Windsor Study in Argumentation.

Gronbeck, B.E. (2007). Theorizing visual argumentation. Three approaches to Jacob Riis. In F.H. van Eemeren, J.A. Blair, C.A. Willard, & B. Garssen (Eds). *Proceedings of the sixth conference of the international society for the study of argumentation* (pp. 541–548). Amsterdam: Sic Sat.

Hahner, L. (2013). The riot kiss. Framing memes as visual argument. *Argumentation and advocacy (49)*, 151–166.

Hahner, L. (2014). Yarn Bombing and argument by aesthetic appropriation. In C. Palczewski (Ed). *Disturbing Argument*. New York: Routledge.

Hitchcock, D.; Verheij, B. (2006). Introduction. In D. Hitchcock & B. Verheij (Eds). *Arguing on a Toulmin Model: New Essays in Argument Analysis and Evaluation* (pp. 1-25). Amsterdam: Springer.

Kišiček, G. (2014). The role of prosodic features in the analysis of multimodal argumentation. In B. Garssen, D, Godden, G. Mitchell & F. Snoek Henkemas (Eds). *Proceedings of the 8th International Conference of the International Society for the Study of Arrgumentation*, (pp. 730-741). Amsterdam: Sic Sat.

Kišiček, Gabrijela (2016). Prosodic features in the analysis of multimodal argumentation. In D. Mohammed, Dima & M. Lewinski (Eds). *Argumentation and reasoned Action: Proceedings of the 1st European Conference on Argumentation, Volume II.* (pp. 629-642). Milton Keynes: Collage Publications.

Kišiček, Gabrijela (2018). Persuasive Power of prosodic features. *Argumentation and Advocacy, 54 (3)*, 124-134.

Kišiček, G. (2019). Auditory arguments – importance of sound in an argumentative discourse. In B. Garssen, D. Godden, G. R. Mitchell, J. H. M. Wagemans (Eds). *Proceedings of the 9th Conference of the International Society for the Study of Argumentation* (pp. 640-651). Amsterdam: Sic Sat.

Kišiček, Gabrijela (2020). Listen Carefully! Fallacious Auditory Arguments. In J. Cook (Ed). *Proceedings of the Ontario Society for the Study of Argumentation Conference, Vol. 12* (pp. 17-32). Windsor: University of Windsor.

Kjeldsen, J. (2007). Visual argumentation in Scandinavian political advertising. A cognitive, contextual and reception-oriented approach. Argumentation and Advocacy (43), 124–132.

Kjeldsen, J. (2013). Strategies of visual argumentation in slideshow presentations: The role of visuals in an Al Gore presentation on climate change. *Argumentation 27(4)*, 425–443.

Kjeldsen, J. (2015). The Study of Visual and Multimodal Argumentation. *Argumentation (29)*, 115–132.

Pineda, R.D., S.K. Sowards (2007). Flag waiving as visual argument: 2006 immigration demonstrations and cultural citizenship. *Argumentation & Advocacy 43(3 & 4)*, 164–174.

Riggoti, E., Greco, S. (2010). Comparing the Argumentum Model of Topics to Other Contemporary Approaches to Argument Schemes: The Procedural and Material Components. *Argumentation (24)*, 489–512.

Riggoti, E., Greco, S. (2019). Inference in Argumentation: A Topics-Based Approach to Argument Schemes. Amsterdam: Springer.

Roberts, K.G. (2007). Visual argument in intercultural contexts: Perspectives on folk/traditional art. *Argumentation & Advocacy 43(3 & 4)*, 152–163.

Roque, G. (2012). Visual argumentation: A further reappraisal. In F.H. van Eemeren, & B. Garssen (Eds). Topical themes in argumentation theory (pp. 273–290). Amsterdam: Springer.

Slade, C. (2002). The real thing: Doing philosophy with the media. New York: Peter Lang.

Torrens, K.M. (1999). Fashion as argument: Nineteenth-century dress reform. *Argumentation and advocacy (36),* 77–87.

Toulmin, S. (1958). *The Uses of Argument.* Cambridge: Cambridge University Press.

Tseronis, A. (2013). Argumentative functions of visuals: Beyond claiming and justifying. In Virtues of argumentation. In D. Mohamed & M. Lewinski (Eds). *Proceedings of the 10th international conference of the Ontario Society for the Study of Argumentation (OSSA)* (pp. 1–17). Windsor: University of Windsor.

Tseronis, A. (2015). Multimodal argumentation in news magazine covers: A case study of front covers putting Greece on the spot of the European economy crisis. *Discourse, Context and Media, (7),* pp. 18-27.

GAMING THE COGNITIVE PRINCIPLE OF RELEVANCE IN SOCIAL MEDIA

MANFRED KRAUS
University of Tübingen
manfred.kraus@uni-tuebingen.de

Abstract

According to the inferential model of communication the meaning of a message is determined by inferences recipients draw on the sender's intentions from the input provided. Arguing is thus both a social and a cognitive process. In this cognitive act the decisive factor is not truthfulness of a message, but the degree to which it meets a recipient's expectations of relevance. According to the Cognitive Principle of Relevance, argumentative input is deemed relevant when it connects with prior information so as to yield conclusions positively relevant to the recipient. The more attractive those conclusions, the more relevant the input will appear; but the greater the effort required in processing, the less relevance will be imputed. Applied to social media, this model suggests that messages posted may parasitically exploit this mechanism and 'game the system' for propagandistic purposes. Since in social media user-created content frequently meets with audiences with low cognitive interest, yet high personal involvement, the principle of relevance may produce paradoxical effects: The more sensational an information is, the more attention it will attract. Messages will also be subject to a process of selection equally guided by relevance: The more contents tally with recipients' preconceived views, the more relevance they will be credited with. Recipients will adopt 'relevant' contents into their 'belief boxes', but also forward them to their own social communities, thus supporting their rapid and widespread proliferation. By these mechanisms, messages posted in social media tend to strengthen rather than challenge preconceived opinions and to create 'filter bubbles' or 'echo chambers'.

1. Introduction

Social media are an undeniable fact in our time. But more than that: They are a social fact in the sense Émile Durkheim defines that term (1895, p. 19: "Est fait social toute manière [...] qui est générale dans l'étendue d'une société donnée tout en ayant une existence propre, indépendante de ses manifestations individuelles"). For, according to the Austrian social media expert Christian Fuchs, the world wide web, in and on which they live, is in itself a social fact. "All media and all software are social," he writes, "in the sense that they are products of social processes. Humans in social relations produce them. [...] Web technologies therefore are social facts" (Fuchs, ³2021, 37). Hence, since social media are social by definition, argumentation in social media must essentially be regarded as social argumentation.

When social media first appeared on the stage, the great expectation across the general public as well as across the community of professional argumentation scholars was that, because of their non-hierarchic structures, exchange of arguments in social media would become supportive of democratic structures and of societal cohesion; yet recent experience over the last few decades has quickly disabused us of such optimism and demonstrated that quite on the contrary they exhibit a dangerously big potential for being disruptive of society. Some scholars, such as Harvard law scholar Cass Sunstein in his books titled *Republic.com* (2001b) and *#Republic* (2017), have therefore been particularly critical of social media and particularly pessimistic about their potentially detrimental effects on the public sphere in an open society.

In the following, I will argue that the danger of being exploited for improper purposes is systemic with argumentation in social media, and try to explain why it is. In my argument, the cognitive dimension of social argumentation will play a crucial role. I will first briefly explain the mechanisms of the inferential model with an emphasis on the factor of relevance and on ways of argument processing, and will then apply it to social media argumentation. This will lead to some observations on the interdependence of cognitive relevance and the attraction of attention and on the role of relevance in argument selection in social media, and to the identification of the main points of vulnerability of the system to improper exploitation.

2. Inferentialism, relevance, and argument processing

For this purpose, I will apply the inferential model of communication and the principle of relevance. Both have been essentially developed by Dan Sperber and Deirdre Wilson in the 1980s and 1990s (Wilson & Sperber, 1981; 1986; 2012; Sperber, 1982; Sperber & Wilson, 1986; 1990; 1995) by way of a constructive criticism of Paul Grice's famous maxims (1975, pp. 45–46; 1989, pp. 26–27), especially those of quality and relation (see Tindale, 2015, pp. 111–118). According to the inferential model, communication does not work as a simple transport of units of information from a communicator to a recipient, as the traditional view would have it, but by way of inferences that recipients may reasonably and legitimately draw about the communicator's communicative intentions from the informational input provided by the communicator. In a similar vein, Robert Brandom defines: "Asserting cannot be understood apart from inferring" (1994, p. 158), and "[t]he function of assertion is making sentences available for use as premises in inferences" (1994, p. 168). Hence, according to the inferential model, arguing is not only a social, but also a cognitive process on the side of the recipient (for a cognition-based model of argument, see also Hample, 1979; 1981; 1985). Relevance theorists further emphasize that in this cognitive act of processing information the decisive factor is not the truthfulness of the message, as Grice had still postulated with respect to his maxim of quality (1975, p. 46; 1989, p. 27), but rather the degree to which it meets a recipient's expectations of relevance (on the difference of audience relevance from premise relevance, see, e.g., Blair, 1989, pp. 77–80; Tindale, 1999, pp. 101–102; 112–113; 2015). Recipients will predominantly process those pieces of information that they deem relevant to themselves and to their preconceived 'belief boxes'.

Yet, according to what Sperber and Wilson call the Cognitive Principle of Relevance, any argumentative input is regarded as relevant to a recipient when it connects with prior background information in such a way as to yield conclusions that are or at least appear relevant to the recipient. As a consequence, since human cognition is naturally guided by a desire for maximization of relevance and minimization of effort, two antagonistic criteria apply: The more attractive those conclusions appear to a recipient, the more relevant the input will be considered to be; on the other hand, the greater the effort required in processing, the less relevance will be ascribed to a message. Tindale in that context speaks of "the production of the maximum amount of contextual effects for the least amount of processing effort" (2015, p. 113). This mechanism of "self-persuasion", as it were, by the recipient of a message in response to attempted persuasive communication has also been aptly described by the so-called Cognitive Response Model propounded by Anthony Greenwald (1968) and by Richard Petty and John T. Cacioppo (Petty & Cacioppo, 1981, pp. 225–254; Cacioppo, Harkins & Petty, 1981).

At this point, so-called dual-process models of argument processing come into play, such as the Elaboration Likelihood Model developed by Richard Petty and John T. Cacioppo (Petty & Cacioppo, 1986b; Petty & Wegener, 1999; O'Keefe, 2008), or similar models described by Dan O'Keefe (1996; 2009). These models imply that only when an argument meets a recipient with high intellectual abilities, high cognitive interest, high personal involvement, and low distraction, it will be predominantly processed according to its argumentative content (which is called the "central route" of processing), whereas with recipients with poorer intellectual abilities and lower cognitive interest (or a higher level of distraction) argumentative content will be likely to be sidestepped in favour of the easier route of more non-intellectual factors of persuasion such as emotional appeal or personal sympathy (called the "peripheral route"), which is not only faster, but also more effortless (see Petty & Cacioppo, 1986a). The audience's attention, however, will of course also be substantially stimulated by imputations of relevance.

3. Applications to social media

These theories and mechanisms are perfectly applicable to the so-called social media. Under social media we understand electronic media such as blogs, social networks, apps, chatrooms, messenger services, photo or video platforms, wikis, search platforms, etc., viz. services such as Facebook, WhatsApp, Twitter, Telegram, Instagram, YouTube, TikTok, Google, Wikipedia, and many more. These are different from the classical mass media by a number of features: Not only are all entirely digitally based, but they also have a potentially global range, operate in real time, are easily accessible and usable to anyone, offer users a fee choice of options and times of accessing, and, last, but most importantly, are characterized by flat hierarchies that lack the traditional one-dimensional hierarchy of sender and audience. According to Douglas Walton (2007, p. 5), these media are dialogical and interactive by their very nature, and hence potentially capable of argumentative exchange. What is more, they are even open to a kind of simultaneous multi-agent communication in which the roles of sender and recipient become interchangeable and progressively blurred. Contents are as a rule user-created instead of being accounted for by a responsible and liable editorial office, and very often not even supervised and monitored by moderators. More often than not it is even easy to the personal identity of the sender.

When we apply the Principle of Cognitive Relevance to these types of media, it emerges that with respect to argumentative content posted therein this mechanism of cognitive relevance is in fact vulnerable to parasitical exploitation for purposes other than proper argumentation. Since exchange of arguments in these media is a multi-agent "game of giving and asking for reasons", as Brandom (2000, p. 197; 2008, p. 112) chooses to say, it is easy to outwit, or, so to speak, 'game the system' for improper purposes such as propaganda. This is particularly precarious in the context of political debate in the public sphere, which is arguably the most sensitive, but also the most prevalent field of argumentation in social media.

4. Cognitive relevance and attracting attention

How does the Principle of Cognitive Relevance operate when applied to arguments posted in social media? In the flat communicative hierarchies of those media, user-created messages often meet with audiences with low cognitive interest on the one hand, yet high personal involvement on the other. Especially in such cases, the principle of relevance can lead to paradoxical effects.

First of all, the mere fact that a message is posted, for the incidental recipient (who may in the first place have come across it by mere chance while surfing the web) already creates an ostensive stimulus that it must have some relevance whatsoever. "Every act of ostensive communication communicates the presumption of its own optimal relevance" (Sperber & Wilson, 1986, p. 158). In the case of high personal involvement with the topic addressed, the recipient's attention will automatically get allured. Consequently, the recipient will try to find the point of relevance and try to relate it in some way to his or her own personal 'belief box'.

Yet it is a well-known fact that factually true messages are very rarely sensational. Hence, from the point of view of the person that posts it, in order to provoke attention, it will be paramount to make the message as thrilling and incredible as possible, albeit by grossly exaggerating facts or, in the worst case, by bluntly making them up, which directs us to the problem of "alternative facts" or "fake news". For the more provocative and outrageous an information is, the more attention it will attract.

Thus, here is the paradox: The more incredible a message is, the more readily it will be believed. This has to do with the mechanism of information processing according to the Elaboration Likelihood Model. For an audience with high personal involvement, but low cognitive interest, the prime choice will be what is called the peripheral route of processing, that is, instead of carefully checking the plausibility of a message along the central route, they will be inclined to accept it on non-rational terms exactly because of its apparent relevance to them and to avoid "superfluous" rational cognitive processing by applying fast and intuitive heuristic mechanisms (see Gigerenzer, 2008). In this way, the Principle of Cognitive Relevance may create a bias for the acceptance of particular arguments by the typical audience of social media.

5. Cognitive relevance and argument selection

Yet in addition to this, messages encountered in the social media will also be subject to a process of selection by recipients, a process that, however, will again be guided (or need we say biased?) by the Principle of Cognitive Relevance: The more the contents of a message tally with the recipient's preconceived world view, the more relevance it will be credited with, whereas messages that challenge a recipient's preconceived opinions and hence require greater argumentative effort in processing tend to be shunned and dismissed as irrelevant. The level of attractivity from one's own point of view on the one hand and the size of required processing effort on the other become the main criteria for imputation of relevance.

Challenging counter-arguments would most certainly have a cognitive effect, too, but it would be a negative one in the eyes of most recipients. As Sperber and Wilson have figured out, an argumentative input will appear relevant to an individual when, and only when, its processing yields *positive* cognitive effects (Sperber & Wilson, 1995, pp. 265–266; Sperber, Cara & Girotto, 1995; Wilson & Sperber, 2002, p. 231). Thus, imputation of relevance has a clear impact on the process of argument selection.

This observation is closely connected with Sperber and Wilson's notion of 'cognitive environments'. "A *cognitive environment* of an individual", they define, "is a set of facts that are manifest to him." But: "It consists of not only all the facts that he is aware of, but also all the facts that he is capable of becoming aware of [...]" (Sperber & Wilson, 1986, p. 39). The final clause clearly also relates to the inferences that, according to the inferential model of communication, a recipient needs to draw from a message input to the sender's intentions. Cognitive environments in the sense of Sperber and Wilson are thus similar to and productive of what others have called 'belief boxes'. Shared cognitive environments, in Sperber and Wilson's view, are an essential requirement for a successful argumentative exchange between two parties (on cognitive environments and audience relevance, see also Tindale, 1999, pp. 101–112; 2015, pp. 144–146). The ultimate aim of an arguer must hence be to make her own argument also part of the addressee's cognitive environment. The supposition that there exists such a shared cognitive environment with the sender of a message may substantially enhance a recipient's willingness to embrace a particular argument or statement as true.

I believe that it is by this mechanism of "sifting" arguments according to their imputed relevance that messages posted in social media show such an obvious tendency to back up rather than challenge or contest the preconceived opinions of their recipients, in other words to create or strengthen closed "filter bubbles" (Möller, 2021), "echo chambers" (Sunstein, 2001a; Edwards, 2013; Barberá, 2020) or "information cocoons"

(Sunstein, 2006, pp. 8–24). Recipients will happily embrace precisely those arguments which they find to be in accordance with their own opinions and 'belief boxes', and hence—so they believe—of relevance to their own situation, and will bluntly reject or sidestep all those that would run counter, and offhandedly denounce them in turn as "fake news". In the minds of uncritical recipients, imputed cognitive relevance, deduced from attention-provoking sensationalism and an insinuated congruence of beliefs, is thus improvidently mistaken for a sign of truth, imputed non-relevance for a sign of falsity. It is this kind of "bias by imputed cognitive relevance" that prevents recipients from taking notice of, let alone heeding any critical information or counter-evidence in the first place and induces them to accept conclusions offhandedly as persuasive qua unchallenged (for a good description of this mechanism from a similar methodical point of view, see Oswald, 2011; 2016). Without any doubt, a certain amount of question-begging is also involved here. But this is most probably how and why it works so smoothly.

It is here that a potential for active parasitic exploitation of the mechanism of cognitive relevance looms large. It may very easily be exploited for instance for propagandistic purposes. Since in many social media it is pretty easy to place in an uncontrolled way and even anonymously any odd sensational message, whether substantiated or not, and to reach therewith a vast global audience in almost no time, prospects are good that a certain percentage of suitably predisposed recipients will be efficiently fooled into believing it and thereby contentedly feeling vindicated. The ultimate aim is to make a particular argument part of the cognitive environments of as many people as possible. As is known well enough, not even abusive ad hominem arguments are excluded from this process.

6. Propagating arguments

But not only will recipients happily embrace any 'relevant' contents and adopt them as facts into their own personal 'belief boxes', but what is more, they will also forward them to their own social communities and thus spread them even more widely. For, owing to the flat hierarchies in social media, users of those media are recipients and communicators at the same time (thereby suspending the hierarchic asymmetry between proponent and recipient that Walton, 2007, p. 109, has basically identified in propagandistic dialogue); by 'sharing' a 'relevant' message with their 'friends' or 'followers', they make for its rapid and widespread proliferation. This has still another unpleasant consequence: In this process of diffusion and propagation the original source of a particular message gets progressively blurred and opaque, while intermediate transmitters act as newsmongers and additionally back it up by investing it with their own authority and trustworthiness.

In this way, circulators of unsubstantiated messages, especially so conspiracy theorists of all sorts, can comfortably hide behind other people's identities and authorities, and trigger a kind of snowball effect that in a flash produces a huge marketplace for any particular argument or statement. Social media parlance in such cases speaks of an argument "going viral". Ironically enough, a paradigmatic case have recently been blunt denials of the Covid pandemic, which "went viral" even more quickly than the virus itself. Denials of climate change or of the legality and regularity of the results of the last U.S. presidential elections have worked the same way. Once such a process has been triggered, it is almost impossible to stop it, since even if counter-arguments and rebuttals are raised instantly and the statement is professionally debunked, it can never be completely eradicated, because, as we all know, the web forgets nothing, and new adherents and partisans will soon arise.

By these mechanisms, too, messages posted in social media as a rule tend to strengthen rather than challenge preconceived opinions among recipients, and hence to contribute to the creation of closed 'filter bubbles' or 'echo chambers'. The true goal of messages posted in social media is thus very often not to communicate factual information, let alone to argue in a rational and reasonable way, but to distribute and propagate a particular input to a maximum audience in as little time and with as little effort as possible.

7. Conclusion

To sum up, by applying an inferential model of communication to argumentative processes going on in social media, we have been trying to show how the Principle of Cognitive Relevance operates as a core element of that model, but is at the same time responsible for paradoxical effects, and have thus been attempting to offer a reasonable explanation for the particular vulnerability of argumentation processes in social media to parasitic misuse and exploitation.

According to Austrian political scientist and psychologist Wolfgang Hofkirchner, cognition, communication and co-operation are the three basic hierarchical modes of sociality with respect to information processing (2013, p. 186). Based on our analysis above, we will ultimately have to conclude that argumentation in the social media, while on the one hand strongly drawing on inferential cognitive processes for gauging argument relevance, precisely on account of these mechanisms very often fails to properly communicate factual information, and for that very reason also evidently falls short of real co-operation between sender and recipient of messages. Paul Grice most certainly might not be amused. If Hofkirchner is right, however, social media, in spite of their shiny label, may not be truly social at all.

References

Barberá, P. (2020). Social media, echo chambers, and political polarization. In N. Persily & J.A. Tucker (Eds.), Social media and democracy. The state of the field and prospects for reform (pp. 34–55). Cambridge: Cambridge University Press.

Blair, J. A. (1989). Premise relevance. In R. Maier (Ed.), Norms in argumentation (pp. 67–83). Dordrecht: Foris Publications.

Brandom, R. (1994). Making it explicit. Reasoning, representing, and discursive commitment. Cambridge, MA: Harvard University Press.

Brandom, R. (2000). Articulating reasons. An introduction to inferentialism. Cambridge, MA: Harvard University Press.

Brandom, R. (2008). Between saying and doing. Towards an analytic pragmatism. Oxford: Oxford University Press.

Cacioppo, J. T., Harkins, S. G., & Petty, R. E. (1981). The nature of attitudes and cognitive responses and their relationships to behavior. In R. E. Petty, T. M. Ostrom, & T. C. Brock (Eds.), Cognitive responses in persuasion (pp. 31–54). Hillsdale, NJ: Erlbaum.

Durkheim, É. (1895). Les règles de la méthode sociologique. Paris: Félix Alcan.

Edwards, A. (2013). (How) do participants in online discussion forums create 'echo chambers'? The inclusion and exclusion of dissenting voices in an online forum about climate change. Journal of Argumentation in Context, 2(1), 127–150.

Fuchs, C. (2021). Social media. A critical introduction. London: SAGE (1st ed. 2014).

Gigerenzer, G. (2008). Why heuristics work. Perspectives on Psychological Science 3(1), 20–29.

Greenwald, A. G. (1968). Cognitive learning, cognitive response to persuasion, and attitude change. In A. G. Greenwald, T. C. Brock, & T. M. Ostrom (Eds.), Psychological foundations of attitudes (pp. 147–170). New York and London: Academic Press.

Grice, H. P. (1975). Logic and conversation. In P. Cole & J. J. Morgan (Eds.), Syntax and semantics 3. Speech acts (pp. 41–58). New York, NY: Academic Press.

Grice, H. P. (1989). Studies in the way of words. Cambridge, MA: Harvard University Press.

Hample, D. (1979). Predicting belief and belief change using a cognitive theory of argument and evidence. Communication Monographs, 46, 142–146.

Hample, D. (1981). The cognitive context of argument. Western Journal of Speech Communication, 45, 148–158.

Hample, D. (1985). Refinements on the cognitive model of argument. Western Journal of Speech Communication, 49, 267–285.

Hofkirchner, W. (2013). Emergent information. A unified theory of information framework. New Jersey, NY et al.: World Scientific.

Möller, J. (2021). Filter bubbles and digital echo chambers. In H. Tumber & S. Waisbord (Eds.), The Routledge companion to media disinformation and populism (pp. 92–100). London: Routledge.

O'Keefe, D. J. (1996). Argumentation studies and dual-process models of persuasion. In J. van Benthem, F. H. van Eemeren, R. Grootendorst, & F. Veltman (Eds.), Logic and argumentation (pp. 61–76). Amsterdam: North-Holland.

O'Keefe, D. J. (2008). Elaboration likelihood model. In W. Donsbach (Ed.), International encyclopedia of communication (vol. 4, pp. 1475–1480). Oxford, UK, and Malden, MA: Wiley-Blackwell.

O'Keefe, D. J. (2009). Theories of persuasion. In R. L. Nabi & M. B. Oliver (Eds.), The SAGE handbook of media processes and effects (pp. 277–278). Los Angeles: SAGE.

Oswald, S. (2011). From interpretation to consent. Arguments, beliefs and meaning. Discourse Studies 13(6), 806–814.

Oswald, S. (2016). Rhetoric and cognition. Pragmatic constraints on argument processing. In M. Padilla Cruz (Ed.), Relevance theory. Recent developments, current challenges and future directions (pp. 261–285). Amsterdam: John Benjamins.

Petty, R. E., & Cacioppo, J. T. (1981). Attitudes and persuasion. Classic and contemporary approaches. Dubuque, IA: William C. Brown.

Petty, R. E., & Cacioppo, J. T. (1986a). Communication and persuasion. Central and peripheral routes to attitude change. New York: Springer-Verlag.

Petty, R. E., & Cacioppo, J. T. (1986b). The elaboration likelihood model of persuasion. In L. Berkowitz (Ed.), Advances in experimental social psychology 19 (pp. 123–205). New York: Academic Press.

Petty, R. E., & Wegener, D. T. (1999). The elaboration likelihood model. Current status and controversies. In S. Chaiken & Y. Trope (Eds.), Dual process theories in social psychology (pp. 41–72). New York: Guilford Press.

Sperber, D. (1982). Mutual knowledge and relevance in theories of comprehension. In N. V. Smith (Ed.), Mutual knowledge (pp. 61–85). London: Academic Press.

Sperber, D., & Wilson, D. (1986). Relevance. Communication and cognition. Cambridge, MA: Harvard University Press.

Sperber, D., & Wilson, D. (1990). Rhetoric and relevance. In D. Wellbery & J. Bender (Eds.), The ends of rhetoric. History, theory, practice (pp. 140–155). Stanford, CA: Stanford University Press.

Sperber, D., & Wilson, D. (1995). Postface. In D. Sperber & D. Wilson, Relevance. Communication and cognition. 2nd ed. (pp. 255–279). Cambridge, MA: Harvard University Press.

Sperber, D., Cara, F., & Girotto, V. (1995). Relevance theory explains the selection task. Cognition 57, 31–95.

Sunstein, C. R. (2001a). Echo chambers. Bush v. Gore. Impeachment, and beyond. Princeton: Princeton University Press.

Sunstein, C. R. (2001b). Republic.com. Princeton: Princeton University Press.

Sunstein, C. R. (2006). Infotopia. How many minds produce knowledge. Oxford: Oxford University Press.

Sunstein, C. R. (2017). #Republic. Divided democracy in the age of social media. Princeton: Princeton University Press.

Tindale, C. W. (1999). Acts of arguing. A rhetorical model of argument. New York: State University of New York Press.

Tindale, C. W. (2015). The philosophy of argument and audience reception. Cambridge: Cambridge University Press.

Walton, D. (2007). Media argumentation. Dialectic, persuasion, and rhetoric. Cambridge: Cambridge University Press.

Wilson, D., & Sperber, D. (1981). On Grice's theory of conversation. In P. Werth (ed.), Conversation and discourse. Structure and interpretation (pp. 155–178). London: Croom Helm.

Wilson, D., & Sperber, D. (1986). On defining relevance. In R. Grandy & R. Warner (Eds.), Philosophical grounds of rationality. Intentions, categories, ends (pp. 243–258). Oxford: Clarendon Press.

Wilson, D., & Sperber, D. (2002). Truthfulness and relevance. Mind, 111, 583–632.

Wilson, D., & Sperber, D. (2012). Meaning and relevance. Cambridge: Cambridge University Press.

EVOCATION OF RELEVANT QUESTIONS: HOW DOES THIS WORK?

LEONARD KUPŚ
Faculty of Psychology and Cognitive Science
Adam Mickiewicz University, Poznań, Poland
leonard.kups@outlook.com

MARIUSZ URBAŃSKI
Faculty of Psychology and Cognitive Science
Adam Mickiewicz University, Poznań, Poland

Abstract
This paper addresses the issue of asking relevant questions given incomplete information. We employ four concepts of evocation of relevant questions and exemplify the workings of our formal approach with British National Corpus data.

1. Introduction: the concept of evocation

There are gaps in our knowledge, and we are aware of this. To fill these gaps, we ask questions (this remark has been put forward already by Aristotle). We do not ask them randomly: if our goal is to address the knowledge gap, we ask questions relevant to the issue at hand. The research presented in this paper models such process of goal-directed question-asking.

We focus on the concept of *evocation* of a question Q by a set X of declarative sentences (Wiśniewski, 2013). Q is evoked by X just in case two conditions are met: (i) truth of all the elements of X warrants the existence of a true direct answer to Q, but (ii) no single direct answer to Q is entailed by X. Evocation is an instrumental concept in addressing the issues of modelling dynamics of information processing. However, the basic version raises some troubling issues related to the possible lack of relevance between Q and X. Consider the question Q_1: "Is two an even number?". As Q_1 is a so-called safe question (that is, there exists a true answer to it), condition (i) is warranted for any X. If, moreover, X consists of elements semantically unrelated to answers to Q_1 and is consistent (as we are going to work here within the limits of inconsistency-sensitive logics, like Classical Propositional Logic, CPL), the condition (ii) is met as well. Thus, for example, Q_1 is evoked by X consisting of a single sentence, "The Moon is the only natural satellite of the Earth" (and also by the empty set, for that matter).

In order to address these issues, we developed a version of evocation rooted in situational semantics, construed set-theoretically (Kupś & Urbański, 2023). Such a basis, more fine-grained than the standard truth-values-based approach, offers the possibility of defining the notion of relevance in a precise way. We defined a family of concepts of evocation that are more sensitive to several semantic and pragmatic phenomena in the natural language use of questions. Most notably, we defined the concepts of relevance and compatibility of sentences and texts, as well as both strong and weak versions of the evocation of relevant and compatible questions.

Our goal in this paper is to demonstrate how various modes of relevant evocation facilitate adequate modelling of question processing in real-life dialogues and argumentation processes and, thus, support comprehension in communication. To this end, we employ examples from the British National Corpus.

We start with the required elements of the underlying logic of questions -- Inferential Erotetic Logic (Wiśniewski, 1995, 2013) and basic definitions of specific concepts -- relevance and compatibility. Then we turn to examples of different modes of evocation. Throughout the paper we assume CPL to be our declarative basis, and we employ the classical Tarskian concept of entailment.

2. Relevance, compatibility, and texts

Let us start with a short and instructive example. Consider the following four sentences:
 α John is 5 years old.
 β Anna is a student.
 γ John is 10 years old.
 δ John is Anna's brother.

First, we assume that the proper names in these examples refer to the same person each time. The sentences α and γ are *relevant* since both concern John, but they give us contradicting information (we assume that one person cannot differ in age at a given time). Thus, those two sentences are *incompatible*. Sentences α and β are *compatible*, but they are *irrelevant* since we do not know what the connection is between John and Anna. Sentences α, β, and δ are both relevant and compatible, and so are sentences β, γ, and δ.

We model these concepts of relevance and compatibility by employing Wiśniewski's situational semantics (Wiśniewski, 1997). The underlying assumptions are: (i) that sentences do not refer to individual situations but to the sets of situations, and (ii) that there exists a non-empty set of all situations, which we are going to call the *universe of situations* $\mathbb{U}$, and that every sentence S is assigned a set of situations $v(S)$, which is a subset of $\mathbb{U}$. The assignment function v maps CPL formulas into Boolean relations between sets of situations in a usual way:

1. for each propositional variable p_i, $V(p_i) \subseteq U$
2. for each CPL formula A, B:
   2.1 $v(\neg A) = U - v(A)$,
   2.2 $v(A \land B) = v(A) \cap v(B)$,
   2.3 $v(A \lor B) = v(A) \cup v(B)$,
   2.4 $v(A \to B) = v(\neg A) \cup v(B)$,
   2.5 $v(A \leftrightarrow B) = (v(\neg A) \cup v(B)) \cap (v(\neg B) \cup v(A))$.

Thus, if S_1 = "John is reading a book" there is a set $v(S_1) \subset \mathbb{U}$ of all the situations in which John is reading a book. If S_2 ="John is not reading a book", then there exists a set $v(S_2) = \mathbb{U} - v(S_1)$ of all the situations in which John is not reading a book. Moreover, the set of situations corresponding to the conjunction $S_1 \land S_2$ (that is, the set $v(S_1) \cap v(S_2)$) is empty.

We need some more auxiliary concepts (cf. Ajdukiewicz, 1974). A term is said to *designate* every and only such object about which it may truly be predicated. The objects designated by a term are called its *designata* and let $\mathbb{D}$ stand for the set of all designata in the given universum. For the sake of simplicity, we assume that each proper name has only one designatum. The set of all designata for a given term is its extension. If the extensions of two terms are equal, they are equivalent, e.g. "London" and "the current capital of the UK" are equivalent terms. Moreover, extensions of terms can be disjoint, include or intersect one another. We may assume that every term t is assigned an extension $E(t)$, which is a subset of $\mathbb{D}$.

Now, on to our crucial concepts. If S_i and S_j are sentences, then:

- S_i and S_j are *compatible* iff the intersection of their corresponding sets of situations is non-empty: $C(S_i, S_j) = 1 \leftrightarrow V(S_i) \cap V(S_j) \neq \emptyset$;

- S_i is relevant to S_j (in symbols $R(S_i, S_j) = 1$) iff there exists a term t_i in a sentence S_i and a term t_j in a sentence S_j such that $E(t_i) \cap E(t_j) \neq \emptyset$.

As we work at a level of propositional logic, we need to make a couple of assumptions concerning the object-level furniture of the described world. Thus, as terms, we understand for now only names and descriptions. Also, for a sentence to be relevant to another, the designata of their terms have to exist in a given universum (this is, in fact, a consequence of the assumed situational semantics).

From such definitions follows, e.g., that if a given sentence is not relevant to another sentence, they are always compatible (cf. α and β above). Relevant sentences may be compatible (α and δ) or not (α and γ), and compatible sentences may be relevant (β and δ) or not (β and γ).

One final notion we need to define is the notion of *text*, which is just a non-empty set of declarative sentences. Thus, if S is a sentence and T is a text:

- S is compatible with T iff the intersection of situations corresponding to all sentences in T and the sentence S is non-empty: $C(S,T) = 1 \leftrightarrow \bigcap_{x \in T} V(x) \cap V(S) \neq \emptyset$;
- If S is a sentence and T is a text, S is relevant to T iff there exists a sentence in T such that S is relevant to that sentence.

Let us consider a further example:
ϵ Anna does not have a brother.
ζ Anna has 4 cats.
and a text $T = \{$John is Anne's brother, John is a student, John has two sisters$\}$. Here ϵ is relevant to but not compatible with T (as it is not compatible with one of its elements), while ζ is both relevant to and compatible with T.

The concepts of relevance and compatibility come hand in hand in accounting for the semantic interplay between sentences and texts. Throughout the rest of this paper, however, we shall focus exclusively on relevance.

As it should be visible now, the concept of relevance may come in flavours, depending on which quantifier we employ in its definition. We shall take advantage of this fact while defining different versions of the evocation of questions in the following sections.

3. Evocation of relevant questions

The basic version of the concept of evocation of question by a set of declarative premises is given by (Wiśniewski, 2013, p. 60): A set of declarative well-formed formulas formulae X *evokes a question* Q (in symbols $E(X,Q)$) iff

1. $X \Rightarrow dQ$ and
2. $\forall A \in dQ: X \not\Rightarrow \{A\}$

The symbol dQ represents the set of all the direct answers to the question Q (see Wisniewski-Lapin chapter). The symbol $\Rightarrow$ refers to multiple-conclusion entailment (mc-entailment): "[...] mc-entailment between [the sets] X and Y holds just in case the truth of all the d-wffs [declarative well-formed formulas] in X warrants the presence of some true d-wff(s) in Y: whenever all the d-wffs in X are true [...], then at least one d-wffs in Y is true". (Wiśniewski, 2013, p. 33). As the classical definition of evocation of questions is prone to troubles we mentioned initially, we augmented it with a relevance perspective (Kupś & Urbański, 2023). Now we will exemplify the workings of our four versions of relevant evocation with the data from the British National Corpus. (Data cited herein have been extracted from the British National Corpus Online service (BNC Consortium, 2007), managed by Oxford University Computing Services on behalf of the BNC Consortium. All rights in the texts cited are reserved.)

In order to present our examples, we need a bit of notation. Let us go back to our text T. It consists of three sentences: δ ="John is Anna's brother", η ="John is a student", θ ="John has two sisters". By α_i-terms (resp. Q_j-terms) we shall designate the set of all the terms (as described above) occurring in the sentence α_i (resp. the question Q_j. Thus: δ-terms ={John, Anna, brother}, η-terms = {John, student}, θ-terms = {John, sister}. Now, consider the question "Does Anna have 4 cats?". Q_1-terms = {Anna, cat}. The affirmative and negative answers to Q_1 are compatible with and relevant to the text T. In the case of the question Q_2 ="Does Anna have a brother?" (Q_2-terms = {Anna, brother}) both the affirmative and the negative answers are relevant to T, but only the positive one is compatible with it.

We shall define four versions of the concept of evocation of relevant questions. The difference between them amounts to answering the following: are all the answers to the evoked questions supposed to be relevant to a given text or just some, and are they supposed to be relevant to all the sentences in this text or just some?

We will employ Q-terms sets as helpful guidance to determine terms in direct answers to questions. It should be noted that this strategy needs to be pursued with some care. In some cases, it works very smoothly (see the example in section 3.1 or the first one in section 3.3); in others, it needs to be augmented with a more thorough semantic analysis (see example in 3.2).

3.1. Strong Evocation of Relevant Questions

A text T *strongly evokes a relevant question Q* (in symbols $E^{\alpha R}(T,Q)$) iff:
1. $T \Rightarrow dQ$ and
2. $\forall A \in dQ: T \not\Rightarrow \{A\}$ and
3. $\forall A \in dQ, \forall x \in T: R(A,x) = 1$.

So, here we require that each direct answer to the evoked question must be relevant to each sentence in the evoking text. Consider the following example (BNC A04 13-18; we shall start with the question, which in the original text is the last element):

Q_{18}^{A04} Is the text or part of the text written so that a reader will benefit in a future encounter with a work of art?
Q_{18}^{A04}-*terms* ={text, art, encounter, part, work, reader}

S_{13}^{A04} There is a massive amount of writing about art, only some of which can immediately be identified by a reader as criticism.
S_{13}^{A04}-*terms* = {art, criticism, amount, reader}
Q_{18}^{A04}-*terms* ∩ S_{13}^{A04}-*terms* = {art, reader}

S_{14}^{A04} Writing by the art critic of a newspaper is self-evidently criticism, in parallel with the writing of music and theatre critics; an exhibition can be treated almost in the same way as a performance.
S_{14}^{A04}-*terms* = {critic, way, theatre, parallel, art, critics, exhibition, performance, criticism, writing, newspaper, music}
Q_{18}^{A04}-*terms* ∩ S_{14}^{A04}-*terms* ={art}

S_{15}^{A04} Articles in magazines are less certainly described as criticism, for their main topics may be personalities or history, and art may be only a small part of the writers' account.
S_{15}^{A04}-*terms* = {topics, art, account, writers, history, criticism, part, magazines, articles, personalities}
Q_{18}^{A04}-*terms* ∩ S_{15}^{A04}-*terms* ={art, part}

S_{16}^{A04} Books and catalogues may contain criticism; but their writers may think of themselves as art historians, philosophers, aestheticians, anthropologists, historians or biographers, and there are many other possibilities; their books may never be identified as art criticism.
S_{16}^{A04}-*terms* ={philosophers, anthropologists, art, writers, catalogues, books, criticism, aestheticians, possibilities, books, biographers, historians}
Q_{18}^{A04}-*terms* ∩ S_{16}^{A04}-*terms* = {art}

S_{17}^{A04} As the following chapters will show, the useful and helpful functions of art criticism will receive preference in the choice of what is quoted or discussed.
S_{17}^{A04}-*terms* = {preference, art, criticism, functions, chapters, choice}
Q_{18}^{A04}-*terms* ∩ S_{17}^{A04}-*terms* ={art}

As Q_{18}^{A04} is a polar question, and thus a safe one (with an obvious proviso concerning truth of its presuppositions, but there is no term like "the present king of France" there), the first condition of the definition of strong relevant evocation is fulfilled. Also, the text $T_{13-17}^{A04} = \{S_{13}^{A04}, S_{14}^{A04}, S_{15}^{A04}, S_{16}^{A04}, S_{17}^{A04}\}$ does not entail any answer to Q_{18}^{A04}, so the second condition is met as well. Finally, since the intersection of Q_{18}^{A04}-*terms* and *S-terms* for each sentence in T_{13-17}^{A04} is non-empty and Q_{18}^{A04} is a polar question, so it is reasonable to assume that direct answers sets of terms will also have non-empty intersections with Q_{18}^{A04}-*terms*. Thus we may conclude that T_{13-17}^{A04} strongly evokes a relevant question Q_{18}^{A04} (in symbols: $E^{\Box R}(T_{13-17}^{A04}, Q_{18}^{A04})$).

3.2. Weak Evocation of Relevant Questions

Now let us mitigate one of the requirements of the third condition of the definition of evocation of relevant questions, thus obtaining its weak version:

A text T weakly evokes a relevant question Q (in symbols $E^R(T,Q)$) iff:
1. $T \Rightarrow dQ$ and
2. $\forall A \in dQ: T \not\Rightarrow \{A\}$ and
3. $\exists A, \forall x \in T: A \in dQ \land R(A,x) = 1$.

Thus in the weak version, we require that at least one direct answer to the evoked question must be relevant to each sentence in the evoking text. Consider the following example (ARR 918-921; again, we shall start with the question which in the original text is the last element; the text in the square brackets is added for context):

Q_{921}^{ARR} But how could a fig tree 'retaliate'?
Q_{921}^{ARR}-terms = {fig, tree, retaliate}
S_{918}^{ARR} They [wasps] lay their eggs in some of the tiny flowers, which the larvae then eat.
S_{918}^{ARR}-terms = {wasps, eggs, flowers, larvae}
Q_{921}^{ARR}-terms ∩ S_{918}^{ARR}-terms = {∅}
S_{919}^{ARR} They pollinate other flowers within the same fig.
S_{919}^{ARR}-terms = {wasps, fig, flowers}
Q_{921}^{ARR}-terms ∩ S_{919}^{ARR}-terms = {fig}
S_{920}^{ARR} 'Defecting', for a wasp, would mean laying eggs in too many of the flowers in a fig and pollinating too few of them.
S_{920}^{ARR}-terms = {wasp, fig, egg, flowers}
Q_{921}^{ARR}-terms ∩ S_{920}^{ARR}-terms = {fig}

We model relevance between sets of direct answers to a question and an initial text. Sets of direct answers to open questions are larger and thus more difficult (possibly impossible) to put forward. Therefore, instead of listing all possible direct answers, we will provide examples of types of direct answers and reason on their basis. By types, we mean the degree of relevance to the initial text.

$\{\lambda, \mu\} \in dQ_{921}^{ARR}$, where λ = "Fig tree can change the shape of its flowers" and μ = "Fig tree can alter its pollen". The direct answer λ is relevant to every sentence in $T_{918-921}^{ARR}$, and μ is not relevant to the S_{918}^{ARR}. Thus $E^R(T_{918-921}^{ARR}, Q_{921}^{ARR})$.

3.3. Partial Strong Evocation of Relevant Questions

Defining the concepts of partial evocation of relevant questions, we play with the quantifiers regarding sentences in the evoking text. In the case of partial strong evocation, we require that each direct answer to the evoked question is relevant to a sentence in the evoking text.

A text T partially evokes a relevant question Q (in symbols $E^{\square \lozenge R}(T, Q)$) iff:

1. $T \Rightarrow dQ$ and
2. $\forall A \in dQ: T \not\Rightarrow \{A\}$ and
3. $\forall A \in dQ, \exists x: x \in T \land R(A, x) = 1$.

Consider the following example (BNC ARM 299-302):
Q_{302}^{ARM} So why do you think Dino is better then you?
Q_{302}^{ARM}-terms = {Dino}
S_{299}^{ARM} I'm not really secretive about new stuff.
S_{299}^{ARM}-terms = {stuff}
Q_{302}^{ARM}-terms $\cap$ S_{299}^{ARM}-terms = $\emptyset$
S_{300}^{ARM} Dino is cool to skate with because he is better then me and he really pushes everything to the maximum.
S_{300}^{ARM}-terms = {Dino, maximum}
Q_{302}^{ARM}-terms $\cap$ S_{300}^{ARM}-terms = {Dino}
S_{301}^{ARM} He pushes me to make stuff whilst you just cruise around and do nose feebles all the time.
S_{301}^{ARM}-terms = {stuff, time, nose, Dino}
Q_{302}^{ARM}-terms $\cap$ S_{301}^{ARM}-terms = $\emptyset$

Here Q_{302}^{ARM} is not a polar question, but any direct answer must refer to the Dino. Thus in this example, the sentence S_{299}^{ARM} does not share a term with any answer Q_{302}^{ARM}, while for all the others elements of the text $T_{299-301}^{ARM}$ there is at least one common term with Q_{302}^{ARM}-term. As the remaining conditions of the partial strong evocation are fulfilled, we may conclude that $E^{\square \lozenge R}(T_{299-301}^{ARM}, Q_{302}^{ARM})$. Notice that if we remove the troublemaking sentence S_{299}^{ARM} from the evoking text, we will obtain just strong evocation: $E^{\square R}(T_{300-301}^{ARM}, Q_{302}^{ARM})$.

Another example of the same type of evocation is to be found in BNC ASH 368-343:

Q_{643}^{ASH} Isn't it about time someone took a long hard look at the laughable regulations on what riders wear and questioned whether they benefit the sport?
$Q_{643}^{ASH}\ terms$ = {look, time, raiders, sport, regulations}
S_{638}^{ASH} In addition, cyclists, runners and skiers have now adopted close fitting garments like our jodhpurs for ease of movement, but they don't top them off with ties or traditional jackets to go flapping around while they have other things to concentrate on.
$S_{638}^{ASH}\text{-}terms$ = {addition, jodhpurs, jackets, movement, ties, skiers, ease, cyclists, things, runners, garments}
$S_{638}^{ASH}\text{-}terms \cap Q_{643}^{ASH}\text{-}terms = \emptyset$
S_{639}^{ASH} Only in endurance riding, polo and cross country do we see riders wearing sensible clothing.
$S_{639}^{ASH}\text{-}terms$ = {clothing, endurance, riders, polo, country}
$S_{639}^{ASH}\text{-}terms \cap Q_{643}^{ASH}\text{-}terms$ = {riders}
S_{640}^{ASH} Riding's governing bodies should be paying more attention to clothing as 'equipment' important for safety and performance instead of for appearance's sake.
$S_{640}^{ASH}\text{-}terms$ = {bodies, appearance, equipment, riding, clothing, attention, safety, performance, sake}
$S_{640}^{ASH}\text{-}terms \cap Q_{643}^{ASH}\text{-}terms = \emptyset$
S_{641}^{ASH} If other sports had riding's attitude to clothing, Nigel Mansell would drive in a leather flying hat and Ian Woosnam would compete in plus-fours, but one is unsafe and the other unnecessary so the motorsport and golf authorities have not tried to cling to them.
$S_{641}^{ASH}\text{-}terms$ ={hat, Ian, Nigel, golf, sports, Mansell, leather, authorities, motorsport, plus-fours, flying, clothing, Woosnam, riding, attitude}
$S_{641}^{ASH}\text{-}terms \cap Q_{643}^{ASH}\text{-}terms = \emptyset$
S_{642}^{ASH} Indeed, in motorsport it's not the colour of your clothes that matters, it's whether they are fireproof, and the regulations stipulate that everything has to comply to the relevant safety standards.
$S_{642}^{ASH}\text{-}terms$ = {clothes, colour, matters, motorsport, standards, safety, regulations}
$S_{642}^{ASH}\text{-}terms \cap Q_{643}^{ASH}\text{-}terms$ = {regulations}

Q_{643}^{ASH} is a polar question; thus again we will assume that direct answers to it include the same terms as the question itself. S_{638}^{ASH}-term, S_{640}^{ASH}-term and S_{641}^{ASH}-term all have empty intersections with Q_{643}^{ASH}-term. Again, the remaining conditions of partial strong evocation are fulfilled. Furthermore, again, the text $T_{638-643}^{ASH}$ contains a subset which strongly evokes Q_{643}^{ASH}; this is the text consisting of S_{639}^{ASH} and S_{642}^{ASH}.

3.4. Partial Weak Evocation of Relevant Questions

Our final version of the evocation of relevant questions is the one that is both partial and weak:

A text T partially evokes a relevant question Q (in symbols $E^{\partial R}(T,Q)$) iff:

1. $T \Rightarrow dQ$ and
2. $\forall A \in dQ: T \not\Rightarrow \{A\}$ and
3. $\exists A, \exists x: A \in dQ \land x \in T \land R(A,x) = 1$.

Here we require that at least one answer to the evoked question needs to be relevant to at least one sentence in the evoking set. Consider an example ASS 218-223:

Q_{223}^{ASS} But how to persuade her to go to Beachy Head?
Q_{223}^{ASS}-terms = {Beachy, Head}
S_{218}^{ASS} They could go down to Beachy Head.
S_{218}^{ASS}-terms = {Beachy, Head}
S_{218}^{ASS}-terms $\cap$ Q_{223}^{ASS}-terms = {Beachy, Head}
S_{219}^{ASS} Wander along the edge of the cliff.
S_{219}^{ASS}-terms = {edge, cliff}
S_{219}^{ASS}-terms $\cap$ Q_{223}^{ASS}-terms = $\emptyset$
S_{220}^{ASS} Some remark, along the lines of 'Oooh look! lines}
S_{220}^{ASS}-terms = {remark, lines}
S_{220}^{ASS}-terms $\cap$ Q_{223}^{ASS}-terms = $\emptyset$
S_{221}^{ASS} Over there, dear!
S_{221}^{ASS}-terms = {dear}
S_{221}^{ASS}-terms $\cap$ Q_{223}^{ASS}-terms = $\emptyset$
S_{222}^{ASS} And then a smart shove in the small of the back.
S_{222}^{ASS}-terms = {back, shove}
S_{222}^{ASS}-terms $\cap$ Q_{223}^{ASS}-terms = $\emptyset$

4. Summary

There is a logical dependency between the defined concepts of evocation of relevant questions: the strong version is logically strongest of them all, the partial weak -- the weakest. The dependencies are pictured in Fig. 1.

The concepts of evocation of relevant questions do adequately address certain phenomena occurring in natural language. However, the

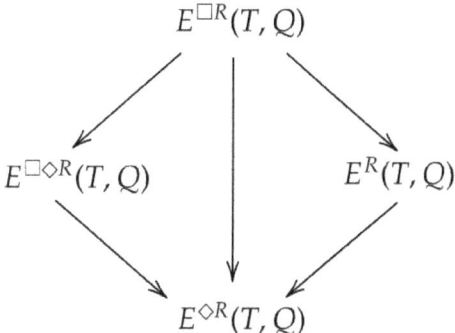

Figure 1. Dependencies between types of evocation of relevant questions.

simplicity of the declarative logical basis employed in this paper comes at a price. We are aware that some version of the predicate calculus, even the First-Order one, would be better suited to capture those phenomena and perhaps would make our examples more convincing. With such an expanded expressive power of the formal language, we could, in particular, offer more comprehensively construed notions of compatibility and relevance to include terms and relations besides term extensions. This is one possible direction of our future studies. Another is more empirically oriented: we are interested in what extent raising questions evoked in various modes is perceived as acceptable or legitimate by language users.

References

Ajdukiewicz, K. (1974). Pragmatic Logic. In Pragmatic Logic. https://doi.org/10.1007/978-94-010-2109-8
BNC Consortium. (2007). The British National Corpus, XML Edition. http://hdl.handle.net/20.500.12024/2554
Kupś, L., & Urbański, M. (2023). Relevance and Evocation of Questions. (Submitted)
Wiśniewski, A. (1995). The Posing of Questions: Logical Foundations of Erotetic Inferences. Kluwer AP.
Wiśniewski, A. (1997). Logika a zbiory sytuacji. In R. Kubicki & P. Zeidler (Eds.), Od logiki do estetyki. Prace dedykowane Profesorowi Włodzimierzowi Ławniczakowi (pp. 13–25). Wydawnictwo Fundacji Humaniora.
Wiśniewski, A. (2013). Questions, Inferences and Scenarios. College Publications.
Wiśniewski, A. Semantics of questions, in: S. Lappin, Ch. Fox (eds.), The Handbook of Contemporary Semantic Theory – second edition, Wiley-Blackwell 2015, pp. 273-313.

THE PARABLES OF JESUS AS ANALOGICAL ARGUMENTATION – A CASE IN POINT: TEN BRIDESMAIDS

NIILO LAHTI
The University of Eastern Finland, School of Theology
niilo.lahti@uef.fi

Abstract

This paper argues that at least some of Jesus's parables should be interpreted as analogical argumentation from a particularistic viewpoint (Juthe, 2016). I shall simultaneously argue against two prominent views in biblical scholarship: 1) the parables are viewed only as allegories or stock metaphors (Zimmermann, 2015) or 2) as generating general rules (Thurén, 2014). I will examine the parable of *Ten Bridesmaids* to assess these three approaches.

1. Introduction

"We will never understand the parables unless we are willing to focus on the function of the analogy ..." (Snodgrass, 2008:517).

This article discusses three main approaches to Jesus's parables in biblical studies considering modern argumentation theory and especially argumentation by analogy. Govier & Ayers (2012:167–168) categorize parables regarding their representativeness and scope of conclusion. They ask whether parables represent *symbols* (i.e., *allegories*) of something else, *instantiations* (i.e., general rules), or *analogies* between singular cases that do not axiomatically utilize general rules.

Symbolism or, as traditionally known in biblical studies, allegory has been promoted ever since the first generations of Christ-believers and it still has advocates also in academic research. E.g., Zimmermann (2015) provides a revitalized version of allegory.

Instantiation is championed by Thurén (e.g., 2014) who is among the few having conducted proper argumentation analyses of the parables with a modern method. Within the debate regarding argumentation by analogy (e.g., Govier, 1989; Guarini, 2004; Shecaira, 2013), Thurén would be labeled as a generalist since he employs the Toulmin model (Toulmin, 2003:91, 93; Marraud, 2022:573).

I shall promote the particularistic analogical approach to the parables, more specifically Juthe's (2016) model within the pragma-dialectical framework (van Eemeren, 2018). I will examine the parable of *Ten Bridesmaids* (Matthew 25:1–13) as a case in point from the three perspectives (2.2.–2.4.). I will argue that at least sometimes the one-to-one approach to analogy enables the most accurate reconstruction (and evaluation) of the parable argumentation. Thus, categorizing Jesus's parables as argumentation needs further refinement.

To clarify the reconstructions of *Ten Bridesmaids*, its argumentative context needs to be sketched.

2. A case in point: Ten Bridesmaids

2.1. Argumentative context[1]

Ten Bridesmaids belongs to an argumentative section of Matthew 24–25 which contains parables and narratives that Jesus uses to argue about eschatological matters and the coming of the divine judge, the Son of Man. In Matthew 23, Jesus condemns the pharisees and Jerusalem, and proclaims desolation to its temple. After Jesus has left the temple, the disciples approach their master. In 24:3, they ask "when will these things be [the temple's destruction], and what will be the sign of your coming and of the end of the age?" Based on their request, their assumed favoritism by Jesus as the judge, and his ensuing argumentation, the pupils seem to wish to ensure their survival at the last judgment. If Jesus granted them a sign, they would be confident to survive. As his followers, the disciples assume to have an advantage over other people.

Jesus's argumentation covers four scenarios where the pupils would survive the judgment. Jesus must cover all four arguments to refute the disciples' implicit standpoint, suggesting coordinative argumentation.

The disciples could survive by meeting the judge's demands only briefly, just before the judgment. They would succeed in this if they knew the sign of the end. Jesus argues, however, that nobody knows the timing of the end (24:36, 39, 42; 25:13) and so useful signs are unavailable (1.1a.1).

Second scenario suggests that if the pupils cannot momentarily meet the demands, they would still survive if they met them continuously until the judgment. Jesus counters this: since they do not know the timing of the end, they must meet three complementary demands which is virtually impossible (1.1b). Especially, the parable of *Thief* (24:43–44) argues that initially the disciples need constant vigilance (1.1b.1a). *Ten Bridesmaids* argues that the followers must also prepare if the judge delays (1.1b.1b;

---

[1] I will reconstruct the argumentative context utilizing the pragma-dialectical system (van Eemeren, 2018) since the particularistic approach that I adopt uses that system. See Lahti (forthcoming) for a more detailed reconstruction.

detailed analysis below). *Talents* (25:14–30) adds risk-taking on top which is relevant despite the timing of the judgment (1.1b.1c).

Third scenario maintains that the disciples survive if failing a demand lacks considerable consequences. Especially in *Good and Bad Servant* (24:45–51), Jesus offers that the punishment is fatal (1.1c).

Lastly, the pupils survive if the judge cancels the punishment. *Ten Bridesmaids* counters the possibility to bypass the punishment by correcting the failure (1.1d.1a; detailed analysis below). *Talents* claims that the judge will refuse excuses to bypass the punishment (1.1d.1b).

The structure of Jesus's counter-argumentation is depicted in Figure 1.

(1)   (It is virtually impossible to survive the Son of Man's judgment)
(1.1a) (You cannot meet the demand(s) momentarily just before the judgment)
1.1a.1 You are unaware when the judgment occurs
(1.1b) (You cannot exert yourselves continuously until the judgment)
1.1b.1a If the judgment occurs early, you need constant vigilance
1.1b.1b Additionally, if the judgment occurs late, you need to be prepared
1.1b.1c Additionally, regardless of the judgement's timing, you need risk-taking
1.1c   The punishment from failing the judgment is fatal
1.1d   The judge will not let you bypass the punishment from failing the judgment
1.1d.1a Punishment cannot be bypassed by late rectifications
1.1d.1b Punishment cannot be bypassed by explanations

Figure 1. Jesus's counter-argumentation against the disciples

Ten Bridesmaids contributes especially to arguments 1.1b and 1.1d. Jesus utilizes the parable to advance two lines of coordinatively connected arguments.

2.2. Symbolism or allegorical interpretation

Often allegorical interpretations stem from contemporary ideological needs. This is understandable already because the Gospels encourage such readings (e.g., Matthew 13:18–23, 37–43). In academic scholarship, Jülicher (1888 & 1889) was the first to heavily criticize allegory, followed by many. E.g., Govier & Ayers (2012:185, 187) and Thurén (2014:11, 19–20) provide more recent critique.

Allegorical interpretations typically bring the elements of comparison into a parable outside of the textual context. In that context, there is usually little that supports the allegorical connection. The elements are not compared argumentatively but those of the story are replaced by outside elements. The original elements act as cover names for the new ones. Thus, as allegories the parables do not argue, i.e., provide reasons to believe (Govier & Ayers, 2012:187).

To be effective, any allegorical interpretation requires that the audience knows the story beforehand. This is seldom true with Jesus's textual audiences. The author inferred from the text, "Matthew," needs to convince a first-time recipient to promote further engagements with the Gospel. This is unlikely to happen with just allegories.

E.g., Zimmermann (2015:201–202, 204) rehabilitates allegory by more reliably discovering a transfer of meaning within a parable by distinguishing between *textual* and *conventional plausibility*. Textual plausibility seeks signals within the text or context, urging to discover a particular transfer of meaning. Conventional plausibility holds that in metaphorical texts, such as parables, certain terms have traditional usages. Zimmermann calls such traditional couplings of metaphorical domains *stock metaphor*. Zimmermann argues that "the greater the conventional plausibility of a metaphor, the lesser textual plausibility is required for the metaphorical transfer and vice versa" (2015:202). These linguistic conventions may come from several traditions, including the Scriptures, early Judaism, rabbinical Judaism, or the Greek linguistic world.

I agree that metaphors usually have pre-existing reference points and meanings. However, the textual plausibility should be prioritized over the conventional one. If a metaphor is used in an argumentation, it should be primarily analyzed how the metaphor is utilized in *that* argumentation. In different-domain-analogies, metaphors in the parable (source) are compared with something in the target via abstract relation. These relations tell the primary use of the metaphor, contrasting any "deeper meaning" (Zimmermann, 2015:201). Sometimes the metaphor is utilized in a novel way instead of traditionally. While the textual and conventional plausibility inform each other, the text is more readily available compared to linguistic conventions. Tracing the correct tradition(s) from numerous options is more conjectural.

Regarding *Ten Bridesmaids*, Zimmermann (2015:278–281, 284) offers these stock metaphors: wedding stands for time of salvation or relationship between God and humans, bridegroom stands for YHWH or Messiah or Christ coming for judgment or God (God traditionally coming at an unknown time implies similarity with the bridegroom), and light (of the torches) stands for God's justice or commandment and the demands placed on the people or good works. Zimmermann does not use these options to defend a particular interpretation of the parable but provides several non-argumentative interpretations from different perspectives. Settling for one interpretation may be difficult when a single element has several stock metaphors.

Table I. Stock metaphors of *Ten Bridesmaids*

Ten Bridesmaids	Stock metaphor
wedding	time of salvation; relationship between God and humans
bridegroom	YHWH; Messiah; Christ coming for judgment; God
light (of the torches)	God's justice; commandment and the demands placed on the people; good works

To doubt these stock metaphors, the story is about the terms the bridegroom is met (25:1) before the wedding. The wedding proper occurs after the parable has ended. Any relations between humans and alleged God happen before and outside the party (25:10–12). Jesus compares the bridegroom to the Son of Man, not primarily any of the stock metaphors. Light is not mentioned in the parable, making the corresponding stock metaphors particularly conjectural.

The argumentative indicators (see van Eemeren, Houtlosser & Snoeck Henkemans, 2007) surrounding the parables and parable-like narrative (*Days of Noah*) in Matthew 24–25 (e.g., 24:37 "for as"; 24:39 "so will be"; 24:44 "therefore you also"; 25:1 "will be like"; 25:14 "for as"; 25:32 "as") showcase the need for considering analogical argumentation instead of allegory or symbolism. The indicator relevant to *Ten Bridesmaids* (25:1) compares the situation with the kingdom of heaven to the whole story (Snodgrass, 2008:509, 515).

2.3. Instantiation or generalistic interpretation

Thurén (e.g., 2014, 2021),[2] using the Toulmin model, and so representing generalism (Toulmin, 2003:91, 93; Marraud, 2022:573), reconstructs a general rule (warrant) instantiated by the parable. He interprets parables to function as backings, supporting the warrant, usually produced by the narrative's ending.

The assumption or attribution of the general rule as an implicit premise manifests some deficiencies in the approach:
1. In a typical reconstruction process, the other relevant elements of the parable that are comparable with the target situation, and that are essential for establishing the assumed rule, are not

---

[2] See Lahti (2021) for a pragma-dialectical analysis of Matt 6:25–34 and the parables therein.

explicated (Govier, 1989:144–145; Govier & Ayers, 2012:187; however, Thurén, 2014:21).
2. The arguer may be held responsible for more than they subscribe to (Govier, 1989:144–146; Govier & Ayers, 2012:178, 187; van Laar, 2014:92, 103).
3. This type of reconstruction has difficulties in addressing changes in the strength of the argumentation (Guarini, 2004:159; see Juthe, 2016:247).
4. The origin of the general rule may be obscure (Govier, 1989:145; Guarini, 2004:155; Marraud, 2020:5; Marraud, 2022:588).

To substantiate the parable's theological application, Thurén (2014:21) often joins the warrant with a general *qal wahomer* principle, typical e.g., in ancient Judaism. With *qal wahomer* an inference is made from minor to major, a reasoning *a fortiori*: if something is true/great/etc. in a minor issue, how much more will it be true/great/etc. in a major issue (King, 2011:76). Sometimes the context and explicit *qal wahomer* proposition enable such a reconstruction (e.g., Matthew 7:11, 12:12, and 12:41–42). Other times, however, the general rule and *qal wahomer* are only assumed (e.g., Thurén, 2014:14–16).

Thurén (2023) has analyzed *Ten Bridesmaids* with the Toulmin model. I hereby provide and discuss his reconstruction which focuses on the parable's ending (Figure 2). The poor girls try to access the wedding despite being late, but the groom provides no mercy or second option.

Backing: *Ten Bridesmaids*
▼
Warrant: No second option or mercy will be provided
▼
Data: You are God's subjects ▶ Claim: You cannot rely on God's mercy after the judgment
(thus, any negligence or mistake can be fatal and must be avoided)

Figure 2. Thurén's reconstruction of *Ten Bridesmaids*

While I agree that the ending argues about the inability to circumvent the result of one's failure, this reconstruction has deficiencies. First, the *data* and *claim* allude to God, whereas the disciples will be judged by the Son of Man. Instead, the disciples and the girls are compared based on being the judged ones. Second, the *warrant* appears too general for Jesus to be held responsible. It is unclear who enforces such a principle. Masters generally? The audience may easily find counterexamples. If instead the argumentation functions as a case-to-case analogy, the reliance on a potentially weak warrant cease (point 2).

While the ending reveals a key argumentation, the narrative implies another essential claim in relation to the preceding parables. In *Thief* and

Good and Bad Servant, Jesus suggests that constant vigilance is needed, especially when the judge arrives early. In *Ten Bridesmaids*, Jesus offers another scenario where the judge, now the bridegroom, comes later than expected. The delay causes all virgins to fall asleep, contrasting Jesus's exhortation to stay awake (25:13). This suggests that constant vigilance becomes impossible when the delay extends long enough. Thus, Jesus implicitly exhorts the disciples to prepare for a possible delay of the Son of Man. In unknown circumstances, the preparation complements constant vigilance, which carries only so far. Moreover, the reconstruction excludes the successful virgins, the reward and punishment depicted in the parable, and their implications to the disciples. The parable's total function needs more than one Toulmin-scheme (point 1).

The reconstruction suffers from an unspecific connection between the *warrant and backing*. If the parable is only a presentational device for the warrant, it is questionable to include it in the structure. If the parable functions as an argument by example, offering just one example in support of the rule is not a strong argument (point 4).

2.4. Analogy or particularistic interpretation

To address the problems of symbolism and generalism, I claim that bare minimum *Ten Bridesmaids* should be analyzed from a particularistic approach to argumentation by analogy. Already due to the argumentative indicator (25:1), the approach should treat the argumentation by analogy as *sui generis* (Juthe, 2019) and irreducible (Juthe, 2016:96–101), and not as a presentational device for another argumentation type (against figurative analogy in van Eemeren & Garssen, 2014 and Garssen, 2021; for criticism, see Juthe, 2016:86; van Poppel, 2020:249; van Poppel, 2021:198).

Jesus invites the disciples to judge the events and outcomes depicted in the story (24:43, 45). If the narrative, manifesting complex argumentation, is deemed reasonable and it resembles their own situation on relevant aspects, based on consistency, the pupils should judge their situation similarly (Juthe, 2016:203). Thus, the approach should enable the reconstruction of complex argumentation structures of the source (parable) and target (Juthe, 2016:255). It should enable the comparison of both the corresponding individual elements (bridegroom – the Son of Man) and the argumentative relations (if the bridegroom delays, then the virgins fall asleep – if the judgment delays, then you cannot stay constantly vigilant). Thus, we can respect most of the story and reveal the issues that may raise doubt or criticism in the argumentation (see Juthe, 2016:69, 123).

Consequently, the approach should illuminate the argumentation occurring on two levels: inside the source and target and then on a meta-argumentative level where the argumentation structures of the source and target are compared and the logical properties are transferred from the source to target (Alhambra, 2022:4–5, 18, 21). To materialize all this, the source and target are reconstructed as parallel argumentation structures (Juthe, 2016:201). Moreover, since the fictional parable and the projected future of the disciples depict hypothetical situations, the arguments should reflect this with "if...then" propositions (Govier & Ayers, 2012:186; Alhambra, 2022:2–4, 7, 13, 15, 21).

Juthe's (2016) model to argumentation by analogy within pragma-dialectics (van Eemeren, 2018) fulfills these requirements, when adjusted with the considerations regarding hypothetical argumentation and meta-argumentation. In Juthe's model, argumentation by analogy is based on a comparison of two relationships: (1) the relationship between the elements of the source that determine the *Assigned-Predicate*, and (2) the one-to-one relationship between those elements and parallel elements in the target. Since in the source those elements determine the Assigned-Predicate, plus since they obtain in the target, the Assigned-Predicate (or a similar one) can be inferred also in the target. Consequently, general principles need not be assumed (Juthe, 2016:81, 124–125).

To understand Jesus's message and to decipher the relevant argumentative elements and relations between the parable and the target situation, we need to reconstruct the argumentation of the parable. The parable simultaneously explains the target situation and argues about it. The narrative introduces starting points by referring to common knowledge and customs and it argues that the story and the disciples' situation are comparable and so should be judged similarly (Figure 3). The situations invite cross-domain mapping.

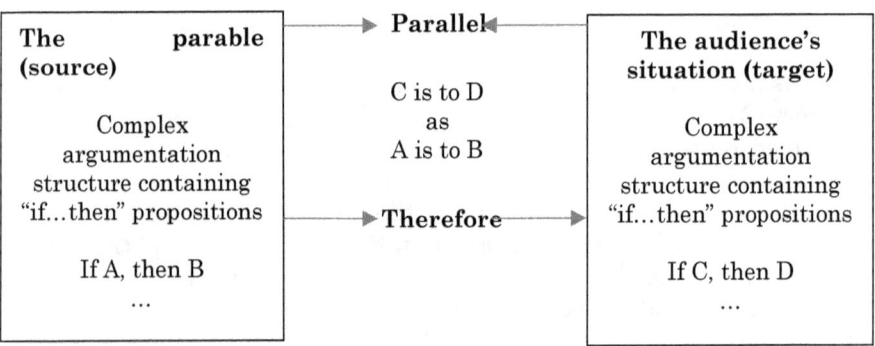

Figure 3. Parallel argumentation between the parable and the audience's situation

Essential for comparing a parable and a "real-world" situation, Juthe distinguishes between *same-domain-analogies* and *different-domain-analogies*:

"A same-domain-analogy is an analogy where not only the relations between the elements of the different objects are the same but also the elements are from the same domain. In a different-domain-analogy the analogy comes only in virtue of having the same relation between the elements of the different objects; the elements of the two objects belong to wholly different domains. ... If a different-domain-analogy is between domains that are very distant, then the elements will be very different and the analogy tends to become a metaphor or parable." (Juthe, 2016:82)

Some elements in the parable may have clear counterparts in the target (bridegroom – the Son of Man: both are judges), however, while others hold more abstract counterparts (having oil – being prepared).

The girls want to enter the wedding with the bridegroom. The bridesmaids having or not having the oil before the groom's announcement determines whether they may enter (1, 1.1b, 1.1c). The bridegroom delays which causes all the girls to fall asleep and the torches of the foolish, that lack the oil, to extinguish (1.1a) (Snodgrass, 2008:516). The foolish virgins try to get the oil after the announcement and before his coming, but they fail (1.1c.1a). Additionally, after finally getting the oil they plea the bridegroom to let them in, but he denies them (1.1c.1b). The failed attempts to rectify the situation before and after the bridegroom's arrival showcase why the oil must be had before the announcement. Considering these motivations and developments, we can reconstruct the storyline's argumentation (Figure 4).

(1)	(If the virgins want to enter the wedding, then they must have the oil before the bridegroom's announcement)
1.1a	If the bridegroom delays, then the virgins' torches extinguish without the oil (5, 8)[3]
1.1b	If the virgins have oil before the bridegroom's announcement, then they can enter the wedding when he comes (4, 6, 10)
1.1c	If the virgins do not have oil before the bridegroom's announcement, then they cannot enter the wedding when he comes (3, 6, 12)
1.1c.1a	If the bridegroom is announced, then the virgins lack time to get the oil before his coming (8–10)
1.1c.1b	If the bridegroom has come, he denies the virgins from entering the wedding despite them having the oil (11–12)

Figure 4. Storyline's argumentation structure

The relevant details and actions in the parable should make sense to the disciples and have counterparts in their situation. In determining the counterparts, it is safe to begin with the elements that are compared with the disciples' projected starting position.

The virgins wait for the bridegroom, who will come and who has authority over them. Similarly, the disciples wait for the Son of Man. The bridegroom decides whether the girls access the wedding, and the Son of Man decides whether the pupils enter the kingdom. Thus, the bridegroom

---

[3] The numbers indicate the verses in Matt 25.

and the Son of Man act as judges and the people in waiting are the judged ones.

The girls' actions are compared with what the disciples can(not) and should (not) do. First, all the virgins fall asleep since the bridegroom delays. This relates to Jesus's exhortation to stay awake. One can initially stay awake but eventually constant vigilance becomes impossible.

The wise virgins survive the delay by having the oil before the groom's announcement. Instead of oil, the disciples, more abstractly, need to be prepared for the delayed judgment. Having the oil and being timely prepared decide whether one passes the judgment. Moreover, the virgins aim for a wedding, whereas the disciples pursue the kingdom. This represents the reward while being denied access depicts the punishment.

Verses 25:8–10 appear to lack specific counterparts for the disciples. After the announcement, the foolish try to get the oil from the wise and from the sellers. If having the oil exemplifies preparedness for the judgment, and since preparedness is not (yet) further specified, it is difficult to gauge how the disciples could take or buy it away from others to themselves. Since the story and the standpoint emphasize places in time, I consider that the primary frame for 25:8–10 is the time between the groom's announcement and his coming. During that time that the poor girls fail to rectify their initial failure. Similarly, it is implied that the disciples cannot rectify theirs before the Son of Man has come to judge.

Perhaps this argument becomes clearer when compared with the parable of *Lightning* (24:27). Thus, if the time between the announcement and the coming is as short as witnessing the start and end of a lightning strike across the sky, it is limited indeed. Nevertheless, I estimate 25:8–9 to be a sore point since the audience might think of colleagues who prepare overabundantly so that there will "be enough for us and you."

We can now map the elements and relations of comparison (Table 2) and reconstruct a parallel argumentation structure of the disciples' situation (Figure 5) considering majority of the story.

Table 2. Mapping the counterpart elements and relations

Ten Bridesmaids	Audience's situation	Relation
bridegroom	Son of Man	judge
virgins	disciples	recipients of judgment
falling asleep	failing at constant vigilance	product of the delay
having oil	being prepared	former exemplifies latter
entering the wedding	entering the kingdom	reward
denied entering the wedding	denied entering the kingdom	punishment
time between announcement and judgment	time between announcement and judgment	too short to rectify initial failure

(1) (If you want to enter the kingdom, then you must be prepared before the Son of Man's judgment is announced)
1.1a If the judgment delays, then you cannot stay constantly vigilant (13)
(1.1b) (If you are prepared before the judgment is announcement, then you can enter the kingdom)
(1.1c) (If you are not prepared before the judgment is announcement, then you cannot enter the kingdom)
(1.1c.1a) (If the judgment is announced, then you lack time to prepare before the judgment)
(1.1c.1b) (If the Son of Man has come to judge, he denies you entering the kingdom even if you are done preparing)

Figure 5. Parallel argumentation structure of the disciples' situation

Compared to prior argumentation in Matthew 24, the novel aspects of *Ten Bridesmaids* are the standpoint that demands preparation and the argument 1.1c.1b about the Son of Man's zero tolerance for failure. Since the audience does not know when the judge comes, it must consider early and late arrivals. Constant vigilance covers the former but not the latter. Thus, preparation creates an additional demand.

The emphasis on branch 1.1c highlights that the punishment following the initial failure cannot be circumvented. This contrasts the disciples' assumption of having an advantageous position at the last judgment.

The crux of the argumentation has become clear. Jesus prophesies that the Son of Man will act as the judge, and he explains via *Ten Bridesmaids* that the judge acts like the bridegroom. Why would this be convincing? Besides the similarity of the two situations, the image of the bridegroom appeals to a common experience about the authority of powerful people over those that must answer them. Yet, the audience may assess whether it is convincing that *the bridegroom* punishes the foolish girls so harshly after their failure and attempt to rectify the situation. Based on the parallel situations, the pupils should assess their fate similarly.

3. Conclusion

Viewing *Ten Bridesmaids* as a case-to-case analogical argumentation has enabled us to appreciate the whole story, its argumentative complexity, and its nuanced contribution in the argumentative context. Moreover, the mapping of counterparts and the reconstruction of parallel structures has enabled a detailed scrutiny of relevant elements and argumentative relations. The level of abstractness must be assessed carefully case-by-case. This strikes a balance between symbolism, where the details get the spotlight, and instantiation, where the abstractness may be overemphasized.[4]

Nevertheless, some parables utilize explicit general rules and allegories may occasionally provide credible secondary interpretations. All three perspectives are viable. We should consider to what extent they accurately reveal the argumentation of Jesus and Matthew and allow us to evaluate their reasonableness.

Finally, I suggest paths for further reflection. First, the particularistic approach requires more testing to determine its suitability to other parables. Nevertheless, I hope to have shown that all parables do not function similarly. Second, Pilgram & van Poppel (2021) have applied an integration of pragma-dialectics and 3D-model of metaphors to analyzing the functions of metaphors in argumentation. Similar analysis may illuminate how Jesus uses certain types of parables based on their (in)directness and (un)conventionality, and their potential and deficiencies in convincing the audience. Combining this approach with Juthe's model promises even more detailed insight of the parables. Third, by perceiving some of the parables as case-to-case analogies, instead of instantiations of general rules, improves their evaluation. Revealing the relevant details, the ensuing relations of the parable, and their counterparts, and the overall meta-argumentativeness enables us to assess whether the relevant critical questions are satisfactorily answered.

References

Alhambra, J. (2022). Argumentation by analogy and weighing of reasons. *Informal Logic*, 42(4), 749–785. https://doi.org/10.22329/il.v42i4.7143

Eemeren, F. H. van. (2018). *Argumentation Theory: A Pragma-Dialectical Approach*. Springer Cham. https://doi.org/10.1007/978-3-319-95381-6

Eemeren, F. H. van & Garssen, B. (2014). Argumentation by analogy in stereotypical argumentative patterns. In R. Ribeiro (Ed.), *Systematic approaches to argument by analogy* (pp. 41–56). Argumentation library 25. Springer Cham. https://doi.org/10.1007/978-3-319-06334-8

Eemeren, F. H. van, Houtlosser, P. & Snoeck Henkemans, A. (Eds.). (2007). *Argumentative indicators in discourse: A pragma-dialectical study*. Argumentation library 12. Springer Dordrecht. https://doi.org/10.1007/978-1-4020-6244-5

Garssen, B. (2021). The maxims of common sense: Strategic manoeuvring with figurative analogies. In R. Boogaart, H. Jansen & M. van Leeuwen (Eds.), *The*

---

[4] While the *Periodic Table of Arguments* (e.g., Wagemans, 2018) offers specific guidelines for reconstructing single argumentations and promotes the ideal of not overloading the argumentation with implicit elements, it is difficult to employ the PTA to a meta-argumentative analogy which constitutes of complex argumentation structures.

language of argumentation (pp. 213–227). Argumentation library 36. Springer Cham. https://doi.org/10.1007/978-3-030-52907-9

Govier, T. (1989). Analogies and Missing Premises. *Informal Logic,* 11(3), 141–152. https://doi.org/10.22329/il.v11i3.2628

Govier, T. & Ayers, L. (2012). Logic and parables: Do these narratives provide arguments? *Informal Logic,* 32(2), 161–189. https://doi.org/10.22329/il.v32i2.3457

Guarini, M. (2004). A defense of non-deductive reconstructions of analogical arguments. *Informal Logic,* 24(2), 153–168. https://doi.org/10.22329/il.v24i2.2141

Jülicher, A. (1888 & 1889). *Die Gleichnisreden Jesu 1 & 2.* J. C. B. Mohr.

Juthe, A. (2016). *Argumentation by analogy: A systematic analytical study of an argument scheme.* [Doctoral dissertation, Universiteit van Amsterdam]. UvA-DARE. https://hdl.handle.net/11245/1.534988

Juthe, A. (2019). A defense of analogy inference as Sui generis. *Logic and Logical Philosophy* [online], 29(2), 259–309. https://doi.org/10.12775/LLP.2019.025

King, L. (2011). Jesus argued like a Jew. *Leaven,* 19/2, Art. 5., 74–79. https://digitalcommons.pepperdine.edu/leaven/vol19/iss2/5

Lahti, N. (2021). Is there a reason to worry? A pragma-dialectical analysis of Matthew 6.25–34. In R. Roitto, C. Shantz & P. Luomanen (Eds.), *Social and cognitive perspectives on the Sermon on the mount* (pp. 205–235). Studies in ancient religion and culture. Equinox.

Lahti, N. (Forthcoming). Argumentation by parable in Matthew 24–25 as *parrhesia. Journal of early Christian history. Special issue: Faith, fear and fearless speech in early Christianity*

Marraud, H. (2020). A modest proposal for classifying the theories of argument. Unpublished. https://www.academia.edu/44808846/A_MODEST_PROPOSAL_FOR_CLASSIFYING_THE_THEORIES_OF_ARGUMENT

Marraud, H. (2022). An unconscious universal in the mind is like an immaterial dinner in the stomach: A debate on logical generalism (1914–1919). *Argumentation,* 36, 569–593. https://doi.org/10.1007/s10503-022-09580-8

Pilgram, R. & van Poppel, L. (2021). The strategic use of metaphor in argumentation. In R. Boogaart, H. Jansen & M. van Leeuwen (Eds.), *The language of argumentation* (pp. 191–212). Argumentation library 36. Springer Cham. https://doi.org/10.1007/978-3-030-52907-9

van Poppel, L. (2020). The relevance of metaphor in argumentation: Uniting pragma-dialectics and deliberate metaphor theory. *Journal of Pragmatics,* 170 (December), 245–252. https://doi.org/10.1016/j.pragma.2020.09.007

van Poppel, L. (2021). The study of metaphor in argumentation theory. Argumentation, 35, 177–208. https://doi.org/10.1007/s10503-020-09523-1

Snodgrass, K. (2008). *Stories with intent: A comprehensive guide to the parables of Jesus.* William B. Eerdmans.

Thurén, L. (2014). *Parables unplugged: Reading the Lukan parables in their rhetorical context.* Fortress Press. https://doi.org/10.2307/j.ctt9m0vdv

Thurén, L. (2021). Parables in the Sermon on the mount: A cognitive and rhetorical perspective. In R. Roitto, C. Shantz & P. Luomanen (Eds.), *Social and cognitive perspectives on the Sermon on the mount* (pp. 174–204). Studies in ancient religion and culture. Equinox.

Thurén, L. (2023). Cracking the code of Jesus's parables with argumentation analysis. *Journal of argumentation in context*, 12(1), 59–76. https://doi.org/10.1075 jaic.22005.thu
Toulmin, S. (2003). *The uses of argument* (2nd ed.). Cambridge University Press. https://doi.org/10.1017/CBO9780511840005
Wagemans, J. H. M. (2018). Analogy, similarity, and the *Periodic Table of Arguments*. Studies in logic, grammar and rhetoric, 55(68), 63–75. https://doi.org/10.2478/slgr-2018-0028
Zimmermann, R. (2015). *Puzzling the parables of Jesus: methods and interpretation*. Fortress Press. https://doi.org/10.2307/j.ctt155j2q7

Employing Argument Mining for Reason-Checking

John Lawrence
Centre for Argument Technology, University of Dundee, UK
j.lawrence@dundee.ac.uk

Jacky Visser
Centre for Argument Technology, University of Dundee, UK

Abstract

This paper describes a software framework, *Skeptic*, aimed at automatically providing pointers for the critical assessment of a persuasive text. That is, with a natural language text as input, the *Skeptic* framework returns a ranked list of questions designed to help readers reason-check fake news and other contentious texts. *Skeptic* combines state-of-the-art argument mining techniques with manually crafted rules to map argumentative features of the text to methods for critical assessment, such as the critical questions of argument schemes, ways of evaluating different types of propositions, and signs of possible biased reasoning.

1. Introduction

While deliberate misinformation, disinformation, and deception are by no means new societal phenomena, the recent rise of fake news (Lazer et al., 2016) and information silos (Flaxman et al., 2016) has become a growing international concern, with politicians, governments and media organisations regularly lamenting the issue. Efforts to combat such disinformation dressed up as genuine news focus too often exclusively on the factual correctness of the claims made. Whilst the truth of purported facts is clearly of crucial importance, there are other, often overlooked, aspects to consider here. It is, after all, very possible to argue from true factual statements to blatantly false or misleading implications by applying skewed, biased, or otherwise defective reasoning. Furthermore, the categorical corrections on factual impropriety delivered by fact-checkers can both alienate readers who believe they are being told what to think (Nyhan & Reifler, 2010) and raise questions around the impartiality of the fact-checkers themselves (Dotson, 2022). For these reasons, attention is increasingly turning to the extension of fact-checking to the broader concept of reason-checking: checking not just factual statements, but whether the entire reasoned discourse is acceptable, relevant, and sufficient (Visser, Lawrence & Reed, 2020).

In the current paper, we propose to address these concerns by employing computational methods for the assessment of argument quality. We deploy a range of software tools aimed at helping a reader critically assess a piece of text, improving their critical thinking abilities, and allowing them to draw their own conclusions as to whether or not they should accept what they are reading. In order to achieve this, we first apply a combination of argument mining techniques to reconstruct argumentative features of the text, such as the argumentation structure, proposition types, and argument schemes. These features are then mapped to potential areas of concern, with the software providing prompts which allow the reader to investigate further.

2. The *Skeptic* Reason-Checking Framework

The *Skeptic* (Visser & Lawrence, 2022) reason-checking framework combines state-of-the-art argument mining techniques with manually defined rules, mapping features of the mined argument structure to potential flaws or weaknesses in the arguments presented. An overview of the framework can be seen in Fig. 1.

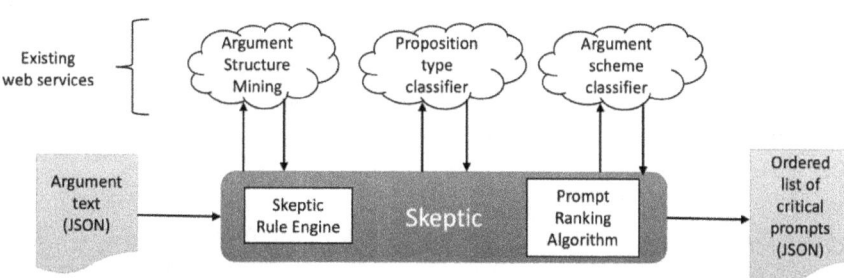

Figure 1. The tasks and levels of complexity in argument mining techniques

The software tool is implemented as a web service[1] that takes an input natural language text and returns a ranked list of questions designed to help readers reason-check the argumentation. The questions are meant to be used as pointers, empowering the readers' critical literacy skills, helping them to draw their own conclusions as to whether or not they

---

[1] Web service is a generic term for a piece of software, the functionality of which is made available over the internet using a common format for communicating with other software tools—in this case, JavaScript Object Notation (JSON).

should accept what they are reading. Actively involving the reader in the reason-checking process should help avoid the instinctive enmity engendered by authoritative fact-checks, while simultaneously broadening the critical spectrum.

The web service maps argumentative features of the persuasive text to methods for critical assessment, such as the critical questions of argument schemes, ways of evaluating different types of propositions, and signs of possible biased reasoning. We employ a pipeline of extant argument technologies (Snaith et al., 2010), all developed to work with the AIF ontology (Chesñevar et al., 2006), using JSON as a common file type to facilitate handover between the different pipeline components. The combined argument mining and classification techniques provide a reconstruction of the argumentative features of the text, such as the structure of the argumentation, the proposition types of premises and conclusions, and the argument schemes instantiated in the text. These features are then mapped to potential areas of concern, which the *Skeptic* framework returns as a ranked list of prompts for readers to investigate further.

Looking at the overall argumentation structure allows us to identify potential areas of bias where only one side of an argument is being exposed. The argumentation structure also allows us to identify the most central propositions in an argument. These are then classified into one of three proposition types: statements of fact, value, or policy (Visser, Lawrence, Reed, Wagemans & Walton, 2020). This classification results in a powerful expansion upon mere fact-checking by broadening the range of proposition types to be checked. Where factual statements can be checked for veracity, policy statements could be checked for consistency or appropriateness, while value statements could be checked for, e.g., popularity. Finally, identified instances of argument schemes are mapped to their associated critical questions (Walton et al., 2008).

The remainder of this paper explores in more detail how these various components work, looking first at the argument mining techniques employed, before moving on to consider how the identified structure is mapped to potential critical prompts and how these can then be ranked by their criticality.

3. Argument Mining

Argument Mining (Lawrence & Reed, 2020) broadly refers to the automatic identification and extraction of argument components and structure in natural language text. Lawrence and Reed (2020) break down the related tasks involved in argument mining as shown in Fig. 2. Starting from the identification of argument components by segmenting and classifying these as part of the argument being made or not (these tasks are sometimes performed simultaneously, sometimes separated and sometimes the latter is omitted completely), tasks move down through levels of increasing complexity: first considering the role of individual clauses (both intrinsic, such as whether the clause is a policy statement, and contextual such as whether the clause is the conclusion to an argument); and, secondly considering argumentative relations from simple premise/conclusion relationships to whether a set of clauses forms a complex argumentative relation, such as an instance of an argument scheme.

The *Skeptic* framework makes use of argument mining components across these various levels of complexity, combining techniques for the identification of overall structure (argument components and their attack/support relationships), intrinsic clausal properties (statements of fact, value, or policy), and the identification of instances of argument schemes. The modular nature of the framework allows for the specific implementation of each of these components to be easily swapped in and out. We detail below the current implementations employed by *Skeptic*.

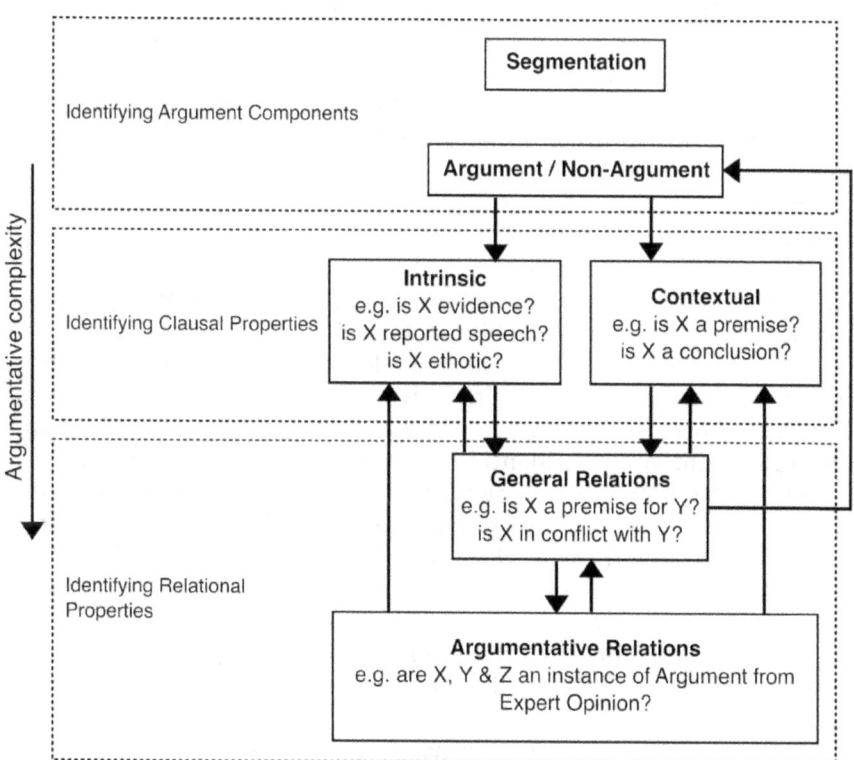

Figure 2. The tasks and levels of complexity in argument mining techniques

3.1. Argument Structure Identification

Argument structure identification is carried out in two phases, first segmenting the text into the components from which the argument is constructed, and then identifying the support and attack relationships which exist between these.

Segmentation involves the extraction of the fragments of text from the original text that will form the constituent parts of the resulting argument structure. Peldszus and Stede (2013) refer to these argument segments as 'Argumentative Discourse Units' (ADUs), and define an ADU as a 'minimal unit of analysis'.

For ADU segmentation we use *Proposition Boundary Learning* (Lawrence & Reed, 2015) a supervised learning approach in which two naïve Bayes classifiers are used, one to determine the first word of a proposition and one to determine the last. The classifiers are trained using a set of manually extracted propositions as training data. The text to be segmented is first split into words and a list of features is then determined for each of these words. The features used cover both intrinsic (the word itself, its length, and Part Of Speech) and contextual (the word/punctuation before and the word/punctuation after). By looking at more general features (length and POS) and contextual features, this approach aims to overcome the variability in specific words that may start (or end) a proposition.

Having trained the classifiers, this same list of features is then determined for each word in the test data, enabling the classifiers to label each word as being 'start' or 'end'. Once the classification has taken place, the individual starts and ends are matched to determine propositions, using their calculated probabilities to resolve situations where a start is not followed by an end (i.e. where the length of the proposition text to be segmented is ambiguous). Using this method, a 32% increase in accuracy is achieved over simply segmenting the text into sentences when compared to argumentative spans identified by a manual analysis process.

Once the ADUs have been determined, the main claims are then identified by considering how semantically similar each ADU is to all of the other ADUs. This method has been shown to provide a reliable indication of which claim is most central within the argumentative structure of a piece of text (Lawrence et al., 2017). Having identified the main claims the supporting reasons for these, and any potential objections, are identified. To do this, all of the other sentences are ranked by their semantic similarity to the main claim, and then checked for indicators of support (e.g., "because") or conflict (e.g., "however") to determine their possible argumentative relations (Lawrence & Reed., 2017).

3.2. Proposition Type Classification

For proposition type classification, we experimented with eight different deep-learning model configurations: four pipelines based on BERT (Devlin et al., 2019), and another four based on DistilBERT (Sanh et al., 2019). The BERT-based models were expected to result in a higher accuracy, while we expected DistilBERT to increase the processing speed of predictions while maintaining most of the high performance levels.

The Scikit Learn (Pedregosa et al., 2011) library for Python was used to implement four different models. Linear SVC and Naive Bayes were selected for their strength in text classification problems. Logistic regression is not normally ideal for text classification problems with more than two classes without modification, but performance was increased by setting the *multiclass* hyperparameter to multinomial. SVC was also tested as an alternative to LinearSVC.

The BERT-SVC pipeline demonstrated the highest performance of all model configurations, with an accuracy of 80% and F1 scores of 84%, 87% and 60% for the classification of *Value*, *Policy* and *Fact*, respectively. As such, this pipeline was chosen as the current proposition type classification model used by *Skeptic*.

3.3. Argument Scheme Classification

In order to identify instances of argument schemes, we use the approach described by Lawrence & Reed (2015), where, rather than aiming to classify complete scheme instances, the individual scheme components are identified and then grouped together into a scheme instance. For example, Walton's *Position to Know* scheme is comprised of an assertion premise, a position to know premise, and a conclusion. Where these elements are identified in close proximity in the text, we can assume that this scheme is being used. By considering the features of the individual types of these components and training individual machine learning classifiers to identify each, accuracies between 0.75 and 0.93 are achieved for identifying at least one component part of a scheme.

4. Identification and Ranking of Critical Prompts

Having identified both the overall argument structure as well as proposition types and instances of argument schemes using the automatic techniques described in Section 3, *Skeptic* utilises manually crafted rules to determine areas of potential weakness in the expressed arguments. These rules are comprised of three categories: structural bias rules, proposition type rules, and scheme-based rules. Examples of each type of rule can be seen in Table I.

Table I. Example rules for generating critical prompts

Rule Type	Condition	Critical Prompt
Structural Bias	A proposition has more than one incoming support, but no incoming conflicts.	Is there any counter-argument against this claim?
Proposition Type	Policy statement	Is this statement consistent with previous proposals?
Proposition Type	Value statement	Does this statement reflect a popularly held value?
Proposition Type	Factual statement	Is this statement actually true?
Scheme-based	Position to Know	Is the source honest, trustworthy, reliable?
Scheme-based	Popular Opinion	What evidence, such as a poll or an appeal to common knowledge, supports that this claim is generally accepted as true?
Scheme-based	Correlation To Cause	Is there any reason to think that the correlation is any more than a coincidence?

Application of these rules results in a comprehensive list of potential critical prompts, addressing potential weaknesses of each proposition and inference scheme contained within the identified argument structure. The presentation of such a complete list has the potential to overwhelm the user, therefore we employ a combination of properties based on network graph theory (Barabási, 2016) to determine the criticality of each component for the structure as a whole. In particular, we calculate the centrality and divisiveness of each proposition and use these to rank their associated critical prompts.

Central issues are those that play a particularly important role in the argumentative structure. In order to calculate centrality scores for each proposition, we adapt eigenvector centrality (used in the Google Pagerank algorithm (Brin & Page, 1998)) Firstly, we create a directed graph, $G = (V, E)$, in which: the vertices (V) are propositions extracted from the corpus; and, an edge exists between two vertices if there is a support or attack relationship connecting them in the identified structure. We then construct the matrix $A = (a_{v,t})$, where $a_{v,t}$ is the weight of the edge between vertex v and vertex t if v and t are connected, and $a_{v,t}=0$ otherwise. The relative centrality score of vertex v can then be defined as shown in

Equation 1 where λ represents the greatest eigenvalue for which a non-zero eigenvector exists.

$$Central(v) = \frac{1}{\lambda}\sum_{t \in G} a_{v,t}\, Central(t)$$

Equation 1. Calculation of centrality.

Divisive issues are those that split opinion and which have points both supporting and attacking them (Konat et al., 2016). The Divisiveness of an issue measures how many others are in conflict with it and the amount of support which the two sides have. In order to calculate this, we now create two directed sub-graphs, one for support and one for conflict: $G_s = (V, E_s)$, in which an edge exists between two vertices if there is a support relationship connecting them; and, $G_c = (V, E_c)$, in which an edge exists between two vertices if there is an attack relationship connecting them. The divisiveness of a vertex v can then be defined as shown in Equation 2, where $deg_s^-(v)$ is the in-degree of vertex v in G_s.

$$Divisive(v) = \sum_{\forall t\ s.t.(t,v) \in E_c\ \vee\ (v,t) \in E_c} deg_s^-(v) * deg_s^-(t)$$

Equation 2. Calculation of divisiveness.

The *Skeptic* framework offers flexibility in how the centrality and divisiveness scores can be combined in order to produce the final criticality ranking. In the current implementation, each proposition is ranked separately according to each measure and then the average ranking taken. An example of the resulting JSON formatted output from the Skeptic framework can be seen in Fig. 3. In this output each critical prompt is comprised of its criticality rank, a reference for the AIF node (either an I-node, corresponding to a proposition, or an S-node corresponding to an argumentative relation of support or conflict) to which it refers, and the text of the question which is being asked in relation to this node.

The output from *Skeptic* offers the possibility for integration in a wide variety of composition and discussion software platforms. An example of how output from the *Skeptic* framework can be integrated into other software tools is shown in Fig. 4. This example shows an interface for text composition in which the user is being asked to consider a critical question relating to the argument from analogy which they are making. Such an interface can allow the user to confirm that they have considered this issue, before then moving on to address further potential points of concern.

```
"Skeptic": {
    "questions": [
        {
            "rank": 1,
            "nodeID": "3_1661780046893607143",
            "question": "Is this statement consistent with previous proposals?"
        },
        {
            "rank": 2,
            "nodeID": "3_1661780046893607143",
            "question": "Is this policy appropriate for this objective?"
        },
        {
            "rank": 3,
            "nodeID": "5_1661780046893607143",
            "question": "How strong is the probability or plausibility that these cited consequences will (may, might, must) occur?"
        },
        {
            "rank": 4,
            "nodeID": "5_1661780046893607143",
            "question": "What evidence, if any, supported the claim that these consequences will (may, might, must) occur if the action is brought about?"
        },
        {
            "rank": 5,
            "nodeID": "5_1661780046893607143",
            "question": "Are there consequences of the opposite value that ought to be taken into account?"
        },
        ...
    ]
}
```

Figure 3. Example output from the *Skeptic* framework.

don't know a shell is in the chamber, is your friend any less dead because you were playing?

Of course, some people will say: "It's just a game!" But like the people who think they can play around with crack or pre-marital sex and not get burned by death, AIDS or pregnancy, the person who thinks they can mess with Dungeons and Dragons without getting burnt is whistling in the dark.

Q: Are there differences between the compared cases that would tend to undermine the force of the similarity cited?

Thus, in my mind, there is no doubt that Dungeons and Dragons and its imitators

Figure 4. An example of how output from the *Skeptic* framework can be integrated into other software tools.

5. Conclusion

The *Skeptic* reason-checking framework combines the identification of argumentative features and mapping these to potential flaws in the expressed reasoning. The software allows the user to enter a piece of text and receive a ranked list of questions that they may wish to consider further. The developed software offers a range of potential applications in, for instance, critical literacy education, tools to improve persuasive writing, and the identification of misinformation and fake news.

As *Skeptic* relies on the output of existing argument mining techniques to recover the underlying argument structure, performance is currently limited by the accuracy of these methods. However, as such techniques continue to improve, we believe that *Skeptic* will offer a solid platform for tools aimed at improving reasoning and critical thinking, with a wide range of potential use-cases.

References

Barabási, A.L. (2016). Network Science. Cambridge University Press.

Brin, S. & Page, L. (1998). The anatomy of a large-scale hypertextual web search engine. *Computer networks and ISDN systems, 30(1-7)*, 107–117.

Chesñevar, C., McGinnis, J., Modgil, S., Rahwan, I., Reed, C., & Simari, G. (2006). Towards an argument interchange format. *The Knowledge Engineering Review, 21(04)*, 293-316.

Devlin, J., Chang, M. W., Lee, K., & Toutanova, K. (2019). BERT: Pre-training of Deep Bidirectional Transformers for Language Understanding. In *Proceedings of the 2019 Conference of the North American Chapter of the Association for Computational Linguistics: Human Language Technologies* (pp. 4171–4186). Association for Computational Linguistics

Dotson T. (2022). Fact-checking may be important, but it won't help Americans learn to disagree better. *https://bit.ly/38OhsVu* Accessed: 2-5-2022.

Flaxman S., Goel S., & Rao J.M. (2016). Filter Bubbles, Echo Chambers, and Online News Consumption. *Public Opinion Quarterly, 80*, 298-320.

Konat, B., Lawrence, J., Park, J., Budzynska, K., & Reed, C. (2016). A corpus of argument networks: Using graph properties to analyse divisive issues. In *Proceedings of the 10th edition of the Language Resources and Evaluation Conference.*

Lawrence, J., & Reed, C. (2015). Combining Argument Mining Techniques. In *Proceedings of the Second Workshop on Argumentation Mining.* Association for Computational Linguistics.

Lawrence, J., & Reed, C. (2017). Using Complex Argumentative Interactions to Reconstruct the Argumentative Structure of Large-Scale Debates. In *Proceedings of the Fourth Workshop on Argumentation Mining.* Association for Computational Linguistics.

Lawrence, J., Park, J., Budzynska, K., Cardie, C., Konat, B., and Reed, C. (2017). Using Argumentative Structure to Interpret Debates in Online Deliberative

Democracy and eRulemaking. *ACM Transactions on Internet Technology, 17(3),* 1-22.

Lawrence, J. & Reed, C. (2020). Argument mining: A survey. *Computational Linguistics, 45(4),* 765–818.

Lazer D. M. J., Baum M. A., Benkler Y., Berinsky A. J., Greenhill K. M., & Menczer F (2016). The science of fake news. *Science, 359(6380),* 1094-1096.

Nyhan, B., & Reifler, J. (2010). When corrections fail: The persistence of political misperceptions. *Political Behavior, 32,* 303–330.

Pedregosa, F., Varoquaux, G., Gramfort, A., Michel, V., Thirion, B., Grisel, O. & Duchesnay, E. (2011). Scikit-learn: Machine learning in Python. *The Journal of machine Learning research, 12,* 2825-2830.

Peldszus, A. and Stede, M. (2013). From argument diagrams to argumentation mining in texts: A survey. *International Journal of Cognitive Informatics and Natural Intelligence (IJCINI), 7(1),* 1–31

Sanh, V., Debut, L., Chaumond, J., & Wolf, T. (2019). DistilBERT, a distilled version of BERT: smaller, faster, cheaper and lighter. *arXiv preprint arXiv:1910.01108.*

Snaith, M., Devereux, J., Lawrence, J., & Reed, C. (2010). Pipelining Argumentation Technologies. In P. Baroni, F. Cerutti, M. Giacomin & G. Simari (Eds.), *Proceedings of the 3rd International Conference on Computational Models of Argument (COMMA 2010)* (pp. 447-454). IOS Press.

Visser, J., Lawrence, J., & Reed, C. (2020). Reason-Checking Fake News. *Communications of the ACM, 63(11),* 38-40.

Visser, J., Lawrence, J., Reed, C., Wagemans, J., & Walton, D. (2020). Annotating argument schemes. *Argumentation, 35,* 101–139.

Visser, J., & Lawrence, J. (2022) The skeptic web service: Utilising argument technologies for reason-checking. *Proceedings of the Ninth International Conference on Computational Models of Argument (COMMA 2022)* (pp. 375-376) IOS Press.

Walton, D., Reed, C., & Macagno, F. (2008) Argumentation Schemes. Cambridge University Press.

NORMAL VS. INTENSE SCRUTINY: DISTINCT MODES OF CRITICAL THINKING?

LAWRENCE LENGBEYER
United Stated Naval Academy
lengbeye@usna.edu

Abstract

We should distinguish between two distinct methodologies of critical thinking, 'Normal Scrutiny' and 'Intense Scrutiny,' whose analytical results are liable to come apart in particular cases. The two modes diverge in their logical rigor and/or sensitivity to utterance context features. Analysis of critical thinking test items raises a concern that standard training in critical thinking, and standardized assessment of it, are aimed only at Normal Scrutiny, thereby discouraging and penalizing the use of Intense Scrutiny despite its great situational utility.

1. The hypothesis & its sources

Critical thinking ("CT") is standardly regarded as a single methodology to be contrasted with uncritical thinking. I suggest, however, that CT comprises at least two distinct cognitive approaches, 'Normal Scrutiny' and 'Intense Scrutiny,' approaches whose analyses are prone to diverge. If this hypothesis is correct, it raises doubts about how we teach and assess CT abilities.

I have formulated this hypothesis as an inference to best explanation from three kinds of observation.

First are the student outcomes obtained at the United States Naval Academy when I taught undergraduate university students a novel argument-mapping-based CT course. The course was constructed by an international team of experts funded by the United States government's IARPA program (Intelligence Advanced Research Projects Activity), then tested at numerous educational sites and with varied student populations in Australia, the United States, the United Kingdom, and Canada (Thomason et al., 2013). In each instance, student CT skills were carefully assessed in a pre-/post-test structure using two respected instruments, the logical reasoning portion of the LSAT (Law School Admission Test; Law School Admission Council, 2007) and the then-recently-developed HCTA

(Halpern Critical Thinking Assessment; Halpern, 2010).[1] The results obtained in my own course were poor, unexpectedly so given the general quality of the students (at least the equal of the student cohorts elsewhere that obtained superior results), their evident effort and high morale, and the apparent understanding that they clearly demonstrated day to day in the mapping activities conducted in our small-enrollment classroom setting.

How to make sense of the unexpected outcomes? There may have been serious deficiencies in my own teaching, and in my perception of student learning. But this seemed unlikely, given my demonstrated reasoning ability, passionate engagement throughout the course, careful attentiveness to teaching in general, and habitual self-critical stance. I then conjectured that (i) the Naval Academy students' test answers shifted from pre-test to post-test by aligning increasingly with what could be termed "Intense Scrutiny," given that during argument mapping exercises I had upheld unusually demanding standards of proper logic and attentiveness to unstated assumptions, an emphasis on rigor that accounted for our slower course progress compared to the sibling courses taught elsewhere; and (ii) this student shift toward employing Intense Scrutiny explained by the disappointing post-test scores, either because the official test answers assess compliance with Normal Scrutiny, such that answers resulting from Intense Scrutiny are often graded as incorrect, and/or because Intense Scrutiny is more time-consuming and hence costly to use on a question-packed timed test.

The second set of observations underlying the hypothesis of competing modes of CT comprises the oft-seen variations in degree of reasoning intensity employed across practical settings in everyday life. We are all familiar with, for instance, the legalistic hair-splitting of many politicians and others who work with the law. Or of teenagers, who can swerve from highly imprecise uses of language to saying something like, "I did say I'd never kiss him ... but I never said I wouldn't let *him* kiss *me*." Some persons have vocational obligations to utilize special levels of rigor in their professional (but not personal) reasonings, such as intelligence analysts (see Kruger et al., 2022), who advise top state officials and thus need to scrutinize the assumptions embedded within their reasoning in scrupulous, almost paranoid ways: e.g., "Are we wrong to assume that those teammates are not actually double agents working for the adversary?" And then there are the conversational contributions of

---

[1] Those tests do not involve argument mapping, but the reasoning skills gained via mapping were expected to transfer to the kinds of verbal problems offered in the LSAT and HCTA. Because two alternate versions of each test were available, it was possible to administer each test twice, a pre-test prior to course instruction and a post-test at course conclusion. The LSAT is particularly highly regarded, and has been used in subsequent studies of argument mapping. See, e.g., Cullen et al., 2018, esp. pp. 3-4.

philosophers, whose esteem for precision in thought and expression marks them off from most of the population in ways that often deserve respect and appreciation, but can earn derision when applied indiscriminately.

We can thus readily see the concrete consequences of adopting differing stances toward verbal discourse that we receive and produce. While practical and ethical appropriateness of stances will turn upon the particulars of situations and participant understandings and expectations—and will often be debatable—it is undeniable that differing stances cause interactants to operate cognitively (and emotionally) with divergent understandings. For example, one who disregards speaker intentions and other pragmatic features of the communicative context, and places exclusive emphasis upon the strict semantics of utterances, will commonly make very different sense of spoken or written language than another person who attends less to the precise linguistic forms used and more to the circumstances of utterance and reception.

Differing stances will bring about divergences in both how text gets comprehended (how the language is understood, how the utterances are parsed, how meanings are assigned, how associations are or are not raised to salience) and how reasoning gets conducted. These two effects can perhaps sometimes merge in practice, but they are conceptually distinguishable. The latter will of course be occasioned in part by the former, as the *inputs* to inference and analysis will thus vary, but also by differing choices of reasoning *process* or method. Are verbally-expressed contents (as construed) to be taken strictly, for example, or is there latitude for charitable re-construals that allow for utterers' casualness in expression, lack of logical sophistication, or outright misstatements, and so don't take them strictly at their words? Do we target what the utterer was *trying* to say, or carefully hew to exactly what she *did* say—and do we vary this choice for the core of her apparent message as opposed to its supporting and peripheral parts, so that we might be stricter about the former than the latter?

Obviously, we have here abundant grounds for suspecting that those asked to engage in critical thinking might well produce analyses of verbal material that do not coincide and are possibly inconsistent. Moreover, the differing analyses can both (or all) be reasonable. Yes, there *may* be response-narrowing features of situations that, in combination with well-known norms of proper 'play' in particular kinds of 'language games,' permit rather objective distinctions to be drawn between competent and faulty performance; but this will not always be the case. Where it is not, as when an unsocialized newcomer engages in a novel language game (or what seems to be a familiar game, but is actually played here according to different norms), then there may be inaccuracy or unfairness in assessments of the newcomer's performance. This will be relevant to how we think about those who undergo CT assessments, especially if the test-taking norms to be followed are not effectively communicated to them, or are even so obscure, complicated, unexamined, or poorly understood as to be incommunicable to them at the time.

The third observation behind the 'two modalities of CT' hypothesis is the not-uncommon admonition that test-takers not "overthink" standardized tests that call upon CT skills (among others), like the American SAT (Scholastic Aptitude Test) for university-seekers.[2] I have seen no careful analysis of what such 'overthinking' consists of or how it leads to underperformance on tests. But some plausible conjectures are that supposed 'overthinkers' are, for purposes of the particular activity of test-taking, resolving ambiguities they encounter by

1. being too attentive to the *semantics* of the items—the prompts and/or answer options—and heedless of their pragmatics;[3]
2. being too attentive to the actual or imagined *pragmatics* of the items, instead of taking the words at face value;
3. being too attentive to the semantics of items' verbal *components*, and neglectful of more *holistic* senses—one version of this being literal-mindedness about figures of speech (e.g., being uncertain whether "sexually suggestive"[4] requires a proposal of sexual interaction);
4. taking terms that lack explicit qualification or condition to be *unqualified* and unconditional (e.g., when an HCTA item, on the topic of inclusion of welfare recipients in reform efforts, offers as one answer option, "Welfare recipients are not to be trusted,"[5] this is construed as across-the-board mistrust, as opposed to mistrust in the context of crafting regulations that will apply to themselves);
5. considering (semantic or pragmatic) *meanings that test-takers are not expected to consider* because knowledge of them typically comes only with greater theoretical background or intellectual sophistication, or because they involve rare or recondite uses of terms;
6. considering (semantic or pragmatic) *meanings that test-takers are not expected to consider* because these meanings are less

---

[2] I shall focus on the overthinking that consists in applying certain kinds of analysis, rather than the sometimes related, but distinct, notions used to point to loss of perspective, self-conscious attention to details of activities that can be done automatically, or excessive self-doubt and dithering. (On the intrusions of thoughts, worries, and other distractions that compete with on-going cognitive tasks, see, e.g., Beilock & Carr, 2005; on counterproductive self-monitoring, see, e.g., Beilock et al., 2004. On the various uses of "thinking too much," see Kaiser et al., 2015.)

[3] See, e.g., the example of "All lives matter," discussed in Chappell, 2023.

[4] HCTA Form B, Question 9 (Halpern, 2012).

[5] HCTA Form B, Question 8 (Halpern, 2012).

paradigmatic and common, or involve references to world features that are less familiar to typical (student) test-takers;
7. imagining/considering unusual instances of, or associations evoked by, phenomena described in test items (e.g., "Well, it could be that he 'asks' via facial expression, so therefore ...") (possibly characterizable as 'lateral thinking').

Is 'overthinking' of these kinds something to be discouraged in education—or encouraged? Penalized in testing—or rewarded? Is not some of this reasoning so stringent and creative that it deserves to wear the name "Intense Scrutiny" as a badge of honor rather than of shame?

2. The divergence in the context of testing: discouraging a valuable practice

It will help us judge these practices if we have specific concrete instances to examine.

One high-value question on the HCTA test (worth 7 of the test's 99 points) asks the test-taker to imagine herself as an instructor at a preschool for children with disabilities:

> You realize that a new employee has been showing signs of an eating disorder. You do not see any signs of his eating disorder during the workday, but you are concerned because you both have to deal with very young children who have special needs and he may not be able to care for the children if he is sick. He has not responded to your hints about his eating disorder problem. As far as you know, no one else knows about his eating disorder. Given these facts, rate each of the following problem statements on a scale of 1 to 7 [(from extremely poor to excellent) based on how good it is as a] ... statement of the problem[:]
> 1. The employee may not be able to properly attend to the children because he has an eating disorder.
> 2. You are the only one who knows he has an eating disorder.
> 3. The employee's parents do not know that he has an eating disorder.
> 4. You need to find a way to give the employee better hints about his eating disorder.
> 5. The employee may be fired if his eating disorder continues.
> 6. The employee may hurt himself if his eating disorder continues.

7. You feel responsible for doing something about the employee's eating disorder. [6]

"The problem" of interest here is open to very varied conceptualizations that seem independent of CT ability, or at least related very uncertainly to it. And is the test-taker to infer from the phrase "the problem" that the test creators believe there to be one (core) problem—such that it would be improper to supply high ratings to *multiple* distinct problem-stating options? And what does it mean to suggest that it's a *problem* to feel responsible (in #7) for acting on a colleague's mental health issue? Is this not an appropriate, even admirable, reaction, rather than a problematic one? Or, is the test-taker intended and expected to use a forgiving, loose interpretive approach and to figure that the test-creators meant option #7 as 'needing to figure out what to do about the colleague's eating disorder'—something that might fairly be thought a problem? Different stances will point test-takers in different directions.

Similar puzzles are presented by this question prompt:[7]

> Psychologists have been debating over whether to include the diagnostic category "self-defeating personality disorder" in their manual of psychological disorders. One psychologist argued that women who are battered (victims of physical abuse) suffer from this disorder, and the reason that they allow themselves to become victims of abuse is because they have this disorder.

The test-taker is then asked, "Which of the following is the best criticism of this argument? (Choose one.)" The official 'right' answer is, "The term 'self-defeating personality disorder' is just another term for a victim of abuse—it does not explain anything." Some students chose instead this 'wrong' answer: "The use of this term suggests that the victim of abuse is somehow responsible for being abused." But surely that is a reasonable answer ... unless one is expected to recognize that "best criticism" as used on this kind of test is asking for the best *'logical'* criticism, that is, the kind of flaw that is part of the standard canon in CT courses—not the best *moral* criticism. Even on logical grounds alone, the 'correct' answer is dubious: had the prompt concerned the claim that *narcissistic personality disorder* is what causes batterers to be physically abusive, surely that would be plausible, and not an empty tautology.

It seems that some CT test items prioritize a kind of thinking (presumably as part of Normal Scrutiny) that uncritically spits back standard textbook lessons. They do not, at least, invite independent, no-holds-barred critical thinking (which "Intense Scrutiny" would include).

---

[6] HCTA Form B, Question 21 (Halpern, 2012). The official answers are (1) good to excellent; (2) very poor to good; (3) extremely poor to very poor; (4) extremely poor to medium; (5) medium to excellent; (6) medium to excellent; (7) medium to excellent.
[7] HCTA Form A, Question 7 (Halpern, 2012).

Items like these give Normal Scrutiny a bad name. They make it seem that CT is capped at a ceiling that discourages (and faults or penalizes) broader and deeper probing of uncertain issues. Arguably, the Normal Scrutiny taught in CT courses consists (primarily?) of an awareness of a set of patterns of logical and illogical reasoning. The emphasis is on training students to recognize certain canonic models of correct reasoning (whether deductive, as with *modus tollens*, or inductive, as with awareness of regression to the mean or small-sample-size distortions), plus the usual misguided deviations (such as the fallacy of affirming the consequent, or the Gambler's Fallacy). Lessons abstract away from the complexities of ordinary discourse.

This might be defensible—it's good for students to gain this limited competence, and maybe too much to ask that they gain more—but we ought not let standard CT pedagogy deceive us into believing that such courses are instilling a practice of true critical thinking. Like other university courses, for instance in applied mathematics or engineering, the typical CT course imparts an array of concepts along with paradigmatic instances that students learn to recognize and manipulate. Learning such identifications and maneuvers can be useful; the course content is not merely narrow gimmicks and tricks. But it seems wrong to encourage what might be called 'academic correctness,' the following of certain authority-prescribed thinking paths that have been emphasized in formal curricula while avoiding more searching kinds of thought.[8] As political correctness directs thinking into certain channels that bypass fields of potential truth and insight, so too might CT lessons ask students to mechanically impose simplified perceptual grids upon encountered instances of reasoning while steering them away from richer though more uncertain (and maybe more incomplete) possibilities of meaning or logic.

This suggestion seems to be supported by the effectiveness of the highly-profitable test-preparation courses that have emerged to aid test-takers. The courses produce results—higher scores, including on CT tests—by informing test-takers that test creators are seeking certain particular kinds of reasoning awareness, kinds that our mass schooling can realistically aim to impart, and not some generalized capacity for handling all reasoning in the wild, let alone for deploying especially careful, i.e. Intense, analysis.

But it seems wrong for tests that purport to assess CT skill to penalize those who stray from the narrow dictates of, say, fallacy-spotting of the usual suspects. True CT is a process and a motivation, and the simple tools we convey in CT courses stand to true CT as, say, basic music theory concepts stand to being musical. Or as the standard rules and patterns of

---

[8] For commentary on the extent to which the SAT actually measures "developed reasoning skills" as opposed to "test-wiseness" and facility with the test "idiom," see the interviews and other materials in PBS, 1999.

grammar, genre, plot, etc. stand to creative writing. Or as mastery of a philosophical glossary stands to being philosophical. In all these cases, practical pressures for standardization of teaching and assessment must not be allowed to limit the ambitions of learners, discouraging their more free-ranging, penetrating, and creative activities.

This analysis supports the longstanding critique that standardized tests confound their assessments of intellectual skills with unacknowledged and undefended assessments of certain sorts of 'cultural' knowledge or expertise. Students who learn what 'academic correctness' expects, and are prepared to provide that instead of thinking more earnestly and independently, are seemingly at a definite advantage.

Another simple illustration: Earlier, I described a test question as asking the test-taker "to imagine herself as an instructor"—employing the female pronoun. When a test-taker encounters such a usage during the test, is the gender to be disregarded, taken to be simply an effort to avoid androcentric linguistic bias? If a test-taker were to suspect, to the contrary, that the choice of "herself" was substantively significant (particularly if gender would affect the choice of answer), ought the test-taker be penalized for not having the academic-cultural savvy to favor the—more commonplace or conventional? more obvious to the academically fluent?—anti-androcentrism explanation? Is favoring the 'academically correct' construal a component of CT, something that ought to affect how someone scores on a CT assessment?

Yet another example from the HCTA:[9]

> Some colleges are considering a new requirement for graduation—that every student must do some meaningful public service in order to graduate.
>
> *For each of the following statements, decide if it is a conclusion (C), a reason (R), or a counterargument (CA) with regard to this issue. ...*
> 1. Students will learn valuable skills through public service.
> 2. Some students will do more harm than good if they are required to perform a service they do not want to do.
> 3. Students should not be required to perform public service.
> 4. Students are already overburdened with homework and other assignments.
> 5. Public service will provide opportunities to improve our community.

The official right answers are R-CA-C-CA-R, thus taking the initial prompt to present an argument whose conclusion is support for the proposal under consideration, i.e., that students should have a public

---

[9] Form A, Question 13 (Halpern, 2012).

service graduation requirement. This is odd, [10] but acceptable. But surely it's no less reasonable (especially given #3's contrary, *negative* framing) to imagine constructing a contrary argument regarding the issue, in which case the correct answers would be CA-R-C-R-CA.

Interestingly, the corresponding item from the alternative HCTA version (Form B) has this prompt: "Some companies are considering a new requirement for all employees—that every employee must attend a diversity seminar with other members of the company." This puts test-takers in an even tighter bind, if their personal political inclinations incline them toward the conclusion "Employees should *not* be required to attend a diversity seminar" (which is indeed again the #3 statement) and thus to go with the alternative, 'wrong' CA-R-C-R-CA designations. So here is a CT test item that not only assesses specific canonic course content rather than good reasoning in general, but is also biased against test-takers of a certain political stripe.

The point of picking apart test questions in this way is to show that usual ways of thinking about, and teaching, CT draw dubious lines dividing the approved modes of making sense of utterances from the disapproved modes. Drawing these lines discourages intellectual practices that are impressively useful in certain practical contexts. The ability to engage in precise, refined, meticulous, imaginative parsings of written or oral texts can be uncommon and prized—in, say, a lawyer crafting contracts, a critic analyzing a work of literature or fine art, a philosopher painstakingly dissecting an argument, an intelligence analyst drawing careful logical inferences about adversaries' intentions, [11] or a comic wit ingeniously unearthing overlooked layers of meaning or implication in what someone says. But the ability's value is not confined to special vocations. Anyone aiming to avoid being hoodwinked by literally true, but deceptive or misleading advertising, for instance, will do well to develop some Intense Scrutiny inclinations and capabilities (see Thomason et al., 2013, pp. 96-99).

So for some purposes, CT is actually *better* when it 'overthinks,' extracting nuanced possibilities of sense and reasoning that ordinarily go overlooked. Intense Scrutiny modes of highly sensitive, possibility-

---

[10] It's dubious, particularly in a test item assessing comprehension of argument components, to treat a statement of fact as constituting an argument.

[11] American intelligence failures regarding Saddam Hussein's intentions for Iraq's nuclear-weapons program were the impetus for IARPA to fund the abovementioned argument-mapping CT course-development project (see Lengbeyer, 2014). Did this failure demonstrate that Normal Scrutiny is inadequate when high-stakes matters become sufficiently complex and even minute reasoning errors can have massive costs? Or was it simply a botched job of Normal Scrutiny, where solid Normal Scrutiny would have sufficed? A more precise delineation between Normal and Intense varieties of CT is needed.

proliferating thinking can have great value in some life contexts, despite being pointless, wasteful, counterproductive, even obnoxious in others.

But it may underestimate the value of Intense Scrutiny to portray it as useful only for specialized purposes calling for generating unusual possibilities of sense and logic. CT is *routinely* "so highly difficult—it's tempting to say 'nearly impossible'—to execute fully effectively, that is, reliably without error or oversight" (Lengbeyer 2014, pp. 14-15).[12] Experience and training are crucial in reasoning, but they afford little protection from error, like a vaccine that reduces the likelihood of infection upon viral contact from 80% to 30%. Given this, Normal Scrutiny often may simply not answer some of the intellectual needs that prompt resort to CT in the first place.

3. Concluding thoughts

Despite the forgoing, it is not difficult to construct a case in favor of the widespread emphasis upon a restricted version of CT as the norm. Perhaps social utility is better served by guiding an entire populace toward a modest shared level of CT performance, rather than encouraging everyone to drill down as far as possible into nuances of meaning and logic. The costs in mental development to the stronger individuals might conceivably be worth the gain in social agreement and harmonious coordination (something we now value more highly, having recently lost much of it).

As teachers, we fear that introducing CT complications will just muddy the waters for our students and muddle their minds, given just how difficult reasoning, and the analysis of reasoning, is for all of us. Utilizing a demanding methodology (like argument mapping) for teaching CT makes clear how unexpectedly challenging it is to reduce ordinary discourse into the forms of logical reasoning. Anything beyond the simplest, most carefully bounded textual passage can be susceptible to multiple plausible analyses of its reasoning. So much of our reasoning is left unarticulated—certainly in everyday discourse, but arguably even in meticulously spelled-out documents—and the missing pieces, the unstated

---

[12] Reasoners must account for "considerations that are relevant yet hidden from us…[,] buried within assumptions that we unwittingly make," yet "possess no systematic procedure for searching the abstract space of ideas where these hidden considerations reside…. Thus the task of uncovering all the possible considerations that might be germane to a given inquiry can be highly challenging. Still more challenging, perhaps, is the task of understanding and clarifying every one …, and integrating each with all the rest …. And, finally, there remain the demands of evaluating all the distinct considerations, of formulating judgments in the absence of clear-cut criteria and then weighting and combining these." Ibid.

assumptions and inference rules or schemata, can be filled in in diverse and mutually inconsistent ways.

Facing these challenges as CT instructors, we understandably rein in our pedagogic ambitions. Intense Scrutiny is difficult (and time consuming) to teach and assess, resists systematization, is variable across reasoners, and can seem ludicrously nitpicky and hence alienating to many students. But it is the real thing. (As are the intricacies of medical anatomy, when compared to the simplistic variants of folk anatomy that suffice for everyday purposes.) Rather than adapt our ideas about valuable reasoning skills to the practicalities of administering formal courses in CT, we ought to adapt our courses (at least non-introductory ones) so that students are encouraged and empowered to do the most careful possible reasoning.

But there remains one loose end to be tied up. What has become of my conjectured explanation for my CT students' lackluster performance on their post-test? Could this, in the end, be accounted for by them having shifted from Normal Scrutiny in the pre-test to Intense Scrutiny in the post-test? I have not performed a complete data analysis, but in the tests that I have regraded to look for these effects, I have detected *no* clear evidence of such a methodological shift.[13]

So perhaps the explanation is that I did invest a substantial proportion of course time in teaching Intense Scrutiny CT, and that this simply subtracted from the time that the students would otherwise have had to master Normal Scrutiny methods and gain more solid grasps of what Normal Scrutiny expects as 'academically correct' thinking. Preliminarily, it seems that my students did *not* actually develop Intense Scrutiny capabilities, or at least were not disposed to employ these during the post-test—whether because they sensed that they could not afford the time needed for exercising these, or because they applied their awareness, developed through long experience in formal education, that the place to indulge in more thorough and speculative reasoning is not on a timed standardized test.

The two-modes hypothesis is thus theoretically plausible, but it awaits confirmation in empirical testing. The need now is for a careful distinguishing between Normal Scrutiny and Intense Scrutiny, in operational terms sufficiently precise and detailed to permit usages of the two CT modes to be discriminated and detected in experimental studies or

---

[13] Before re-examining the student answer papers, I constructed an HCTA alternative answer key that granted test-takers leeway to attempt more-probing readings of test items—readings that could plausibly have resulted in reasonable deviations from the standard set of prescribed answers. I then regraded the student answers, hoping to find substantial performance jumps for some students. But I detected no significant effect of this kind.

observational investigations like corpus analyses.[14] Despite the oceans of ink already spilled discussing CT, there remains much about it that we still do not understand.

References

Beilock, S. L., & Carr, T. H. (2005). When high-powered people fail: Working memory and "choking under pressure" in math. *Psychological Science, 16,* 101–105.

Beilock, S. L., Kulp, C. A., Holt, L. E., & Carr, T. H. (2004). More fragility of performance: Choking under pressure in mathematical problem solving. *Journal of Experimental Psychology: General 133,* 584-600.

Chappell, R. Y. (2023). Text, subtext, and miscommunication. *Good Thoughts* blog, Jan. 15, 2023. https://rychappell.substack.com/p/text-subtext-and-miscommunication (Feb. 17, 2023).

Cullen, S., Fan, J., van der Brugge, E., & Elga, A. (2018). Improving analytical reasoning and argument understanding: a quasi-experimental field study of argument visualization. *NPJ Science of Learning, 3,* 21. https://doi.org/10.1038/s41539-018-0038-5.

Halpern, D. F. (2010). *Halpern Critical Thinking Assessment.* Vienna: Schuhfried.

Halpern, D. F. (2012). *Halpern Critical Thinking Assessment: Test manual.* Mödling, Austria: Schuhfried.

Kaiser, B. N., Haroz, E. E., Kohrt, B. A., Bolton, P. A., Bass, J. K., Hinton, D. E. (2015). "Thinking too much": A systematic review of a common idiom of distress. *Social Science and Medicine 147,* 170-183.

Kruger, A., Thorburn, L., & van Gelder, T. (2022). Using argument mapping to improve clarity and rigour in written intelligence products. *Intelligence and National Security, 37.* https://doi.org/10.1080/02684527.2022.2026584.

Law School Admission Council. (2007). *The Next 10 Actual, Official LSAT Prep Tests.* Newtown, PA: Law School Admission Council.

Lengbeyer, L. (2014). Critical thinking in the intelligence community: The promise of argument mapping," *INQUIRY: Critical Thinking Across the Disciplines 29,* 14-34.

PBS (Public Broadcasting Service). (1999). Secrets of the SAT. *Frontline.* https://www.pbs.org/wgbh/pages/frontline/shows/sats/ (Feb. 12, 2023).

Plumer, G. (2000). A review of the LSAT using literature on legal reasoning. *Law School Admission Council Computerized Testing Report, 97(8),* 1-19.

Thomason, N. R., Adajian, T., Barnett, A. E., Boucher, S., van der Brugge, E., Campbell, J., Knorpp, W., Lempert, R., Lengbeyer, L., Mandel, D. R., Rider, Y., van Gelder, T., & Wilkins, J. (2013). *A012. Critical Thinking Final Report* to

---

[14] It might also be enlightening to search the research repositories of the College Board (which runs the SAT) and the Law School Admissions Council (which runs the LSAT), to see if they have recorded observations that accord with a Normal-vs.-Intense distinction. (See Plumer, 2000, for an example of the research commissioned by the LSAC.) They might, for a start, have taken note of test-takers who perform well on the harder questions while falling down on the easier ones—plausible candidates for being consistently Intense critical thinkers.

IARPA (n66001-12-c-2004). Tech. Rep. Melbourne, Australia: University of Melbourne.

AUDIENCE: A CENTRAL CONCEPT IN SOCIAL ARGUMENTATION

JIAXING LI
Nankai University
1120200792@mail.nankai.edu.cn

Abstract

Social argumentation, as a type of argumentation, is inherently interactive and always aims to have a specific effect on the targeted people in a certain community. By analyzing the characteristics of social argumentation, this paper focuses attention on the concept of audience, which plays the role of participant, moderator, and evaluator. As for this central concept, contemporary argumentation researches engage in addressing issues related to audience identity and how can arguer get access to the audience. Perelman's universal audience and Tindale's cognitive environment are two key concepts that provide the theoretical basis and forward paths for audience issues in argumentation. Through the analysis and elaboration of the universal audience and cognitive environment, this paper points out that Perelman's universal audience is an exploration of universality, which also concludes from specific contexts. Tindale's cognitive environment, on the other hand, reflects the particularity and relevance of argumentation, revealing the mechanisms by which argumentation can be effectively conducted between the arguer and the audience. In the treatment of social argumentation, the combination of the two can provide a sound solution for the treatment of audiences in social argumentation.

Keywords: audience; universal audience; cognitive environment; social argumentation

1. Introduction

Argumentation, as the main form of communication, occurs in all areas of people's lives and plays a powerful role in their lives. Social argumentation, activities of presenting and receiving reasons, is closely connected with human cognition and thinking, which ensures the effective conduct of social lives. In the professional sphere, social argumentation generally is used for persuasion. In politics, social argumentation is often used as a common way of asserting one's position, expressing an opinion, and appealing to people to support or defend a policy or position. In the medical field, a doctor or specialist uses social argumentation to make other specialists aware of his treatment plan and to persuade their colleagues and patients to accept and cooperate with the treatment.

In social argumentation, the purpose of arguers is not to pursue a premise-conclusion structure but to reach agreements in certain contexts. van Eemeren (1996) claims that argumentation is "a verbal and social activity of reason aimed at increasing (or decreasing) the acceptability of a controversial standpoint for the listener or reader, by putting forward a constellation of propositions intended to justify (or refute) the standpoint before a rational judge" (5). Novaes (2021) gives her definition in the Stanford Encyclopedia of Philosophy, "Argumentation can be defined as the communicative activity of producing and exchanging reasons to support claims or defend positions, especially in situations of doubt or disagreement". As a real-world-based argument, social argumentation is goal-oriented, emphasizing the proof of a position and its practical efficacy on the object of the argument, and it does not involve a single object. From a broader perspective, since humans are social animals, all arguments related to humans can be seen as social arguments. Tindale (2015) focuses on the social nature of argumentation, envisioning argumentation as "something through which we assert ourselves and measure the assertions of others; it is part of the fabric of the social world (3)". "Argumentation involves the practices of using arguments to interact with, explore, understand, and (sometimes) resolve matters that are important to us (2)." Indeed, we are in an environment surrounded by all kinds of social argumentation. From political speeches to television commercials or war propaganda, it can appeal to emotions, mobilize political action, influence public opinion, affect market products, and even enable dictators in power (Walton, 2007,5).

Social argumentation serves as a framework for human interaction, facilitating the communication of ideas and expression of intentions, which in turn influence belief modification and behavioral adoption. Interaction, process, and goal orientation are typical features of social argumentation. Integral to these characteristics is the concept of the audience to whom social argumentation is directed. It is the interaction with the audience that enables social argumentation to deliver information to influence, persuade, or resolve disputes. Thus, since the middle of the last century, with the recognition that formal evolutionary structures are inadequate for analyzing everyday argumentation, numerous scholars have embarked on an informal exploration, and Perelman and Olbrechts-Tyteca's (1969) masterpiece, *The New Rhetoric*, has pointed the way for contemporary argumentation research. In their theory, argumentation and audience are inseparable.The success of argumentation hinges on securing the audience's compliance, with its effectiveness determined by the audience's characteristics, which in turn shape the arguer's behavior and the approach to argumentation.

With the shift from formalism to pragmatism and empiricism, the concept of audience has become an important concern in current argumentation theory research. Pragma-dialectics developed the "strategic manipulation" theory in its late stage, which involves the consideration of potential audiences at various stages of argument construction for better resolving differences of opinion (van Eemeren, 2002). Johnson (2013) also emphasizes the vital role of audiences and argues that audiences are relevant to argumentation and have the function of evaluating evidence, making arguments more accurate in a practical sense.

2. The role of the audience in social argumentation

As a key role in argumentation, the audience plays the role of participant, moderator, and evaluator in social argumentation, which not only shows the characteristics of the audience but also reveals the rhetorical mechanism of argumentation. Social argumentation is a communicative activity, which consists of an arguer, an audience, and a social environment. The audience bears the brunt of being a direct participant in the argument. As a thinking community, the audience is not just a static listening presence presented to the arguer, but is always an active participant in any communicative exchange. Reception theory describes how readers must cognitively fill in the gaps and open places in any text. (Holub, 1984) As the argument proceeds, the audience is invited to participate and get in touch with the arguer so that the purpose of the argument can be achieved. It is because of audience participation that the process of social argumentation can take place. In addition to being passive participants, audiences provide much of the discursive content, embodying active participation, as they are central to the context. (Tindale, 2015)

Audiences also play the role of moderators in social argumentation. With the rapid development of technology, people now have more access to express themselves to the public, which raises the issue of poor argumentation with bias, discrimination, or other undesirable information. At the same time, seeking to explain audience behavior and changing it to suit one's needs has become one of the many strategies of social and political struggle. In this case, the audience can regulate the argumentation by giving feedback rather than maintaining the dominant position of the arguer during a dynamic social argument. Audience also plays positive roles in argumentation, like Walton (2015) argues that according to constructive theory, as learners inquire into complex problems through arguments and construct counterarguments, they develop their argumentative knowledge.

The third role of the audience in social argumentation is that of the evaluator, which can be found in Aristotle and Perelman. In *Rhetoric*, Aristotle proposes three ways of persuasion, considering ethos, pathos, and logos. In the *New Rhetoric*, Perelman initially combines argumentation with the audience to form a theory of argumentation that focuses on audience theory. Through the analysis of the structure of everyday arguments, this theory is to study inferential techniques that are used to enable or enhance the compliance of ideas to controversial propositions that seek the agreement of others. Audience, as the argumentative target of the arguer in social argumentation, is the direct manifestation and evaluator of the effectiveness of argumentation. The persuasive and influential orientation of social argumentation makes the audience, in turn, an instrument of evaluation. The audience is located in the real world and social argumentation aims to influence the attitudes, beliefs, and actions of audience located in the real world.

Thus, the audience becomes a key element in the study of social argumentation, and a proper understanding of the audience contributes to the achievement of the purpose of argumentation. The audience, instead of being a passive and static being, is an active participant invited into the argument to exchange ideas with the arguer to achieve the purpose of the argument, so it is natural to consider the audience in the context of the dynamic situation of society. If there is no audience, it is just a human action, far from the goal of social argumentation. Moreover, audiences can provide useful feedback or manipulate social argumentation on certain special occasions. When evaluating social argumentation, we give concerns to communicative validity, and the most direct evaluator is the audience.

3. Current Status and Problems of Audience

In current argumentation research, more and more scholars have paid attention to the critical role of the audience in argumentation. The above analysis also provides a detailed discussion for the central position of the audience in social argumentation, who plays roles of participants, moderators, and evaluators. In social argumentation, the audience also reacts to the construction of argumentation by the arguer. Thus, a correct understanding of the audience helps construct the argument and realize the argumentative goal.

However, the audience, as a dynamic and diverse concept, is difficult to capture. Since ancient Greece, numerous scholars have paid different attention to this vital topic of rhetorical research. In the ancient Greek period, rhetorical argumentation was used in speeches to influence people in courts and assemblies or to praise and glorify certain people. The Sophists, who prided themselves on being teachers of the art of oratory, also placed particular emphasis on the skill of overcoming opponents and the ability to argue eloquently. One of Plato's contributions to classical rhetorical theory was to argue that oratory should be adapted to the audience. Socrates also emphasized that stylistic discourse should be adapted to the response of the audience. Aristotle divided rhetoric into three parts: the orator, the subject, and the audience, with special emphasis on the role of the audience, and contributed three types of audience-centered persuasion, which are ethos, pathos and logos.

With the contemporary focus on audiences, questions have arisen. Some scholars are confused about the audience's identification and the connection between the arguer and the audience. Johnson (2013) mentions his confusion about how the arguer gains knowledge of the audience, saying that the rhetorical approach is fundamental but still "problematic." Other scholars have raised questions about audience-centered approaches to argumentation research, typically including: What is an audience? Is an audience static or dynamic, fixed or changing? What types of audiences are there? How can the arguer get to know the audience, and how can he or she relate to it? How can the audience be handled appropriately in an argument?

The answers to the above questions begin with a key figure: Chaïm Perelman. Based on classical rhetorical theory, Perelman combined argumentation and audience to form a unique theory of rhetorical argumentation, and his ideas were mainly reflected in *The New Rhetoric*, in which audience, as a key concept, also became an important topic in the study of argumentation. Perelman (1969) inherits and extends Aristotle's theory of audience by describing the audience as "the collection of people whom the speaker wishes to influence by his or her argumentation," which solves the problem of the non-physical existence of an audience. In this sense, the idea that the audience is a construction in the mind of the speaker, rather than a limited number of people who are presented to the arguer according to his purpose, sheds light on the confusion of how to recognize the identity of the audience. However, as a construction of the arguer, this type of audience cannot be arbitrary and fixed. Instead, it is an audience made in reality that "conforms to the reality of the intended audience (20)" while also changing and evolving dynamically as the argument proceeds. At the same time, the goal of argumentation is to seek the audience's approval, and only by carving out the target audience in advance can the construction and development of argumentation be more focused. In addition, Perelman attempts to develop a universal audience as a criterion for argument evaluation in response to the complexity of the audience. Universal audiences and particular audiences constitute the main categories of audiences, and audiences are also considered valuable tools for evaluating arguments.

The contemporary rhetorical theorist, Tindale, gives many unique insights into the development of the audience concept in argumentation theory. Tindale absorbed the advantages of Aristotle's and Perelman's theories, developing his ideas about audiences through a comprehensive consideration of reality. Tindale emphasizes that the identity of the audience is particularly important because the issues of persuasion and evaluation of arguments either depend on it or can be attributed to it to some extent. (Tindale, 2013) However, audience identity is complex because it has a dynamic character and is intrinsic to the dynamics of social communication. Audiences may be actual, potential, virtual, simple, complex, particular, or universal. Faced with the complexity of the audience and how to achieve successful argumentation between the arguer and the complex audience, Tindale (2015) developed the concept of cognitive environment and proposed an approach to argumentation personalization. He pointed out the relative rationality resulting from the universal audience and the non-realistic type of existence of universal rationality. Then he developed the universal audience by illustrating the functional role of the criterion of objectivity of this concept, combining epistemology and value, and proposed the concept of cognitive environment to serve as a tool of replacement. Since different groups live in different social environments, their cognitive environment varies. The validity of argumentation is achieved through the shared cognitive environment of the speaker and the listener. At the same time, the cognitive environment provides the cognitive and ethical criteria for the judgment of argumentation.

The above views of Perelman and Tindale on audiences allow us to make a further generalization about audiences. An audience is a construction by the arguer based on the reality of the argument. In social argumentation, there are various categories of the audience involved, as different arguments are addressed to different audiences. The question of the identity of the audience and how to achieve a connection with the audience are two major controversial issues in the construction of arguments. When we are trying to give solutions to these questions, at first, we should consider rationality in social argumentation. Universal audience and cognitive environment provide the theoretical basis and development path for dealing with the audience issue in argumentation. In the following section, these two concepts are further elaborated to explore a comprehensive solution for dealing with audiences in social argumentation.

4. Universal audience and cognitive environment

This consideration of Perelman's audience is related to his philosophical quest for objectivity. Perelman is aware of the serious problems that real audiences face concerning value arguments: rhetoric based strictly on the beliefs of particular groups is biased, narrow, and parochial, and arguments that win particular audiences' compliance are often not accepted by the majority of rational people. Moreover, arguments aimed at particular audiences are vulnerable to rebuttal by opponents. For this reason, he proposes a universal audience to avoid the problems associated with arguments aimed at special audiences. Perelman emphasizes that the starting point for understanding the nature and importance of the universal audience is the "universally accepted opinion" claimed in the Topics by Aristotle and that the arguments of rhetors should be based on the generally accepted opinions, to ensure objectivity and rationality of the argument.

On the other hand, Perelman's universal audience is proposed as a solution in search of the complexity of real audiences. In the analysis of argumentation, the separate consideration of various types of audiences and the treatment of composite audiences face complex operational problems. Perelman, therefore, explores the universals and commonalities in audiences to find workable solutions for such audiences. For an audience whose specific characteristics are difficult to grasp, the common universality is the individual with the capacity of reason. This universality is universal in the largest and broadest sense, like his assertion "all reasonable people". For an audience whose identity is not so complex as to be captured, the universality is closely related to the specific cultural context. Thus the universal audience is given the meaning of universality in different cultural contexts in later development. This is not inconsistent or self-contradictory, but rather an exploration of universality based on different contexts. Thus, Perelman's universal audience is, by its very nature, reasonable and aims at exploring and generalizing the universality of the audience. Based on such a concept of universality, the arguer is better able to construct and evaluate arguments. Perelman emphasizes in the preface to his book that " It is the idea of self-evidence as characteristic of reason, which we must assail if we are to make place for a theory of argumentation that will acknowledge the use of reason in directing our actions and influencing those of others." (1969,3) Thus, Perelman's universal audience is not in pursuit of Hegelian absolute rationality, but rather a consideration of the universality and objectivity of reality in an Aristotelian generalization of empirical reason.

By interpreting the universal audience, we can find that the concept of universal audience has two effects. First, as a classification of the audience, it represents a way of dealing with the audience in argumentation and provides a path for the arguer to choose an audience. Starting from the concept of universal audience, the orator decides the type of audience that best suits the situation, and in choosing claims and arguments, the orator is choosing the universal or particular audience which fits the argument ; second, the universal audience provides criteria and standards for distinguishing good and bad arguments, a purpose that is more relevant to philosophical arguments. For philosophical arguments, the universal audience approves of arguments they consider good and rejects arguments they consider bad, so it provides a sense of rationality. Perelman does not consider truth and validity in argumentation to be absolute, asserting that argumentation should provide different interpretations of reality and that to make the rationality of philosophical claims consistent with a pluralistic philosophical system, we must recognize that the appeal to rationality is an appeal to the compliance of the audience rather than to a particular truth.

Tindale's exploration of the cognitive environment gives an appropriate solution to how the arguer is informed about the audience and how to connect with it. Tindale (2015) states that the foundation of argumentation is a shared cognitive environment. By providing a common space for communication, the cognitive environment effectively connects the arguer to the audience. Cognitive environments can separate us, but when mutual cognitive environments are modified to allow maximum access, they become the conditions for our connection.

Derived from Sperber and Wilson (1986), the cognitive environment is the basic concept and core of the relevant theory and is an alternative concept to "mutual knowledge" or "shared information." It is the projection of the "physical environment" in mind and thought, the "space" generated by Aristotle's "Topoi" type of reasoning. A person's total cognitive environment is the set of all facts that he can perceive or infer: all facts that are obvious to him. The key elements also include all facts that can be acquired through cognitive activities in the physical environment, cognitive abilities (perceptual or inferential), knowledge, and memory. It is these cognitive abilities, perception, inference, and memory, combined with the physical environment that allow people to acquire, store, and use the information according to their knowledge and context, leading to activities such as expression, communication, and exchange.

The formulation of Tindale's cognitive environment concept explains how effective communication between the arguer and the audience can occur. The audience is closely related to the argumentative situation, and a shared cognitive environment is important for effective argumentation. It surrounds the whole argument, which has great significance for the action of the argument. The shared cognitive environment is the basis and medium of interaction between the arguer and the audience. People have a common cognitive environment when the same facts and assumptions manifest themselves in different people's cognitive environments and can be used as references for negotiated interpretations. Through the collision and integration of cognitive environments, the arguer makes the audience resonate with them in terms of ideas or conceptual values. The cognitive environment of the audience is influenced, which leads to the realization of the argumentative effect. For example, if an elementary school teacher explains the Pythagorean theorem from a professional mathematical perspective to students, complex formulas and abstract concepts will confuse the students, hindering their identification because, without a shared cognitive environment, they find no place to join, think, and believe.

Argumentation is an attempt by the arguer to change the cognitive environment of the audience, thus changing the beliefs of the audience and even the effect of the argumentation. In a dynamic social argumentation process, the unfolding of the argument is a continuous exchange of information between the arguer and the audience, building shared information and thus a shared cognitive environment. As the argument unfolds, the audience becomes more and more specific, and the argument becomes more and more relevant and persuasive. The overlap of cognitive environments reflects the sharing of knowledge in human communicative activity. In seeking to create an identity or commonality with the audience, the arguer can, by appealing to emotions, put the audience in a state of mind where they are most open to the arguments on offer, while also trying to use the shared cognitive environment to stimulate the public's emotions to achieve an argumentative effect.

5. Treatment of audience in social argumentation

Through the above analysis, Perelman's universal audience reflects his consideration of the universality of the audience and the objectivity of the argument. Tindale's cognitive environment, on the other hand, is an effective tool for the particularity of argumentation in different environments. Although Tindale's concept of cognitive environment is proposed to explain the mechanism of how effective communication between the arguer and the audience can be achieved. However, the cognitive environment is only a tool, not an audience. And this concept faces challenges in practical application as well. Firstly, constructing a cognitive environment is inherently complex. The uncertainty in pinpointing and comprehending distinct cognitive aspects within the audience, which complicates the articulation of argumentative elements. Furthermore, the cognitive environment framework encounters constraints in addressing indirect audiences, such as historical audiences who are beyond direct contact or immediate feedback. Although the cognitive environment can establish norms for specific argumentative contexts, it exhibits comparative limitations in maintaining argumentative objectivity, especially when contrasted with the concept of the universal audience. Therefore, in consideration of the audience, it is important to return to Perelman's distinction between universal audience and particular audience. Such a universal audience reveals the consideration of the universality of the audience in the pursuit of effective argumentation, thus ensuring the rationality and validity of the argument.

In constructing and evaluating social argumentation, ignoring universality and objectivity can lead to inconsistency and confusion. The pursuit of objectivity can also lead to a mismatch and disconnect with reality, so objectivity and specificity should coexist. Given the complexity and dynamics of audience identity in social argumentation, different considerations should be given to different situations. Among them, the cognitive environment and the universal audience are embedded in the construction and evaluation of argumentation as the theoretical essence of universality and specificity.

In the construction of the social argument, we should first construct it based on the universal audience to realize the objectivity of the argument. According to our interpretation of Perelman's universal audience above, the construction of an audience can be divided into two cases. In the first case, if we have direct or indirect access to a particular audience, the construction of the audience will be precise and easy, and the analysis of the audience will extract its universality to form a universal audience in a particular context so that the argument can be developed on this basis. In the other case, if we do not have access to a particular audience, we should construct it based on a broad universal audience. Consider the universality of the possible audiences related to reality to make generalizations and abstractions, based on which the construction of a universal audience is carried out, to select arguments and ensure the effective development of the argument.

In concrete practice, the audience can first be simplified in the initial stages of constructing it. The universal audience in a given social argument situation is not all rational beings, but the universality of the audience in a specific context. As Perelman argues, there is no need to think of the audience in a complex way at the very beginning. Since social argumentation is goal-oriented and directed, it is possible to restrict the audience to objectives that are only relevant to the topic at the initial stage of argument construction. This simplified audience is a presentation of Perelman's universal audience and is also a particular construction with the consideration of specific context and topic. The broad universality abstracted in the construction of the audience for social argumentation thus ensures the most fundamental objectivity of the argument, its universality, and at the same time applies to a variety of situations where there is no specific knowledge of the audience before the argument. On this basis, the initial construction is carried out and the arguments invoked should be generally accepted opinions, such as knowledge and axioms.

When addressing specific social arguments, building on the initial simplified version of the audience described above, the arguer also needs to consider the rationality and validity of specific contexts, so that the issue of audience identity can be further addressed concerning specific situations. The main purpose of audience research is to understand the interaction between rhetorical situations, the characteristics of discourse, audience reception of rhetoric, and negotiation (Kjeldsen, 2018). The legitimacy of the argumentative material, the progress of the argument, and the actual effectiveness of the argument can only be assessed if the argument is put back into the actual argumentative situation. For this, we expand the argument with the help of the cognitive environment as an effective tool. Social argumentation is a cognitive activity in which people acquire, process, digest, and even evaluate arguments through their minds. Information is a crucial presence and one effective way to acquire the knowledge of an audience is to gather information through practical means, both direct and indirect. Specifically, this can be done through research, experiences, and experiments to obtain audience information based on which a shared cognitive environment can be constructed to target the argument. At the same time, the construction of the argument should take into account the cognitive environment of the target audience in terms of the challenges they may raise, and refine the construction of the argument by responding to these challenges.

6. Conclusion

As a purposeful, contextualized, and interactive communicative activity, social argumentation should be studied mainly from a rhetorical perspective to achieve a proper grasp of argumentation. The audience becomes the central concept in social argumentation, as the construction of the argument in the mind of the arguer and the context of reality, the audience penetrates the whole process of argumentation, and the audience's approval reflects the effect of the argument and determines how the argument is constructed.

In the construction of social argumentation, Perelman's universal audience gives us insights. Based on the complexity of the composite audience and the pursuit of Aristotle's universally accepted opinion, Perelman proposes the concept of universal audience and gives an appropriate solution for the treatment of the audience. Perelman's universal audience represents the pursuit of universality, which ensures the rational roots of the argument. Therefore, universalizing the audience at the initial stage of audience construction helps to clarify the audience categories and ensure the rational development of the argument. Simplifying the audience as rational people related to the topic is a broad generalization of the audience's universality in a specific context.

Meanwhile, the cognitive environment is a practical tool for considering audiences in social argumentation to achieve better communicative outcomes and is another key concept and important component of social argumentation. Social argumentation is an interaction that relies on our cognitive abilities, an expression that people use to communicate their ideas and intentions, and it significantly impacts our beliefs and behaviors. As the subject of argumentation, the audience acquires, bears, and accesses social argumentation through cognitive activity. Through the connection of the cognitive environment, the arguer and the audience achieve the exchange and interaction of ideas, which provides a way to achieve the goals of social argumentation. The cognitive environment bridges the gap between the arguer and the audience and directly reflects the specificity and relevance of a particular social argument. Thus, the combination of the cognitive environment and the universal audience can provide a perfect solution for the effective construction of the argument in the identification and construction of the audience in social argumentation.

References

Holub, R.C. (1989). *Reception Theory: A Critical Introduction*. Repr. London: Routledge.
Johnson, R.H. (2013). The Role of Audience in Argumentation from the Perspective of Informal Logic. *Philosophy & Rhetoric*, 46(4):533-549.
Kjeldsen, J. E. (Ed.). (2018). *Rhetorical Audience Studies and Reception of Rhetoric*. Cham: Springer International Publishing.
Novaes, C.D. (2021): 'Argument and Argumentation'. The Stanford Encyclopedia of Philosophy, Edward N. Zalta (ed.). https://plato.stanford.edu/entries/argument/.
Perelman, C., & Olbrechts-Tyteca, L. (1969). *The New Rhetoric. A treatise on argumentation*. Notre Dame-London: University of Notre Dame Press.
Sperber, D.,& and Wilson, D. (1986). *Relevance: Communication and Cognition*. Cambridge, MA: Harvard University Press.
Tindale, C. W. (2013). Rhetorical Argumentation and the Nature of Audience: Toward an Understanding of Audience - Issues in Argumentation. *Philosophy & Rhetoric*, 46 (4): 508-532.
——(2015).*The Philosophy of Argument and Audience Reception*. Cambridge: Cambridge Press.
van Eemeren, F. H. , Grootendorst, R., & Snoeck-Henkemans, F. (1996). *Fundamentals of argumentation theory: A handbook of historical backgrounds and contemporary developments*. Mahwah, NJ: Lawrence Erlbaum.

——(2002). Strategic Maneuvering in argumentative discourse: Maintaining a delicate balance. In F. H. van Eemeren & P. Houtlosser (Eds.) *Dialectic and Rhetoric: The Warp and Woof of Argumentation Analysis.* (pp. 131–159). Dordrecht: Kluwer.

Walton, D. N. (2007). *Media Argumentation: Dialectic, Persuasion and Rhetoric.* Cambridge: Cambridge University Press.

—— (2015). Argument Evaluation and Evidence (Vol. 23). Springer.

ON THE REDUCIBILITY AND THE IRREDUCIBILITY OF ANALOGICAL ARGUMENTS

YAN-LIN LIAO
Department of Philosophy, Sun Yat-sen University, China
liaoyanlin@mail.sysu.edu.cn

Abstract

This paper focuses on whether analogical arguments should be reducible to non-analogical arguments, namely the dispute between the reducibility and the irreducibility of analogical arguments. First, this paper argues that the existing defense for the irreducibility of analogical arguments is unsuccessful. To defend the irreducibility of analogical arguments, this paper contends that there are two types of analogical arguments: (a) Analogical arguments containing conductive arguments and principle-based arguments; (b) Analogical arguments containing IBE and principle-based arguments. In this way, the irreducibility of analogical arguments can be defended.

1. Introduction

Analogical arguments (analogies) are usually viewed as one of the most important types of arguments. Analogies have different functions, such as argumentative, explanatory, and descriptive functions, but this paper intends to focus on the argumentative function (i.e., analogical arguments). In this paper, "analogy (analogies)" refers to analogical arguments unless otherwise noted. There are different structures of analogical arguments proposed by scholars with different backgrounds (e.g., Hesse, 1966; Walton et al., 2008; Bartha, 2010). Walton, Reed, and Macagno presented a simple argument scheme of analogical arguments, which can probably be regarded as the fundamental structure of analogy (2008, p. 56):

Major Premise: Generally, case C1 is similar to case C2.
Minor Premise: Proposition A is true (false) in case C1.
Conclusion: Proposition A is true (false) in Case C2.

By convention, C1 is called "the analogue," C2 is called "the primary subject," and this paper calls the major premise "the similarities proposition." As a necessary premise, the similarities proposition represents the logical role of similarities between two cases in analogical arguments. However, some theorists argue that the similarities proposition only has an epistemic function but not a logical function so that the similarities proposition can be eliminated (e.g., Agassi, 1988; Waller, 2001; Kaptein, 2005). They think that the similarities of analogy only lead people to figure out certain underlying principles (i.e., the epistemic function), and the similarities of analogy do not provide support for the conclusion (i.e., the logical function). When the similarities proposition is eliminated, analogical arguments will be reducible to non-analogical arguments. To defend the irreducibility of analogies, some theorists argue that the underlying principle leads all analogical arguments into deductive arguments. In this paper, I will try to give a preliminary critical analysis of the dispute between the reducibility and irreducibility of analogical arguments and provide a new defense for the latter.

2. The Reducibility of Analogical Arguments

Agassi thinks that the fascination of analogy is its "heuristic," which means a technique to "jog the intellect to make wild hypotheses." The similarities of analogy sometimes are vague and indefinite, but he contends that they can "stimulate one's thinking (1988, pp. 403-404)." In this regard, Waller comes up with an argument by taking the violinist analogy in ethics as an example.

Is it morally wrong to prohibit a woman who is pregnant due to rape from having an abortion? For people in certain cultures, the issue of abortion might be highly controversial. To help people make judgments, we can first consider the so-called violinist analogy. The violinist analogy can be briefly stated: One morning, you wake up to find that you have been kidnapped and are in a hospital now. You are lying in a bed next to a famous but seriously ill violinist, and your two bodies are connected. The violinist is suffering from acute kidney failure, and if her blood is not purified, the toxins in her blood will soon kill her. Music fanatics somehow find out that you are currently single and that your blood type matches the violinist, so they kidnap you and connect your body to the violinist. Now your kidneys have a double responsibility: to purify your blood and the blood from the violinist. This process will last about nine months and will cause inconvenience but no harm to your body. If you disconnect from the violinist now, she will die. Given this, the fanatics will force you to stay connected to the violinist to save her life (Waller, 2001, p. 201).

Is what the fanatics will do morally wrong? Most probably think so because forcing someone to save the other is simply unacceptable. This is a much easier decision than deciding on the abortion case. Importantly, if you think such compulsion is morally wrong, you should conclude that it is immoral to prohibit a pregnant woman due to rape from having an abortion *for the same reason*. This is an analogical argument in ethics. Nevertheless, how exactly does the argument work? Waller reconstructs the argument as follows (ibid.):

1. We both agree with case a.
2. The most plausible reason for believing a is the acceptance of principle C.
3. C implies b (b is a case that fits under principle C)
4. Therefore, consistency requires the acceptance of b.

According to Waller's above structure, "case a" (the analogue) refers to the violinist case, and "b" refers to the abortion case. Premises 2 and 3 show that case a's function lies in stimulating the arguer to find principle C that implies b. For the violinist case, Waller thinks that the principle might be that "we do not have an obligation to save or sustain a life when we have done nothing to take on the obligation (2001, p. 202)." The arguer can imply that a woman who is pregnant due to rape does not have an obligation to continue her pregnancy (i.e., it is immoral to force her to keep her pregnancy) based on the principle. Obviously, there is no similarities proposition in Waller's structure. The similarities between the violinist and abortion cases give the arguer a *clue* to discover the underlying principle, but they do not become a premise in Waller's structure. Thus, the similarities proposition is *eliminated* in Waller's structure. In Waller's reconstruction, analogical arguments can be reducible to non-analogical arguments.

The main advantage of the reducibility of analogical arguments is that it makes argument analysis and evaluation more manageable. The similarities between cases are vague, indefinite, or "unanalyzed (Gamboa, 2008, p. 233)," so the argument analysis and evaluation would have to rely on "rough intuitions (Waller, 2001, p. 210)." On the contrary, the underlying principle is clear and definite so that the evaluator has a better understanding of the inferential link of the argument. In particular, the advantage of Waller's structure would be highlighted when we need to compare an analogical argument and its *counter-analogy*, such as Thompson's violinist analogy versus Fischer's analogy of the starveling (ibid., p. 209). Compared to rough intuitions between these two conflicting arguments, it is much more understandable and manageable for the evaluator to analyze and compare two underlying principles. Therefore, the similarities between cases are not qualified to bear the logical function in analogical arguments while the underlying principle is up to the task.

The reducibility of analogical arguments presents a severe challenge to the *logical foundation* of analogy. If the similarities proposition loses its logical role in analogical arguments, analogical arguments will not be a *distinctive* type of argument but be reducible to non-analogical arguments, namely principle-based arguments.

3. The Irreducibility of Analogical Arguments

In defense of the irreducibility of analogical arguments, on the contrary, theorists try to attack the legitimacy of the underlying principle. I think the most important attack is that the reducibility of analogical arguments leads all analogical arguments into deductive arguments, thereby making analogical arguments not capture varying degrees of argument strength (Govier, 2002, p.156; Guarini, 2004, pp. 156-161). Shecaira (2013, pp. 427-429) argues that Guarini misunderstands the move from 1 to 2 in Waller's structure (i.e., a deductive account of analogical arguments) as the strength of analogical arguments can vary according to the plausibility of 2 in Waller's structure. Following Toulmin's conception of inference and a Toulmin-inspired account of argument evaluation, Bermejo-Luque (2012) contends that certain analogical arguments can be "deductive but defeasible."

I agree with Shecaira's insights, but he seems to overlook another sense of argument strength (call it "the narrow sense of argument strength"), which refers to the inference strength only measuring the supporting degree from the premises and the conclusion. The argument strength Shecaira uses refers to the broad sense of argument strength, which consists of the inference strength and the acceptability of the premises. Nevertheless, Bermejo-Luque's argument concerns the narrow sense of argument strength. He tries very hard to justify a seemingly self-defeating claim that an argument can be deductive but defeasible by applying Toulmin's framework. His argument is undoubtedly interesting and inspiring, but it assumes a clear distinction between the warrant and the implicit premise. That will likely involve another controversy in the relationship and distinction between warrants and premises. It would be beyond the scope of this paper to discuss the controversy.

Unlike Shecaira and Bermejo-Luque, I contend that the reducibility of analogical arguments does not necessarily lead to a deductive account of analogical arguments. So, how is it possible that a principle-based argument is not a deductive argument? To answer this question, I need to introduce a distinction between *strict* and *defeasible* modus ponens proposed by Verheij (2000, as cited in Walton, 2005):

Strict Modus Ponens (SMP)
As a universal rule not subject to exceptions, if A then B.
A is true.
Conclusion: B is true.

Defeasible Modus Ponens (DMP)
As a rule subject to exceptions, if A then B.
A holds as true.
It is not the case so far that there is a known exception to the rule that if A then B.
Conclusion: B holds tentatively, but subject to withdrawal should an exception arise.

Interestingly, theorists (including theorists in favor of the reducibility of analogical arguments) generally characterize the structure of analogical arguments by applying SMP, which is a deductively valid form of argument. Due to the deductive validity, SMP does not allow for degrees of the narrow sense of argument strength. DMP, however, is a plausible but invalid form of argument, which allows for degrees of the narrow sense of argument strength. The introduction of DMP provides theorists with an alternative: If the evaluator discerns that the arguer intends to come up with a principle with exceptions, then the evaluator should characterize the analogical argument by applying DMP. If the evaluator discerns that the arguer intends to come up with a principle without exceptions, then SMP should be applied. According to the principle of charity, the evaluator should characterize the analogical argument by applying DMP if she tries hard and still gets into trouble in clarifying the arguer's intention. In this way, the principle-based argument is not necessarily deductive valid. As a result, the reducibility of analogical arguments allows for the degrees of argument strength (in both the narrow and broad senses). It follows that the existing defense of the irreducibility of analogical arguments has *not* succeeded yet.

4. A New Defense for the Irreducibility

In previous sections, I have given a critical account of the dispute between the reducibility and irreducibility of analogical arguments and shown that the defense of the latter has not succeeded. It means that we need a new defense for the irreducibility of analogical arguments if we want to rescue the distinctiveness of analogical arguments.

Gamboa (2008) provides an essential clue for defending the irreducibility of analogical arguments. He agrees that many so-called analogical arguments merely have "the deceptive analogical form" to be reducible to non-analogical arguments (i.e., principle-based arguments). Nevertheless, he argues that "legitimate resemblance-based analogical inferences" are possible (ibid., p. 234). The main idea of his argument is that the similarities between cases cannot be eliminated and have their logical function in some cases. To elaborate the argument, he takes a biological experiment as an example, which can be briefly summarized as follows (ibid., pp. 235-238):

Animal experiments are very common in biological research. For instance, scientists usually use animal models to investigate possible effects on humans caused by external factors (e.g., environments, drugs, etc.). In a scientific study on human male reproductive fertility, scientists investigate possible effects on humans caused by environmental toxins and environmental estrogens by using animal models. According to the reducibility of analogical arguments, the analogical argument of the animal model here can be reducible to a non-analogical argument, namely a statistical inductive argument:

1. Z% of sampled mammals exposed to test substance developed fertility-related properties P.
∴ Z% of mammals exposed to test substance develop fertility-related properties P.
2. Z% of mammals exposed to test substance develop fertility-related properties P.
3. Human males are mammals.
∴ Z% of human males exposed to test substances will develop fertility-related properties P.

In doing this, the similarities between mice and human males are reducible to a principle that Z% of mammals exposed to test substances develop fertility-related properties P (i.e., premise 2). Gamboa, however, thinks that the above reconstruction *misinterprets* the animal models in science because it captures only the common features (e.g., mammals) but *ignores* the differences. In the study of reproductive fertility, mice and humans have important differences in reproductive systems: the shapes of sperm, the fertility levels, the weight, the efficiency of sperm production, and so on. In a few words, the common features and differences between cases need to be considered in analogical arguments of animal models. It means that the similarities proposition representing that case a is similar to case b cannot be eliminated and replaced by the underlying principle only capturing the common features between cases. In this way, Gamboa concludes that the similarities proposition has the logical function so that the analogical arguments are a distinctive type of argument (ibid., pp. 235-241).

I think that Gamboa's above argument is quite inspiring because it makes important progress against the reducibility of analogical arguments. Unfortunately, it seems *insufficient* to defend against it fully. One potential objection could be raised as follows. The similarities proposition in Gamboa-style analogical arguments (i.e., the proposition representing the common features and differences between cases) can still be eliminated as they only stimulate the scientist to investigate the principle, such as:

"If mice exposed to test substance t have a toxic effect e on their reproductive systems, then human males exposed to test substance f(t) will have the same toxic effect e on their reproductive system."

In the above principle, "t" represents the dose of test substances for mice; "e" represents the degree of toxic effect; "f(t)" represents the dose of test substances for human males, which is a function with t as the independent variable. In this way, Gamboa-style analogical arguments can still be reducible to non-analogical arguments:

1. If mice exposed to test substances t have a toxic effect e on their reproductive systems, then human males exposed to test substances f(t) will have the same toxic effect e on their reproductive systems.
2. Mice exposed to test substances t have a toxic effect e on their reproductive systems.
3. Therefore, human males exposed to test substances f(t) will have the same toxic effect e on their reproductive systems.

Gamboa seems *not* to consider such a potential objection. To prevent Gamboa-style analogical arguments from being eliminated the common features and differences between cases, I contend that Gamboa-style analogical arguments are, in effect, *linear* arguments containing two different arguments: *conductive arguments* and *principle-based arguments*. Shecaira (2013) suggests that analogical arguments can be viewed as complexes containing the inferences to the best explanation (IBE) and the deductive inferences. My contention here is partly inspired by Shecaira's idea that the analogical argument could be viewed as a complex argument comprising different subarguments, but my contention is substantially different from his. I will revisit his idea later. Principle-based arguments here refer to the above non-analogical arguments. Conductive arguments here refer to the inference to the principle, which can endow the common features and differences between cases with a logical function rather than just an epistemic one.

Conductive arguments, also known as conduction or pro-con arguments, were first raised by Wellman (1971) and developed by theorists in recent years (Blair & Johnson, 2011). Conductive arguments consist of pro-reasons (PR, i.e., considerations supporting the conclusion), counter-considerations (CC, i.e., considerations undermining the conclusion), and the on-balance premise (OBP, i.e., the premise indicating positive considerations outweigh the negative considerations). PR, CC, and OBP have their own unique logical function, so conductive arguments are a distinctive type of argument. The structure of conductive arguments can be shown as follows (Hansen, 2011; Zenker, 2011):

PR: Pro-reasons 1, 2, 3...
CC: Counter-considerations 1,2,3...
OBP: PR outweigh CC
Conclusion

There are criticisms that CC have only the rhetorical function but not the logical one (e.g., Xie, 2017), and OBE cannot be viewed as a premise in the structure of conductive arguments (e.g., Possin, 2010, as cited in Hansen, 2011). Although some scholars have convincingly responded to these criticisms (e.g., Hansen, 2011), I will not discuss the controversy any further because this is beyond the scope of this paper. In Gamboa-style arguments, the inference to the principle can be characterized by conductive arguments. Specifically, the common features can be viewed as pro-reasons, the differences can be viewed as counter-considerations, and the on-balance premise can represent the weighing mechanism between them. In the case of the toxic effect on the human male's reproductive system, the structure of the conductive argument is formulated as follows:

PR: In terms of the toxic effect on the reproductive system, the relevant common features between mice and human males are germ cell development and spermatogenesis.
CC: In terms of the toxic effect on the reproductive system, the relevant differences between mice and human males are the shapes of sperm and the fertility levels.
OBP: In terms of the toxic effect on the reproductive system, the relevant common features between mice and human males outweigh the relevant differences.
Conclusion: If mice exposed to test substance t have a toxic effect e on their reproductive systems, then human males exposed to test substance f(t) will have the same toxic effect e on their reproductive systems.[1]

With the structure of conductive arguments, the logical function of the similarities between cases is demonstrated in two aspects: (a) The common features support the conclusion while the differences undermine the conclusion; (b) The common features support the conclusion outweigh the differences undermine it. In this way, the similarities proposition cannot be eliminated and is not reducible to the principle. Instead, the link between the similarities proposition and the principle is an inferential link, namely the conductive argument.

---

[1] In the case of the toxic effect on the human male's reproductive system, Gamboa (2008, pp. 236-237) lists many relevant common features and differences between mice and human males, but I just mention a few of them for simplicity.

I have argued that Gamboa-style analogical arguments can be reconstructed as linear arguments containing conductive and principle-based arguments. Gamboa-style analogical arguments can defend against the reducibility of analogical arguments, so they are arguably a distinctive type of argument. However, it is noteworthy that Gamboa-style analogical arguments might not apply to case-to-case arguments in ethics and law that only care about the underlying principle inspired by the common features between cases (e.g., the violinist case). In other words, the *interpretative power* of Gamboa-style analogical arguments seems to be very limited—only those case-to-case arguments concerning the weighing mechanism between common features and differences can be seen as analogical arguments (e.g., animal models in scientific reasoning). Interestingly, the deficiency in the interpretative power of Gamboa-style analogical arguments can be complemented by Shecaira-style analogical arguments (Shecaira, 2013, p. 429):

1. It is true that a.
2. The most plausible (i.e., the best) reason for believing a is principle C.
3. Therefore, it is true that C.
4. C implies b.
5. Therefore, it is true that b.

Shecaira (2013) argues that analogical arguments in ethics and law can be regarded as complexes containing IBE (i.e., the move from 1 to 3) and deductive arguments (i.e., the move from 4 to 5). In Shecaira-style analogical arguments, the logical function of the similarities proposition can be justified by IBE. The similarities proposition here refers to the relevant common feature between cases. The logical function of these relevant common features is to help the arguer to get the "best" principle by narrowing down the list of potential principles. In this way, the similarities proposition cannot be eliminated and is not reducible to the principle. Also, the interpretative power of Shecaira-style analogical arguments is satisfactory—case-to-case arguments that are concerned about the principle behind the common features can be seen as analogical arguments. It means that those analogical arguments people are most familiar with (e.g., analogical arguments in ethics, law, and daily life) can be seen as a distinctive type of argument. This is what Shecaira's insights contribute to the defense of the irreducibility of analogical arguments. As I argued in the previous section, however, the arguments from principle need not be deductive because of the introduction of DMP. Thus, a modified version of Shecaira-style analogical arguments can be formulated as follows: Analogical arguments in ethics and law can be regarded as complexes containing IBE and principle-based arguments (rather than deductive arguments).

5. Conclusion

The main goal of this paper is to give a preliminary critical account of the dispute between the reducibility and the irreducibility of analogical arguments and provide a new defense for the latter. The existing defense for the irreducibility of analogical arguments is unsuccessful. The new defense for it can be summarized as follows. There are two types of analogical arguments: (a) Analogical arguments containing conductive arguments and principle-based arguments, which are concerned with the weighing mechanism between the common features and the differences (e.g., animal models in scientific reasoning); (b) Analogical arguments containing IBE and principle-based arguments, which are concerned about the underlying principle behind the common features (e.g., case-to-case arguments in ethics and law). These two types of analogical arguments are not reducible to non-analogical arguments, so the irreducibility of analogical arguments can be defended.[2]

References

Agassi, J. (1988). Analogies Hard and Soft. In: D.H. Helman (Ed.), *Analogical Reasoning* (pp. 401–419). Dordrecht: Kluver Academic Publishing.
Bartha, P. (2010). *By Parallel Reasoning: The Construction and Evaluation of Analogical Arguments*. Oxford: Oxford University Press.
Bermejo-Luque, L. (2012). A Unitary Schema for Arguments by Analogy. *Informal Logic, 32(1)*, 1-24.
Blair, J. A., & R. H. Johnson (Eds.). (2011). *Conductive arguments: An Overlooked Type of Defeasible Reasoning*. London: College Publications.
Gamboa, S. (2008). In Defense of Analogical Reasoning. *Informal Logic, 28(3)*, 229–241.
Govier, T. (2002). Should a priori Analogies be Regarded as Deductive Arguments? *Informal Logic, 22(2)*, 155–157.
Govier, T. (1989). Analogies and Missing Premises. *Informal Logic, 11(3)*, 141–152.
Guarini, M. (2004). A Defense of Non-deductive Reconstructions of Analogical Arguments. *Informal Logic, 24(2)*, 153–168.

---

[2] I want to thank the ECA 2022 chair Prof. Fabio Paglieri for allowing me to present this research work at the conference online. This paper is a revised and extended version of a preliminary paper in Chinese that I published in Ziran Bianzhengfa Tongxun in 2022. This work was supported by the Philosophy and Social Science Foundation of Guangdong Province (GD22YZX03) and the China Postdoctoral Science Foundation (2021M693691).

Hansen, H. V. (2011). Notes on balance-of-consideration arguments. In J. A. Blair & R. H. Johnson (Eds.), *Conductive Arguments: An Overlooked Type of Defeasible Reasoning* (pp. 31-51). London: College Publications.

Hesse, M. (1966). *Models and Analogies in Science*. South Bend, IN: University of Notre Dame Press.

Kaptein, H. (2005). Legal Progress Through Pragma-Dialectics? Prospect Beyond Analogy and E Contrario. *Argumentation, pp. 19*, 497–507.

Liao, Y. (2022). Leibi shi yizhong dulide lunzhengleixing ma? [Is Analogy a Distinct Type of Argument?], *Ziran Bianzhengfa Tongxun, 44(11)*: 1-8.

Shecaira, F. (2013). Analogical Arguments in Ethics and Law: A Defence of a Deductivist Analysis. *Informal Logic, 33(3)*, 406–437.

Waller, B. (2001). Classifying and Analyzing Analogies. *Informal Logic, 21(3)*, 199–218.

Walton, D., C. Reed and F. Macagno (2008). *Argumentation Schemes*. New York: Cambridge University Press.

Walton, D. (2005). Justification of Argumentation Schemes. *Australian Journal of Logic, 3*, 1-13.

Wellman, C. (1971). *Challenge and response: Justification in Ethics*. Carbondale, IL: Southern Illinois University Press.

Xie, Y. (2017). Conductive argument as a mode of strategic maneuvering. *Informal Logic, 37(1)*, 2–22.

Zenker, F. (2011). An Attempt at Unifying Natural Language Argument Structures. In J. A. Blair & R. H. Johnson (Eds.), *Conductive Arguments: An Overlooked Type of Defeasible Reasoning* (pp. 74–85). London: College Publications.

Resolving Open-Textured Rules with Templated Interpretive Arguments

John Licato
Advancing Machine and Human Reasoning (AMHR) Lab
Department of Computer Science and Engineering
University of South Florida
licato@usf.edu

Logan Fields
Advancing Machine and Human Reasoning (AMHR) Lab
Department of Computer Science and Engineering
University of South Florida
ldfields@usf.edu

Zaid Marji
Advancing Machine and Human Reasoning (AMHR) Lab
Department of Computer Science and Engineering
University of South Florida
zaidm@usf.edu

Abstract

Open-textured terms in written rules are typically settled through interpretive argumentation. Ongoing work has attempted to catalogue the schemes used in such interpretive argumentation. But how can the use of these schemes affect the way in which people actually use and reason over the proper interpretations of open-textured terms? Using the interpretive argument-eliciting game Aporia as our framework, we carried out an empirical study to answer this question. Differing from previous work, we did not allow participants to argue for interpretations arbitrarily, but to only use arguments that fit with a given set of interpretive argument templates. Finally, we analyze the results captured by this new dataset, specifically focusing on practical implications for the development of interpretation-capable artificial reasoners.

1. Introduction

When people delegate decision-making authority to others, that delegation is often accompanied with guidelines in the form of *open-textured rules*— i.e., rules containing open- textured terms, such that the exact interpretations of those terms can be delegated to the discretion of boots-on-the-ground decision-makers. In this way, the intent of the rule-writer can be formalized such that it provides guidance but is not overly constraining, to avoid placing unnecessary constraints on the adaptability of the rule-interpreters. For example, a social media company may instruct its employees to "not allow users to use our website in a manner intended to artificially amplify or suppress information." Should the situation arise, the employees with the responsibility of executing this directive must determine the proper interpretation of the open-textured term 'artificially amplify'. Suppose a university decides to post an announcement and has it re-tweeted through the Twitter accounts of its multiple academic departments, colleges, alumni groups, etc. Would this action constitute artificial amplification in the sense meant by the rule? Furthermore, suppose that the employee is not human at all, but an automated Twitter moderation-bot. How should an answer to the preceding question be searched for and justified, in such a way that is satisfactory to the rule-writers? Such a situation may not have been anticipated by the rule-writers at the time of rule writing, and so the attempt to state the rule in a form general enough to sufficiently cover unanticipated scenarios can result in open-textured language.

For the most part, the use of open-textured rules to guide human behavior works because of the assumption that when the rule-interpreters interpret the rules, it will likely be done in a manner that is sufficiently similar to the interpretation that would be used by the rule-writer under similar circumstances. But this assumption cannot be made when the rule-writers are human beings and the rule-interpreters are artificially intelligent reasoners. It is therefore crucial that serious research be carried out into developing automated interpretive reasoning techniques that can give us some guarantee of human likeness.

One appealing approach is to require interpretations of open-textured rules to be justified with a graph of human-like interpretive arguments, properly organized according to some norms of argument strength and combination (possible frameworks for which are well-known to the AI in argumentation community). If reasonable effort is placed into generating the strongest possible interpretive arguments for and against a possible interpretation (given constraints of time, computational power, and available information), comparing them to each other, and combining them, the resulting interpretation would have some claim to being the most rational, or at least satisficing. It is our belief that such an approach to automated interpretive reasoning is significantly preferable to current approaches in AI; specifically, the so-called "explainable AI" paradigm which seeks to explain how an AI black box reaches its conclusion, rather than seeking to justify that conclusion with normative argumentation. This is the approach argued by the MDIA view, which states that rule-following AI should act in accordance with the interpretation best supported by *minimally defeasible interpretive arguments*—i.e., those interpretive arguments which are such that there are a minimal number and quality of interpretive counterarguments that can be levied against them (Licato, 2021).

Taking some variant of the MDIA position as our base assumption,[1] we seek to advance progress towards interpretation-capable AI: First, we describe Aporia, a new tool for eliciting and comparing naturalistic interpretive arguments. Second, we present a new dataset, collected using Aporia, with the caveat that all interpretive arguments used are required to utilize a fixed set of interpretive argument templates. Finally, we analyze the results captured by this new dataset, specifically focusing on practical implications for the development of interpretation-capable artificial reasoners.

2. Experimental Setup

Aporia (Marji et al., 2021) is a gamified framework to elicit naturalistic interpretive arguments and comparisons of interpretive arguments. It begins with an open-textured rule (e.g., "No vehicles are allowed in the park") and an intentionally ambiguous scenario (e.g, "A non-working antique car, meant for an artistic exhibit, is pushed into the park"). Three players—Player 1, Player 2, and the judge—are then asked to determine whether the rule applies. Player 1 is asked to first decide on a position: either that the action described in the scenario is a violation of the rule, or that the rule does not apply. Player 1 must then argue their position. Next, Player 2 is given time to counter Player 1's arguments. Finally, the judge is asked to determine whether Player 1 was more convincing, or whether Player 2 cast sufficient doubt onto Player 1's argument that it cannot be accepted, and to provide their rationale for their decision.

---

[1] We will not mount a full defense of the MDIA position, instead relying on the defense in [1].

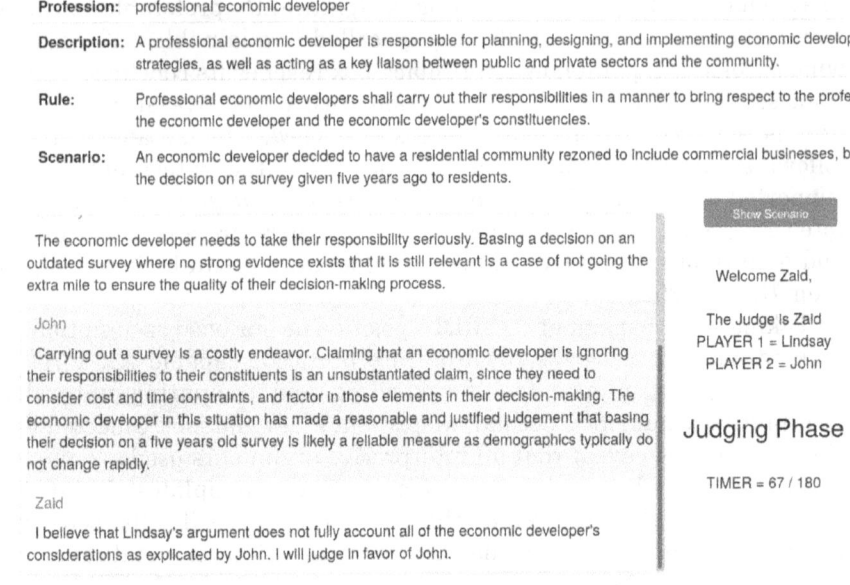

Figure I. Example of Aporia Gameplay

Previously, Aporia has been used to collect a dataset of played games, seeded with rules and scenarios originally collected from codes of ethics of various professional societies (Licato et al., 2019). However, this work only included one judgment per argument pair, and thus does not allow us to rigorously study the range of judgments used to compare interpretive reasoning. Secondly, it did not enforce the use of interpretive argument templates, thus making annotation and comparison of interpretive argument types difficult.

We therefore adopted a typology of interpretive argumentation schemes, primarily adapted from Walton et al. (2021) and simplified to be: (1) more accessible to laypeople (i.e., non-experts in legal reasoning), (2) less fine-grained, and (3) de-emphasize arguments rooted in specific legal practices, so as to be more broadly applicable to general interpretive reasoning of rules. The resulting argument types we employed are:

1. **Argument from Authoritative Source:** An authoritative source (an official definition, authoritative document, or authority figure / expert) defines a term a certain way, which then requires a certain interpretation. **Critical questions:** Does the authoritative source actually say what is claimed? Is the authority's area of expertise or jurisdiction relevant to the rule? Is there another authority that has an alternative interpretation?

2. **Argument from Higher Reason or Principle:** There is a higher principle of reasoning, ethics, or law that requires us to interpret the rule a certain way.

Critical questions: Is the claimed higher principle actually used in practice? Does it actually govern our interpretations? Is there another example where the claimed higher principle is clearly violated?

3. **Argument from Practicality / Consequence:** It is more effective, efficient, or otherwise practical to interpret the rule in a certain way when considering certain consequences of alternate interpretations.
Critical questions: Would this interpretation lead to a bending of the rule that would render the rule useless? Is the claimed consequence actually going to happen with any certainty, or does the new level of certainty justify the new interpretation?

4. **Argument from History:** The historical interpretation of a term is, or has evolved in, a specific way that constrains how we can interpret it today.
Critical questions: Is the claimed historical interpretation actually accurate? Is there a reason to believe that the historical interpretation is outdated and no longer in use?

5. **Argument from Public Understanding:** The public, commonsense, reasonable, or de- fault understanding of a term is such that it requires a certain interpretation.
Critical questions: Is the public understanding of the term claimed here actually correct, or is the arguer mistaken? Are there other understandings of the term that lead to different interpretations, but which are just as public, commonsense, reasonable, or default?

6. **Argument by Analogy / Similarity:** If the action being considered (A1) is similar to another previous action that the rule does or does not apply to (A2), then we can use that to determine whether the rule applies to A1.
Critical questions: What kind of similarity does A1 and A2 have? Is it actually relevant to the rule? Are there important differences or dissimilarities between A1 and A2? Is there another similar case (A3) that is also similar to A1, except that the rule applies differently to A2 than it did to A3?

7. **Argument from Word Choice / Exclusion:** If the rule is phrased in a very specific way that leaves out certain words or terms, then there is a reason for it. And that reason requires us to interpret the rule in a certain way.
Critical questions: Is the omission explainable by some other reason? E.g., is it possible that the rule writers did not think it was necessary to include because it was to be assumed, or inferred?

We then created six rule-scenario pairs. All rules were taken directly from Twitter's rules of conduct in March 2022 (prior to current substantial management changes in that organization)[2] and scenarios were designed to intentionally exploit the open-texturedness of the rules:

1. **Rule:** You may not promote violence against, threaten, or harass other people on the basis of race, ethnicity, national origin, caste, sexual orientation, gender, gender identity, religious affiliation, age, disability, or serious disease.
   Scenario: You retweet a position held by violent extremists, without commentary, and frequently retweet similar opinions from other people. Your account has a large number of followers.
2. **Rule:** You may not promote or encourage suicide or self-harm.
   Scenario: You never seriously encourage someone to commit suicide or self-harm, and would go out of your way to prevent someone from doing those things, but you publicly joke and laugh about suicide-related jokes.
3. **Rule:** You may not engage in the targeted harassment of someone, or incite other people to do so. This includes wishing or hoping that someone experiences physical harm.
   Scenario: A group of reddit users work together to solve a crime. They put together evidence and conclude that person X is guilty, posting images, personal information, and other details publicly to support their investigation and claims. However, X is later proven to be innocent. Did the group participants harass X?
4. **Rule:** You may not use Twitter's services in a manner intended to artificially amplify or suppress information.
   Scenario: You work for the University's social media team, and when a faculty member earns a new grant, you post it on the social media pages of the department, college, university, alumni group, faculty group, and ask friends and colleagues to like and re-share the posts.
5. **Rule:** You may not impersonate individuals, groups, or organizations to mislead, confuse, or deceive others, nor use a fake identity in a manner that disrupts the experience of others on Twitter.
   Scenario: You create an account called "Donald Drumpf" meant to mock things the former president does, by simply re-posting things he actually said as tweets, without additional context.
6. **Rule:** You may not threaten violence against an individual or a group of people. We also prohibit the glorification of violence.
   Scenario: You frequently post positive memes, thoughts, etc. honoring the "greatest generation," who are named this because they fought in, and won, WWII.

Thirty undergraduate students at the University of South Florida were assigned as follows: thirty games were created, each one employing one of

---

[2] https://help.twitter.com/en/rules-and-policies/twitter-rules

the rule-scenario pairs above, such that each rule-scenario pair was assigned to exactly five games. Each student was first assigned to one game, asked to choose a position to defend, and create two arguments in support of their position. Each argument was required to utilize one of the templates, instructions for which were provided to them in randomized order. The students were then shuffled and assigned to another game, and asked to address the first student's arguments—not to simply devise arguments for the opposite position, but to focus on undercutting or undermining Player 1's argument (thus differing slightly from Marji et al. (2021), which instead encouraged general counterarguments). Students were then re-assigned again, and asked to judge which arguments were more convincing for a third game, to explain why, and rate their confidence in their decision on a scale from 1 to 5 (5 being the highest confidence level).

This last step was repeated three more times, so that each game had four unique judgments (in order to ensure each judgment was made independently, each judge was unable to see the previous judges' decisions). Every player was assigned such that for every one of their six rounds (initial argument, counterargument, and four judging rounds), they were assigned to a game with a different rule-scenario pair.

Table I. Evaluations of whether argument schemes were correctly applied.

Template name	Classified as		
	1	2	3
Public Understanding	0	3	11
Word choice or exclusion	1	6	7
Practicality or consequence	2	2	9
Authoritative source	0	3	1
Analogy or Similarity	0	1	3
History	0	0	3
Higher reason or principle	1	0	4
Total	4	15	38

In order to determine how well argument schemes were correctly applied, three Advancing Machine and Human Reasoning Lab researchers (the present paper's authors: one faculty member and two PhD students), each with several years of experience in working with interpretive argumentation, independently assessed the arguments provided by participants. They were instructed to categorize each argument into one of the following categories: (1) the annotator could not understand the structure of the argument and figure out how it fit into its supposed argument scheme after reading it at least three times; (2) the annotator identified the structure of the argument being made, but it was of a different scheme than the one it was claimed to be; and (3) the structure of the argument being made is closest to the argument scheme it was claimed to be. The resulting evaluations are listed in Table I. Note that due to technical errors, three arguments were excluded from the analysis, thus the bottom row of Table I totals to 57 rather than 60 (these excluded arguments are also excluded from all statistics we report for the remainder of this paper).

All arguments for all stages were randomly assigned to the annotators so that each argument was categorized by two annotators independently. If the annotators for any argument disagreed on the proper category, then the third annotator would independently categorize it. In such cases, the category chosen by two out of the three annotators would be counted as the correct one. In no cases did the three annotators select three different categories.

3. Analysis

We selected our rule-scenario pairs in order to encourage alternate interpretations, so that we could focus on the argumentation used to justify interpretations. Since Player 1 in each game was allowed to choose which position they would defend (*violation* or *non-violation*), the relative number of times players chose each position for each rule-scenario pair is a way of estimating how well they invite competing interpretations. According to this measure, we were partially successful: rule-scenario pairs 3, 4, and 6 were as close to perfectly balanced as possible (3 players chose *violation*, and 2 chose *non-violation*, or vice-versa). All players chose *non-violation* for rule-scenario pairs 1 and 5, and all chose *violation* for pair 2.

Table II. Frequency of argument types used

Template name (Linguistic, Systemic, or Teleological-Evaluative)	Used as 1st arg	Used as 2nd arg	Total
Public Understanding (L)	10	4	14
Word choice or exclusion (L)	6	8	14
Practicality or consequence (S)	7	6	15
Authoritative source (S)	3	1	4
Analogy or Similarity (S)	1	3	4
History (S)	0	2	2
Higher reason or principle (TE)	1	4	5

MacCormick and Summers (1991) suggested that interpretive arguments be considered and deployed with linguistic arguments coming first, systemic arguments next, and teleological-evaluative arguments last. Table II categorizes and sorts our argument schemes in this order, in order to easily see whether arguers naturally selected arguments in this order. Indeed, linguistic arguments were most commonly chosen as the first argument type, with systemic arguments the second most common choice for first arguments. Given that our arguers were not formally trained in legal reasoning or formal argumentation, it is interesting to see MacCormick and Summers' ordering reflected (albeit weakly) in the arguers' preferences.

We further set out to examine whether the evaluations of argument quality, as made by our participants acting in the role of *judge*, were consistent. We consider a simple majority vote of judges assigned to a game as deciding whether Player 1 or 2 was the winner of that game. Across all 30 games, we see that Player 1 was the winner 53.6% of the time, Player 2 won 32.1% of the time, and there was a tie 14.3% of the time, suggesting that although there is a preference for Player 1, the advantage that Player 1 gets from being able to go first is not non-existent.

A similar pattern manifests when we aggregate the individual judge decisions together, regardless of game: 58.7% and 41.3% of judge decisions were for Player 1 and 2, respectively. Furthermore, when judges selected Player 1 as the winner, they had an average confidence of 4.25 (out of 5). However, when selecting Player 2, their average confidence dropped to 3.76. This may reflect a hesitation to go against the first arguments they read (those of Player 1), consistent with known anchoring and ordering effects in argumentation. However, our study design did not allow for us to vary the order in which players' arguments were read, since Player 2's arguments were always responses to those of Player 1.

In order to determine whether shallow heuristics might have been employed to determine argument quality, we calculated the word length of arguments and used Spearman correlation to compare them to the proportion of judges that voted for them. This effect was significant ($r=0.534$, $p < 0.005$), suggesting that arguments that had more words were considered more persuasive.

Furthermore, we calculated the correlation between the average confidence of judges in their decisions (recall that they self-rated their confidence from 1 to 5) and the level of agreement between judges on that game (defined as the proportion of judges who voted for the majority vote, or zero if there was a tie). The correlation was small ($r=0.347$, $p<0.08$), weakly suggesting that when a game's two players produced competing arguments that were difficult to decide between (in the sense that a sample of judges will disagree as to which is more convincing), judges were able to anticipate this controversiality and it lowered their confidence in their ratings.

4. Conclusion and Future Work

It is not fully known how controlled argumentation dialogue environments, such as *Aporia*, affect the types of argumentation used. We expect that, at least among non-experts, unrestricted argumentation environments will result in arguments that are more difficult to classify as one of the known interpretive argument types. The work presented in this paper offers a data point that can be used as a baseline against which to compare future studies.

In particular, it would be interesting to see how experience with using interpretive argumentation changes the way in which they are used. Table I summarized our evaluations on whether argument schemes were applied correctly. In our study, arguments were rated as category 3 (meaning they were applied correctly) 66.1% of the time, a value we attribute to the short amount of time that arguers were given to familiarize themselves with the argument schemes (roughly 10 minutes).

An under-studied aspect of interpretive reasoning is how experts and non-experts compare and evaluate interpretive arguments. Our present work offers some empirical insight into this---e.g., we showed that argument length correlates with judges' perceptions of argument quality, and the confidence of judges also negatively correlates with how controversial they think their decision will be. But are there deeper relationships between the rationales used by judges in evaluating interpretive arguments and their ultimate decisions? Since the present study required judges to provide short justifications of their decisions, we anticipate the data we collected can be used to further study this question in future work.

References

Licato, J., Marji, Z., & Abraham, S. (2019). Scenarios and Recommendations for Ethical Interpretive AI. In *Proceedings of the AAAI 2019 Fall Symposium on Human-Centered AI*, Arlington, VA, 2019.

Licato, J. (2021). How Should AI Interpret Rules? A Defense of Minimally Defeasible Interpretive Argumentation. *arXiv e-prints*.

Marji, Z. and Licato, J. (2021). Aporia: The argumentation game. In *Proceedings of The Third Workshop on Argument Strength (ArgStrength 2021)*.

Walton, D., Macagno, F., and Sartor, G. (2021). *Statutory Interpretation: Pragmatics and Argumentation*. Cambridge University Press.

MacCormick, D. and Summers, R. (1991). *Interpreting Statutes: A Comparative Study*. Routledge.

LIST OF CONTRIBUTORS TO ALL VOLUMES
(IN ALPHABETICAL ORDER BY SURNAME)

SCOTT AIKIN
Vanderbilt University

JOSÉ ALHAMBRA
Autonomous University of Madrid

JOSE M. ALONSO-MORAL
Universidade de Santiago de Compostela, Spain

KATIE ATKINSON
University of Liverpool

SHARON BAILIN
Faculty of Education, Simon Fraser University, Vancouver Canada

MARK BATTERSBY
Department of Philosophy, Capilano University, Vancouver Canada

TREVOR BENCH-CAPON
University of Liverpool

SARAH BIGI
University of the Sacred Heart, Milan, Italy

PETAR BODLOVIĆ
NOVA Institute of Philosophy (ArgLab),
FCSH, Nova University of Lisbon

MIEKE BOON
University of Twente

ELENA CABRIO
Université CôteD'Azur, CNRS, Inria, I3S

JOHN CASEY
Northeastern Illinois University

ALEJANDRO CATALA
Universidade de Santiago de Compostela, Spain

LIST OF CONTRIBUTORS

DORIANA CIMMINO
Independent researcher

DANIEL COHEN
Colby College

FEDERICA COMINETTI
Università dell'Aquila

CLAUDIA COPPOLA
Università Roma Tre, La Sapienza Università di Roma

MARÍA INÉS CORBALÁN
ArgLab-IFILNOVA, NOVA Universidade de Lisboa

HÉDI VIRÁG CSORDÁS
Assistant Lecturer at Budapest University of Technology and Economics

GIULIA D'AGOSTINO
Università della Svizzera italiana (USI)

DANIEL DE OLIVEIRA FERNANDES
University of Fribourg, Switzerland

EMMANUELLE DIETZ
Airbus Central R&T, Germany

ÁLVARO DOMÍNGUEZ-ARMAS
NOVA Institute of Philosophy, NOVA University of Lisbon

GONEN DORI-HACOHEN
University of Massachusetts Amherst

IOVAN DREHE
Technical University of Cluj-Napoca

LUCIJA DUDA
University of Manchester

MICHEL DUFOUR
University Sorbonne-Nouvelle

ALINA DURRANI
University of Massachusetts Amherst

LIST OF CONTRIBUTORS

CATARINA DUTILH NOVAES
Department of Philosophy, Vrije Universiteit Amsterdam

ISABELA FAIRCLOUGH
University of Central Lancashire

LOGAN FIELDS
Advancing Machine and Human Reasoning (AMHR) Lab, University of South Florida

JOSÉ ÁNGEL GASCÓN
Departamento de Filosofía, Universidad de Murcia

GIULIA GIUNTA
University of Neuchâtel

GEOFFREY C. GODDU
University of Richmond

SARA GRECO
USI-Università della Svizzera italiana

MARCELLO GUARINI
University of Windsor, Canada

PASCAL GYGAX
University of Fribourg, Switzerland

DALE HAMPLE
Western Illinois University

AMALIA HARO MARCHAL
University of Granada

ANNETTE HAUTLI-JANISZ
University of Passau

BITA HESHMATI
University of Groningen

MIKA HIETANEN
Lund University

MICHAEL J. HOPPMANN
Northeastern University, Boston

LIST OF CONTRIBUTORS

BROOKE HUBSCH
The Pennsylvania State University

BETH INNOCENTI
University of Kansas

CHIARA JERMINI-MARTINEZ SORIA
Università della Svizzera italiana

ANTONIS KAKAS
Dept. Computer Science, University of Cyprus, Cyprus

ALEXANDRA KARAKAS
Assistant Lecturer at Budapest University of Technology and Economics

IRYNA KHOMENKO
Taras Shevchenko National University of Kyiv

ZLATA KIKTEVA
University of Passau

KONRAD KILJAN
University of Warsaw
Laboratory of The New Ethos, Warsaw University of Technology

GABRIJELA KIŠIČEK
University of Zagreb

MARCIN KOSZOWY
Laboratory of The New Ethos, Warsaw University of Technology

ADAMOS KOUMI
Dept. Computer Science, University of Cyprus, Cyprus

MANFRED KRAUS
University of Tübingen

LEONARD KUPŚ
Faculty of Psychology and Cognitive Science
Adam Mickiewicz University, Poznań, Poland

NIILO LAHTI
The University of Eastern Finland, School of Theology

JOHN LAWRENCE
Centre for Argument Technology, University of Dundee, UK

LAWRENCE LENGBEYER
United Stated Naval Academy

MARCIN LEWIŃSKI
Nova Institute of Philosophy
Nova University Lisbon, Portugal

JIAXING LI
Nankai University

YAN-LIN LIAO
Department of Philosophy, Sun Yat-sen University, China

JOHN LICATO
Advancing Machine and Human Reasoning (AMHR) Lab, University of South Florida

DAVIDE LIGA
University of Luxembourg

EDOARDO LOMBARDI VALLAURI
Università Roma Tre

COSTANZA LUCCHINI
Università della Svizzera italiana (USI)

CHRISTOPH LUMER
University of Siena, Italy

GIORGIA MANNAIOLI
Università Roma Tre, La Sapienza Università di Roma

MAURIZIO MANZIN
University of Trento

ZAID MARJI
Advancing Machine and Human Reasoning (AMHR) Lab, University of South Florida

HUBERT MARRAUD
Universidad Autónoma de Madrid (Spain)

LIST OF CONTRIBUTORS

SANTIAGO MARRO
Université CôteD'Azur, CNRS, Inria, I3S

VIVIANA MASIA
Università Roma Tre

DAVIDE MAZZI
University of Modena and Reggio Emilia (Italy)

GUIDO MELCHIOR
University of Graz

CHIARA MERCURI
Università della Svizzera Italiana

DIMA MOHAMMED
Institute of Philosophy, Faculty of Social and Human Sciences, NOVA University of Lisbon, Portugal

ELENA MUSI
University of Liverpool

HENRI MÜTSCHELE
Heinrich- Heinrich Heine University Düsseldorf, Germany

ZI-HAN NIU
Department of Philosophy, Sun Yat-sen University, China

PAULA OLMOS
Universidad Autónoma de Madrid

MARIANA OROZCO
University of Twente

RAHMI ORUÇ
Ibn Haldun University, Comparative Literature, ArguMunazara Research Center

STEVE OSWALD
University of Fribourg, Switzerland

WENQI OUYANG
Department of Philosophy, Sun Yat-sen University, China

FABIO PAGLIERI
Istituto di Scienze e Tecnologie della Cognizione, Consiglio Nazionale delle Ricerche (ISTC-CNR), Italy

ROOSMARYN PILGRAM
Leiden University Centre for Linguistics

FEDERICO PUPPO
University of Trento

MENNO H. REIJVEN
University of Amsterdam

THÉOPHILE ROBINEAU
Université Paris Cité

ANDREA ROCCI
Università della Svizzera italiana (USI)

MARIA GRAZIA ROSSI
Institute of Philosophy, Faculty of Social and Human Sciences, NOVA University of Lisbon, Portugal

LUCIA SALVATO
Università Cattolica del Sacro Cuore

CRISTIÁN SANTIBÁNEZ,
Universidad Católica de la Santísima Concepción

MENASHE SCHWED
Ashkelon Academic College, Israel

BLAKE D. SCOTT
Institute of Philosophy, KU Leuven

HARVEY SIEGEL
University of Miami

ILIA STEPIN
Universidade de Santiago de Compostela, Spain

JÁNOS TANÁCS
Department of Argumentation Theory and Marketing, ELTE, Budapest

LIST OF CONTRIBUTORS

GIULIA TERZIAN
ArgLab-IFILNOVA, NOVA Universidade de Lisboa

CHRISTOPHER W. TINDALE
Department of Philosophy, University of Windsor, Ontario, Canada

SERENA TOMASI
University of Trento

MARIUSZ URBAŃSKI
Faculty of Psychology and Cognitive Science
Adam Mickiewicz University, Poznań, Poland

MEHMET ALÌ ÜZELGÜN
IFILNOVA, Universidade Nova de Lisboa
CIES-ISCTE, Instituto Universitário de Lisboa

CHARLOTTE VAN DER VOORT
Leiden University Centre for Linguistics
JAN ALBERT VAN LAAR
University of Groningen

LOTTE VAN POPPEL
Center for Language and Cognition Groningen

SERENA VILLATA
Université CôteD'Azur, CNRS, Inria, I3S

JACKY VISSER
Centre for Argument Technology, University of Dundee, UK

JEAN H.M. WAGEMANS
University of Amsterdam

HAILONG WANG
Wuhan University

JIANFENG WANG
Fujian Normal University, China

MARK WEINSTEIN
Montclair State University

HARALD R. WOHLRAPP
Universität Hamburg

List of Contributors

MING-HUI XIONG
Guanghua Law School, Zhejiang University, China

OLENA YASKORSKA-SHAH
Università della Svizzera italiana (USI)

SHIYANG YU
College of Philosophy, Nankai University

GÁBOR Á. ZEMPLÉN
Eötvös Loránd University-Faculty of Economics, ELTE, Budapest

FRANK ZENKER
College of Philosophy, Nankai University, Tianjin, China

www.ingramcontent.com/pod-product-compliance
Lightning Source LLC
Chambersburg PA
CBHW070715160426
43192CB00009B/1197